Polymer
Liquid Crystals

This is a volume in the
Materials Science and Technology series.
Editors: *A. S. Nowick and G. G. Libowitz*

A complete list of the books in the series appears at the end of the volume.

POLYMER LIQUID CRYSTALS

Edited by

A. Ciferri

Istituto di Chimica Industriale
University of Genoa
Genoa, Italy

W. R. Krigbaum

Paul M. Gross Chemical Laboratory
Duke University
Durham, North Carolina

Robert B. Meyer

Martin Fisher School of Physics
Brandeis University
Waltham, Massachusetts

1982

ACADEMIC PRESS
A Subsidiary of Harcourt Brace Jovanovich, Publishers

New York   London
Paris   San Diego   San Francisco   São Paulo   Sydney   Tokyo   Toronto

ACADEMIC PRESS, INC.
111 Fifth Avenue, New York, New York 10003

United Kingdom Edition published by
ACADEMIC PRESS, INC. (LONDON) LTD.
24/28 Oval Road, London NW1 7DX

Library of Congress Cataloging in Publication Data
Main entry under title:

Polymer liquid crystals.

Includes index.
1. Liquid crystals. 2. Polymers and polymerization.
I. Ciferri, A. II. Krigbaum, W. R. (William Richard),
Date. III. Meyer, R. B.
QD923.P64   548'.9   82-6766
ISBN 0-12-174680-1   AACR2

PRINTED IN THE UNITED STATES OF AMERICA

82 83 84 85   9 8 7 6 5 4 3 2 1

Contents

5   Mechanical Properties of Nematic Polymers

P. G. de Gennes

6   Macroscopic Phenomena in Nematic Polymers

Robert B. Meyer

7   Techniques for the Evaluation of Material Constants in
    Lyotropic Systems and the Study of Pretransitional Phenomena
    in Polymeric Liquid Crystals

Donald B. DuPré

8   Instabilities in Low Molecular Weight Nematic and Cholesteric
    Liquid Crystals

S. Chandrasekhar and U. D. Kini

III. Two Illustrative Experiments on Nematic Flow 211
IV. Hydrodynamic Instabilities in Nematics 215
V. Thermal Instabilities in Nematics 230
VI. Instabilities in Cholesteric Liquid Crystals 236
Appendix. Material Constants for MBBA and HBAB 241
References 242

9 Rheo-Optical Studies of Polymer Liquid Crystalline Solutions

Tadahiro Asada

I. Introduction 248
II. Structure of Solutions of Liquid Crystalline Polymers 248
III. Experimental Methods 250
IV. General Optical Considerations for the Polarized Light Method 253
V. Processes of Structure Formation in Cholesteric Liquid Crystals Studied by Spectrophotometric Methods 256
VI. Deformation and Flow Mechanisms for Some Typical Polymeric Liquid Crystalline Solutions 265
VII. Concluding Remarks 271
References 272

10 The Effects of External Fields on Polymeric Nematic and Cholesteric Mesophases

W. R. Krigbaum

I. Introduction 275
II. Magnetic-Field Effects 277
III. Electric-Field Effects 285
IV. Viscous Flow 295
References 305

11 Liquid Crystal Display Devices

G. Baur

I. Introduction 309
II. Design of Displays 309
III. Optimization of Displays 312
IV. Back Light Layout 323
References 326

12   Recent Advances in High-Strength Fibers and Composites

D. C. Prevorsek

List of Contributors

Numbers in parentheses indicate the pages on which the authors' contributions begin.

Tadahiro Asada (247), Department of Polymer Chemistry, Kyoto University, Kyoto, 606 Japan

G. Baur (309), Fraunhofer-Institut für Angewandte Festkörperphysik, D-7800 Freiburg, Federal Republic of Germany

S. Chandrasekhar (201), Raman Research Institute, Bangalore 560 080, India

A. Ciferri (63), Istituto di Chimica Industriale, University of Genoa, Genoa, Italy

P. G. de Gennes (115), Collège de France, Paris, France

Donald B. DuPré (165), Department of Chemistry, University of Louisville, Louisville, Kentucky 40292

H. Finkelmann (35), Physikalisch-Chemisches Institut, Technische Universität Clausthal, 3392 Clausthal-Zellerfeld, Federal Republic of Germany

P. J. Flory (103), Stanford University, Stanford, California 94305

G. W. Gray (1), Department of Chemistry, University of Hull, Hull, HU6 7RX, North Humberside, England

U. D. Kini (201), Raman Research Institute, Bangalore 560 080, India

W. R. Krigbaum (275), Paul M. Gross Chemical Laboratory, Duke University, Durham, North Carolina 27706

Robert B. Meyer (133), Martin Fisher School of Physics, Brandeis University, Waltham, Massachusetts 02254

D. C. Prevorsek (329), Allied Corporation, Morristown, New Jersey 07960

Preface

This book was developed from lectures presented during the seminar "Polymer Liquid Crystals: Science and Technology," which was held at Santa Margherita Ligure, Italy, May 19–23, 1981.

This work is indicative of the evolution that has occurred in the field of highly oriented polymers in the past decade. The first report of ultrahigh-modulus, high-strength fibers spun from nematic dopes appeared in the patent literature in 1972. Interest in the application of these polymers as composite materials spawned vigorous research activity to investigate the structure and properties of high-modulus polymers. As this research evolved, the connection between these new mesogenic polymers and low molecular weight liquid crystals became more evident. Prior to these developments, the liquid crystal interests of polymer scientists had been confined to a few lyotropic systems such as the homopolypeptides. The synthesis of new lyotropic and thermotropic polymers, and the search for appropriate tools for the investigation of these polymeric mesophases, stimulated interest in the experimental techniques and theoretical foundations that had been developed for low molecular weight liquid crystals. Also, those active in the latter field were intrigued by the possibility that polymeric mesophases might reveal new phenomena requiring the extension of the theory that had been developed for low molecular weight mesogens. However, a potentially fruitful interchange of ideas was hindered by the problems common to all interdisciplinary endeavors: the application of techniques unfamiliar to one group to materials only partially known by the other group; and by the need to develop a mutually understandable language to facilitate the discourse.

This project was planned with the objective of encouraging communication and interaction between these two groups of scientists. The objective of each chapter is to present a review of the current status of a particular field, addressed to scientists of the other discipline. The chapters are arranged under the headings: molecular basis of liquid crystallinity, orientation of liquid crystals, and applications.

Some problems in communication undoubtedly remain in this area. Nevertheless, this book is an attempt to remove such barriers, and to stimulate an interaction that may have far ranging scientific and practical results.

1

Relationship between Chemical Structure and Properties for Low Molecular Weight Liquid Crystals

G. W. Gray

Department of Chemistry
University of Hull
Hull, North Humberside
England

I. Introduction

The study of liquid crystal (LC) properties in polymer systems holds out much promise for the future, not only because such systems will provide us with new knowledge about both polymers and liquid crystals, but also from the standpoint of their technological importance. Activity in the field of liquid crystal polymers is growing apace, and this volume is itself a testimony to the escalating interest in such systems.

As pointed out by Finkelmann *et al.* (1978a,b,c) and by Finkelmann (1980) in his recent review of thermotropic liquid crystalline polymers, numerous instances are to be found in the literature of polymer systems exhibiting an

POLYMER LIQUID CRYSTALS

anisotropic structure in their glassy state but no liquid crystal phase above the glass transition temperature. Only more recently has it been possible to obtain LC phases similar to those formed by low molecular weight (MW) systems by heating macromolecular polymeric materials.

A. Thermotropic LC Polymers

Two main principles have been used in synthesizing such LC polymers:

(1)   By linking together suitable mesogenic monomers, a backbone is constituted. The units are linked together through suitable functional groups A and B located at the ends of the mesogenic monomer; these interact, usually through a condensation reaction, to form the polymer

The linkages between B and A and the mesogen cores (▭) may be direct, giving a rigid type of backbone, or of a flexible nature, e.g., an alkylene chain, giving a polymer backbone with alternating rigid (mesogenic) and flexible segments. Both types are called LC main chain polymers, and each preserves the original LC moieties. As Finkelmann (1980) says, this suggests the idea that the mesogenicity of the monomer can be preserved in the polymer.

(2)   By using mesogenic monomers carrying a terminal group that can give addition polymerization, a backbone can be formed that carries the mesogenic units as pendant side chains:

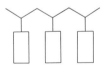

Again, the mesogenic units can be linked directly to the backbone or via flexible spacer units, e.g., alkylene chains. This turns out to be the crucial point. Direct linkage (Blumstein and Hsu, 1978; Shibaev and Platé, 1977), except in a few cases (Perplies et al., 1976; Cser and Nyitrai, 1976; Shibaev et al., 1976), gives only glasses with an anisotropy of structure that is lost at the glass transition. Coupled with steric interactions between the side groups, the tendency toward a statistical distribution of chain conformations hinders the ordered arrangement of the pendant groups and LC formation in the polymer is suppressed. Decoupling of the side groups by using a flexible spacer allows the main chain motions to occur without disturbance of the anisotropic arrangement of the side chains. The polymer then may exhibit LC properties and such polymers are called LC side chain polymers (or sometimes comb polymers).

In the case of LC main chain polymers it has been demonstrated that the thermal stability of the LC phase diminishes with increasing flexibility of the polymer, and this parallels the behavior of low MW LC systems, in relation to which similar examples can be quoted, e.g., reduction of a *trans*-stilbene **I** to the phenethyl compound **II** greatly decreases T_{N-1} (*the nematic–isotropic liquid transition temperature*).

(**I**)   (**II**)

The loss of rigidity in **II** and its ability to adopt bent molecular conformations readily explains the lower tendency of the system to give anisotropic fluid states.

Similarly, in the case of LC side chain polymers, strong analogies with low MW LC systems occur. Both smectic and nematic polymers may be obtained, and, as is commonly found with low MW systems, the tendency to increased smectic character grows as terminal alkyl chains are increased in length and as longer spacer units are used, e.g., as *m* and/or *n* increase in the structure **III**

(**III**)

Moreover, although attempts to obtain cholesteric polymers by use of a chiral center in each mesogenic side group have resulted till now in only smectic products, use of chiral comonomers (i.e., diluting the number of chiral pendant groups) does lead to formation of cholesteric properties. This is analogous to the principle of producing longer pitch chiral nematics (cholesterics) by using mixtures of low MW nematogens and cholesterogens.

Such observations led Finkelmann (1980) to conclude that, provided the polymer main chain does not too strongly influence the mesogenic side groups, i.e., when a sufficient length of spacer group is used, formation of the polymer LC phase will essentially follow the principles for conventional low MW LC systems.

B. Low Molecular Weight LC Systems

This encourages us to inspect (for use in the field of LC polymers) the empirical relationships accumulated over the years between LC phase behavior and molecular structure of low MW systems. The body of data is, however, a very large one to review, and in seeking to achieve some selectivity of what might profitably be considered, two further factors may be borne in mind:

(1)   Again quoting from Finkelmann (1980): "If nematic polymers are obtained, in most cases the corresponding monomers exhibit either no mesophase or a monotropic nematic phase. Nematic monomers give mainly smectic polymers." These comments would suggest that an increase in LC order occurs on passing from monomer to polymer and that we should be looking for structural features in low MW systems that only just favor the formation of nematic phases.

(2)   Of the two polymer types—LC main chain and LC side chain polymers—the latter are probably of most interest to the LC materials chemist, holding out as they do the promise of alternative LC substrates for electro-optical displays—through independent reorientation by electric fields of side groups of say positive $\Delta\varepsilon$—and of cast films of cholesteric polymer in the glass state in which the locked-in cholesteric phase gives the selective reflection of a fixed wavelength of linearly polarized light. Alternatively, films of the *cholesteric* LC polymer phase will still respond to external stimuli: electric fields or temperature.

In fact, the bulk of materials work in the LC field is not today concerned with obtaining materials with high T_{N-I} values. A surfeit of these already exists. What is required are materials with moderate, borderline, or even latent LC characteristics, but coupled with other physical features that give rise, in admixture with higher T_{N-I} systems, to superior electro-optical or cholesteric responses. It seems therefore that much of the data currently being collected on these types of systems may be of greatest relevance in the polymer field. Therefore, although we may begin by looking at some broader factors in relation to molecular structure and T_{N-I} values, we should perhaps rather quickly concentrate attention on structural features relating to modern technically important LC materials of low MW and their properties in a wider sense than simply nematic thermal stability. In all cases we would judge too that avoidance of smectic behavior might be very important.

II. Structural Considerations of Low MW LC Systems

A. Aromatic Systems[†]

Looking back to the period before the 1970s, the great majority of LC material were aromatic in character and of general structure **IV**

(**IV**)

where

(a) X and Y represent a range of terminal substituents such as alkyl, alkoxy, and cyano.

(b) A—B represents a linking unit in the core structure, e.g.,

$$-CH=N-, \quad -N=N-, \quad -N=N-, \quad -CO\cdot O-$$
$$\downarrow$$
$$O$$

(c) a and b have small integral values.

It was quickly realized that any substituent X or Y that does not broaden the molecule too much is superior to an H at the end of the molecule, and that groups such as cyano and alkoxy are more favorable than others such as alkyl or halogen in promoting high T_{N-I} values. A typical nematic terminal group efficiency order is

$$CN > OCH_3 > NO_2 > Cl > Br > N(CH_3)_2 > CH_3 > H$$

The nature of the central linkage is also of great importance and linking units containing multiple bonds that maintain the rigidity and linearity of the molecules are most satisfactory in preserving high T_{N-I} values. Thus in simple systems **IV** with a and b = 1, the CH_2—CH_2 flexible unit is a poor one and gives rise to very weak or nonexistent nematic tendencies. The ester function contains no multiple bonds in the chain of atoms actually linking the rings, but conjugative interactions within the ester function and with the rings do lead to some double bond character and a stiffer structure than might be expected. Esters are in fact fairly planar systems and quite strongly nematogenic. A typical central group nematic efficiency order is

$$trans\text{-}CH=CH > N=N > CH=N > C\equiv C > N=N > CO\cdot O > none$$
$$\downarrow \qquad \qquad \downarrow$$
$$O \qquad \qquad O$$

[†] The broad principles discussed in this section have been discussed and reviewed by Gray (1974, 1976, 1979).

In accordance with the general requirement of an elongated and fairly rigid molecular structure, the following points were also established:

(1)   Any increase in a and/or b rapidly elevates the T_{N-I}.

(2)   Replacement of one of the 1,4-disubstituted benzene rings by a 2,6-disubstituted naphthalene ring giving **V** gives T_{N-I} values intermediate between those for structure **IV** with $a = 1$ and $a = 2$, consistent with the molecule having been elongated but broadened.

(**V**)

(3)   Substitution of any of the lateral H atoms of the rings in **IV** by a bulkier substituent strongly depresses T_{N-I}, the effect being proportional to substituent size, i.e., $H < F < Cl \approx CH_3 < Br < I \approx NO_2$, but independent of the permanent ring–X dipole. The magnitude of the decrease for a given substituent increases with decreasing length of the elongated molecule.

The systems studied in most detail (Gray and Worrall, 1959) were acids of structure **VI**, which exhibited nematic and smectic C phases.[†] The smectic C

(**VI**)

phases were affected differently, and the $T_{S_C-N\,or\,I}$ values were influenced in the order

$$H < F < Cl < Br \approx NO_2 < CH_3 < I$$

suggesting that the ring–X dipole is unimportant in relation to nematic properties but does counteract the broadening effect in relation to S_C properties, i.e., lateral dipoles act to stabilize S_C phases (see also p. 29).

(4)   Substitution of a lateral H atom in a protected position in the side of a molecule can enhance both N and S_C properties. For example, in a naphthalene system whose molecular perimeter is defined by the dashed line, replacement of the H atom by a substituent X does not broaden the molecule

---

[†] In a smectic C phase, the long molecular axes are inclined to the layer planes and the molecular centers are disordered in the planes of the smectic layers.

unless X is large. Thus space is more effectively filled by X, and higher T_{N-I} and T_{S_C-N} values are observed with X = F, Cl, and Br than with X = H; X = I does somewhat diminish T_{N-I}, but T_{S_C-N} is again higher.

(5) Conversely, lateral substitution causing a steric effect that results in a (further) twisting of the molecular structure, e.g., about the inter-ring bond in a biphenyl derivative

causes very large decreases in nematic and particularly smectic thermal stabilities. For example, a small fluoro substituent in a nonsterically affecting position in a molecule of considerable length, e.g., **VI**, causes decreases of as little as 0.5° and 9° in smectic and nematic thermal stability, respectively, but in a sterically affecting situation, e.g., in **VII**, these decreases become 71°

(**VII**)

and 49°, respectively. Note that the smectic properties are the more strongly repressed.

(6) Branching of the terminal alkyl chain also diminishes the liquid crystal tendencies; the effect is most pronounced when chain branching is at the carbon next to the aromatic ring and diminishes as the branching group is moved to the chain terminus. For example, for structure **VIII** T_{N-I} is

(**VIII**)

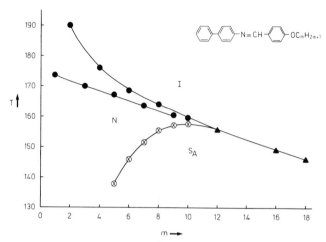

Fig. 1. Plot of transition temperatures against the number of carbon atoms (m) in the alkyl chain for the series of 4-p-n-alkoxybenzylideneaminobiphenyls. (From data in Gray et al., 1955.)

136.5°C. When a hydrogen is replaced separately by a CH_3 group at C_1, C_2, C_3, and C_4, the T_{N-I} values become <20, 112, 103, and 119.5°C, respectively. Smectic A[†] properties are affected less markedly by such chain branching, and certain more ordered smectic phases such as B and E may have their thermal stabilities enhanced.

(7)   The 1,4-disubstituted benzene rings in the core structures of mesogens can be replaced by heterocyclic rings such as pyridyl or pyrimidinyl. The effects are variable from system to system. In some cases T_{N-I} values are enhanced, but the general effect is toward more pronounced smectic behavior.

(8)   Extension of an n-alkyl chain occupying a terminal position in the molecule of a mesogen gives rise to regular trends in the liquid crystal transition temperatures. T_{N-I} values alternate regularly (Fig. 1) but with a diminishing amplitude as the chain grows in length, and it was noted that when the series has high T_{N-I} values (>100°C), the T_{N-I} values lie on two curves, which fall more steeply at first and then level out. With T_{N-I} values that are low (<100°C), the two T_{N-I} curves rise steeply at first and then level out. With series with T_{N-I} values about 100°C, the T_{N-I} curves are often fairly flat. Smectic properties were consistently found to become more marked as the chain lengthened, the S–N temperatures usually lying on a rising curve that eventually leveled off to merge with the T_{N-I} curve and give

[†] A smectic A phase is the orthogonal analog of a smectic C. For a brief review of molecular organization in the various smectic polymorphic phases see Coates and Gray (1976).

direct smectic–isotropic transitions. Most of these observations related to S_A or S_C phases. Figure 1 shows a typical example, and Gray (1976) describes other cases.

The regular trends in the T_{N-I} curves have been interpreted in the following terms. At low temperatures the chain extends with some statistical preference for a regular zigzag conformation, maintaining the best molecular shape for the nematic structure. Thus the anisotropy of molecular polarizability ($\Delta\alpha$) increases as the chain lengthens, but the increase is greater on passing from even to odd than from odd to even alkyl chains. Therefore the T_{N-I} curves rise, and the values for the odd homologs lie on the upper of the two curves. Note that for ethers, where the O atom replaces the first CH_2 group next to the ring, the even members are on the upper curve.

At high temperatures thermal effects increase the statistical probability that the alkyl chain will adopt other than the extended all-trans conformation. To allow for this an expression for T_{N-I} was proposed (de Jeu et al., 1973) such that

$$T_{N-I} = 2A/(4.54k - 2B)$$

where A is proportional to the anisotropy of molecular polarizability ($\Delta\alpha$), B is in some way proportional to the effective molecular length–breadth ratio and k is, of course, Boltzmann's constant.

With longer chains and high temperatures, the decrease in B may more than compensate for any increase in A, and T_{N-I} will fall as the chain length grows.

At low temperatures, B and A may both increase, giving, as said above, an increase in T_{N-I} with chain length.

Such considerations could be applied to a large number of homologous systems, although when smectic properties occurred in the series, they were in most cases limited to the S_A and S_C types, and in a smaller number of cases to the S_B and S_E types. As a result, alternating transition temperatures and smooth curve relationships became accepted for a range of transition types, and, knowing the behavior of two or three members of a series, it was possible to make reasonable predictions for the transition temperatures of other members.

The effects of a range of lateral and terminal substituents were also fairly predictable, and indeed an additivity of effect when more than one lateral substituent was present appeared to operate (Byron et al., 1965; Van Meter, 1973; Van Meter and Klandermann, 1973; Young et al., 1972a,b; Young and Green, 1972, 1974). Likewise, extension of molecular length and chain branching had consistent effects. It seemed then that the empirical rules, most of which followed the concepts of Maier and Saupe (1958a,b, 1960) theory, i.e., that T_{N-I} is determined by dispersion forces, allowed resonable predictions to be made when assessing the probable behavior of new mesogens.

In fact, the synthesis of the commercially important 4-alkyl-4'-cyano-biphenyls, and 4-alkoxy-4'-cyanobiphenyls in the early 1970s was undertaken because it could be predicted that they would be interesting systems (Gray et al., 1973a, 1974, 1975; Gray, 1975). It will be remembered that the stimulus for the development of these systems was the demand of the display device industry for good, stable nematic LC materials of strongly positive dielectric anisotropy ($\Delta\varepsilon$). Our approach was as follows.

First, with reference to structure **IV**, we wished to do away with the central A—B linkage, from which it seemed that all the problems—sensitivity to UV, oxidation, hydrolysis, isomerization, etc.—with earlier materials arose. Since low mp materials were wanted, the structure had to be kept simple, i.e., a and b in **IV** would desirably be 1. These considerations led us therefore to the simple biphenyl system **IX** with no central unit other than

(IX)

the inter-ring C–C bond. Since such a system could have a very low $T_{N–I}$, we fixed Y as CN, (i) to elevate $T_{N–I}$, and (ii) to make $\Delta\varepsilon$ strongly positive.

From the relative behaviors of compounds **IV** with a —CH=N— linkage and those (derivatives of p-terphenyl, in fact) with no central linking unit, we knew that removal of the —CH=N— unit reduced $T_{N–I}$ by about 50°. Knowing the properties of the compounds

we were in a position to expect that the compounds **X** would have $T_{N–I}$ values of about 60–70°C, and that the compounds **XI** would have $T_{N–I}$

(X)   **(XI)**

values of about 30–40°C. The mps of the compounds could not, of course, be predicted with any certainty. This is a built-in problem in materials research concerning LC behavior, and may also be a problem in the design of practically useful LC polymers.

In fact, the compound **X** with R = n-C_5H_{11} had $T_{C–N} = 53°C$ and $T_{N–I} = 67.5°C$ and the compound **XI** with R = n-C_5H_{11} had $T_{C–N} = 22.5°C$, $T_{N–I} = 35°C$.

The commercial success of these materials and related homologs for use in twisted nematic displays resulting from this analytical approach to the situation therefore raised confidence in the ability to predict LC behavior. Indeed, taking the story a little further, when chiral analogs of these systems were needed, so that cholesteric analogs (Gray *et al.*, 1973b; Gray and McDonnell, 1975, 1976) with the same desirable properties could be obtained, the $T_{Ch–I}$ values of systems of general structure

$$CH_3CH_2CH(CH_3)(CH_2)_n \text{—} \langle \bigcirc \rangle \text{—} (\langle \bigcirc \rangle)_b \text{—CN}$$

where $n = 1, 2, 3, \ldots$, and $b = 1$ or 2 (and the related alkoxy compounds), could be predicted with reasonable success.

Incidentally, this study, coupled with work by Billard and colleagues (personal communication), gave rise to the SOL/SED (ROD/REL) rule (Gray and McDonnell, 1977), which allows one to predict the handedness of the cholesteric helix (L or D) in relation to the asymmetry of the chiral center (S or R) and its position (odd or even) relative to the ring (O or E). Hence the compounds

$$\left. \begin{array}{l} CH_3CH_2CH(CH_3)CH_2 \\ \text{and } CH_3CH_2CH(CH_3)CH_2O \end{array} \right\} \text{—} \langle \bigcirc \rangle \text{—} \langle \bigcirc \rangle \text{—CN}$$

give cholesteric helices with opposite handedness, D and L, respectively, for the S configurations.

However, the biphenyl systems **X** and **XI** did give rise to one disturbing feature. The plots of $T_{N–I}$ against chain length for both series were expected to be rising curves. Instead, the curves fell quite steeply from CH_3 and C_2H_5 or CH_3O and C_2H_5O to minima at 3- or 4-carbon chains and then rose, finally leveling off for longer chains (see Fig. 2). To explain this, suggestions have been made (Gray, 1979; Gray and Mosley, 1976) that, with chains longer than four carbons, there may arise a statistically significant population of conformers (which may be energetically unfavorable if we are considering the isotropic liquid or the gas phase) with more elongated structures that are possibly therefore energetically favorable to the ordered nematic state. For instance, with $R = C_4H_9$ and C_5H_{11} we may have

$$\text{C}\overset{\text{C}}{\underset{\text{C}_3}{\text{—}}}\overset{\text{C}}{\underset{2}{\text{—C}}}\text{—}\langle \bigcirc \rangle \text{—}\langle \bigcirc \rangle \text{—CN} \quad \text{and}$$

$$\text{C—C}\underset{\text{C}_3\text{—C}_2}{\overset{}{\diagdown}}\text{C—}\langle \bigcirc \rangle\text{—}\langle \bigcirc \rangle\text{—CN}$$

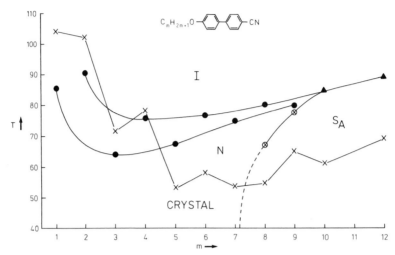

Fig. 2. Plot of transition temperatures against the number of carbon atoms (m) in the alkyl chain for the series of 4-n-alkoxy-4'-cyanobiphenyls. (From data in Gray, 1978.)

by a rotation about the 2–3 bond in the pentyl homolog. This gives a better on-axis development of the chain than does the more usual, extended all-trans conformation.

Because of their commerical importance, the cyanobiphenyls have been the subject of extensive physical studies, and it has emerged that there is a strong tendency to a loose molecular pairing (Leadbetter *et al.*, 1975, 1977) in these systems, such that an antiparallel distribution of the CN dipoles occurs. Each cyano group in a pair is located close to the opposite end of its neighbor.

This association is dominant in the S_A phases of these compounds, which have a lamellar spacing $d \approx 1.4 \times l$ (where l = molecular length). However, x-ray studies confirm that the pairing persists in the nematic state and even in the isotropic liquid. The effective molecular length is therefore enhanced and this may explain the strong promotion of T_{N-I} by terminal CN groups. The effect is general for a variety of terminally cyano substituted mesogens.

Considering polymers, if the side groups have terminal CN groups, the antiparallel pairing of cyano groups attached to different polymer backbones may be less easily achieved, or, once achieved, may give rise more readily to smectic ordering. The influence of terminal CN groups in low MW LC systems in promoting T_{N-I} and giving weak smectic behavior (no S phases are observed in systems **X** and **XI** until the C_8 members) may therefore be somewhat different in the polymer LC environment.

B. Alicyclic Systems

1. Structural Aspects

The fairly low T_{N-I} values of the cyanobiphenyls **X** and **XI** gave little scope for structural modification by lateral substitution of the molecules, although this was possible in the case of the related terphenyl systems, used as high T_{N-I} additives.

The first major development of the biphenyl class of mesogen came therefore with the preparation of the cyclohexane analogs (Eidenschink *et al.*, 1977) of structure **XII** where *a* may be 1 (the PCH compounds) or 2 (the BiCH compounds).

(XII)

The idea behind this work stemmed from studies by Deutscher *et al.*, (1974), who replaced one of the benzene rings in simple dialkyl benzoate esters by a trans-1,4-disubstituted cyclohexane ring to obtain **XIII**.

(XIII)

Surprisingly, these had higher T_{N-I} values than the analogous benzoate esters. This was contrary to the expectation that T_{N-I} would fall through the resulting decrease in anisotropy of the molecular polarizability ($\Delta\alpha$).

However, compounds **XII** with *a* = 1 also showed this trend and had higher T_{N-I} values than the biphenyl analogs **XI**, e.g., R = C_5H_{11}, T_{N-I} (biphenyl) = 35°C and T_{N-I} (PCH) = 55°C. The higher T_{N-I} BiCH materials however had lower (20–27°C) T_{N-I} values than the corresponding terphenyls. These results have since been confirmed (Gray, 1981; Carr *et al.*, 1981; Eidenschink, 1979) for a variety of systems with varying structures, and it is now accepted that replacement of a benzene ring in a nematogen by a trans-1,4-disubstituted cyclohexane ring will enhance T_{N-I}, unless this is already rather high ($\gtrsim 220°C$), when the change will diminish T_{N-I},

This disadvantageous effect of the cyclohexane ring at high temperatures may be due to the flexibility of the ring and its ability to adopt skewed or twisted conformations that are higher in energy and less conducive to nematic order.

With regard to polymers, the advantages or disadvantages of using cyclohexane rings would also depend on T_{N-I}. However, there are other features of using cyclohexane rings that may be important. The viscosity of low MW LC systems is lowered, e.g.,

$$\textbf{XI};\quad R = C_5H_{11},\quad \eta_{20°}\text{ (nematic)} = 32\quad cP$$
$$\textbf{XII }(a = 1);\quad R = C_5H_{11},\quad \eta_{20°}\text{ (nematic)} = 21\quad cP$$

Also, the birefringence (Δn) is lowered from about 0.25 to 0.14, although so also is $\Delta\varepsilon$ (11 to 10). The tendency to molecular pairing still occurs, but in the PCH series no smectic properties occur up to and including the C_9 member. The shapes of the T_{N-I} curves are very different for the PCH series. Both curves rise very steeply from the members with one and two carbon atoms and eventually level off (Fig. 3). Thus a knowledge of the relative T_{N-I} values of **XI** and **XII** with $a = 1$ and $R = CH_3$, $T_{N-I} = +45 \pm 3$ and $-25°C$, respectively, would lead to quite wrong conclusions about the relative T_{N-I} values of the higher homologs, such as $R = C_5H_{11}$, $T_{N-I} = 35$ and $55°C$, respectively.

Alicyclic systems are, as a consequence, attracting a great deal of attention. In my own group, we have directed attention to the more rigid, axially symmetric, 1,4-disubstituted bicyclo[2.2.2]octane (BCO) ring system. Since this is quite bulky, and Dewar and co-workers (Dewar and Goldberg,

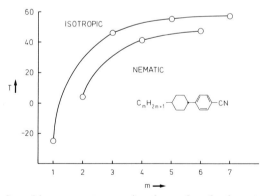

Fig. 3. Plot of transition temperatures against the number of carbon atoms (m) in the alkyl chain for the series of *trans*-1-(4'-cyanophenyl)-4-*n*-alkylcyclohexanes. (From data in Eidenschink *et al.*, 1977.)

1970a,b; Dewar *et al.*, 1974; Dewar and Riddle, 1975; Dewar and Griffin, 1975) had shown that replacement of any of the benzene rings (particularly the end rings) in the diesters **XIV** by a BCO ring depressed T_{N-I}, we chose

(XIV)

the alkoxy rather than the alkyl derivatives and therefore first prepared the compound

We found (Gray and McDonnell, 1978) that this was a most disappointing material. It had an mp of $\sim 50°C$ and a virtual T_{N-I} value below room temperature. This did not augur well for the alkyl analogs, but we did make compound **XV**, which in fact melts at 60°C and has $T_{N-I} = 100°C$. This

(XV)

was confirmed by preparing other homologs and the terphenyl (3-ring) analogs (Gray and Kelly, 1981a,b). Except for the first two homologs of the series **XV**, for which the relative T_{N-I} values are again determined by the shapes (similar to those for the PCH series) of the T_{N-I} curves, the BCO ring is more favorable for the maintenance of nematic order against the disruptive effects of heating than is a benzene or a cyclohexane ring. This is now an established situation for eleven different mesogenic systems (Gray, 1981; Carr *et al.*, 1981).

The cyano-substituted BCO systems again have a strong tendency to molecular pairing, and there therefore seems to be no major structural difference on which to base the explanation of the high T_{N-I} values. Again therefore we have a behavior contrary to that anticipated on the basis of $\Delta\alpha$, and indeed experimental results by Dunmur and Tomes (1980) and by Ibrahim and Haase (1981) confirm the intuitive view of the organic chemist that $\Delta\alpha$ in fact increases in the order

$$\text{cyclohexane} \longrightarrow \text{BCO} \longrightarrow \text{benzene}$$

as do the isotropic polarizabilities.

Such results have not yet been fully explained, nor have the observations that (a) the CCH compounds **XVI** (Eidenschink *et al.*, 1978) have T_{N-I}

values between those of the PCHs and the BCO compounds, and (b) the T_{N-I} values of the compounds **XVII** (Eidenschink, 1979) and **XVIII** (Gray

(XVI)

(XVII)

(XVIII)

and Kelly, 1982a) are much lower (-20 and $50°C$, respectively) than those of the analogs with the rings in the opposite sequence.

The pairing situation in the compounds **XVI** is certainly different (Leadbetter, 1982); only the two cyano groups overlap, and the greater length of the "pair" may explain the higher T_{N-I} values. The pairing behavior of **XVII** and **XVIII** is not known, but would need to be very different than that for the CCHs, e.g., an end-to-end antiparallel arrangement, to explain the lower T_{N-I} values.

In the cases of compounds **XI**, **XII**, and **XV** it has been suggested (Gray, 1981; Carr *et al.*, 1981) that in the paired arrangement repulsive interactions between an alicyclic ring and a nearby CN or other dipolar group are less than in a fully aromatic system. Also it has been proposed (Gray, 1981; Carr *et al.*, 1981) that the bulkier cyclohexane and BCO rings more effectively fill space and give a better packing arrangement. In this context, Leadbetter and Mehta (1981) have proposed that, although $d \approx 1.4l$ for all three systems, the degree of core overlap for **XII** and **XV** may be less, but is coupled with an interdigitation of the end chains, which it is proposed does not occur for the compounds **XI**. This difference in chain packing could then be connected with the radically different shapes of the T_{N-I} curves for the system **XI** and the systems **XII** and **XV**, and the packing arrangements may in turn determine the order of the T_{N-I} values. Alternatively, Eidenschink (1979, 1980) has developed the novel idea that the molecules may be regarded as masses of coupling emitters and receivers of energy and he treats them as antennae. His arguments, not easy to follow in places, claim to rationalize the above uncertainties relating to T_{N-I} values.

Whatever the reasons may be, the T_{N-I} order (except in some cases for the first two homologs of a series

$$BCO > cyclohexane > benzene$$

applies to a wide variety of molecule types (Gray, 1981; Carr *et al.*, 1981). Therefore the T_{N-I} order must be explained in terms other than those involving the cyano group. Examples of these other systems are:

$$R-\underset{}{\boxed{A}}-CO\cdot O-\hexagon\begin{cases}CN\ or\\R'\ or\\OR'\end{cases}$$

$$R-\underset{}{\boxed{A}}-CH_2O-\hexagon\begin{cases}CN\ or\\R'\ or\\OR'\end{cases}$$

$$R-\underset{}{\boxed{A}}-\hexagon\begin{cases}R'\ or\\OR'\end{cases}$$

$$R-\underset{}{\boxed{A}}-CH_2CH_2-\hexagon\begin{cases}CN\ or\\R'\ or\\OR'\end{cases}$$

This is important because in modern LC electro-optical displays multiplexed driving of the display is highly desirable. This requires that the nematic LC have a sharper threshold and a lower temperature dependence of the threshold voltage than can be achieved with available cyano-substituted nematogens of strong positive $\Delta\varepsilon$. The accepted technique uses mixtures of cyano compounds with other materials (azoxy compounds or esters) of lower positive or even weakly negative $\Delta\varepsilon$. Of these, esters with terminal alkyl or alkyl and alkoxy groups are most effective. It is proposed that the ester molecules interfere with the molecular pairing and give rise to beneficially low values of $\Delta\varepsilon/\varepsilon_\perp$ and of k_{33}/k_{11} (the ratio of bend and splay elastic constants) (Bradshaw and Raynes, 1981). Moreover, the esters (Gray and Kelly, 1981c)

$$R-\boxed{}-CO\cdot O-\hexagon-OR$$

give the lowest value $(0.13\%/°C)$ of $(1/V)(dV/dT)$ recorded for any additive of this type (Constant and Raynes, 1981).

Since molecular pairing is undesirable for multiplexed driving of displays, if this does not occur in cyano-substituted comb polymers, this may be a desirable feature for such high MW LC systems. Alternatively, the concept of copolymers with alternating cyano- and noncyano- terminally-substituted side groups seems an attractive thought.

Incorporation of cyclohexane and BCO rings in the side groups in polymers is therefore of interest from several viewpoints, and we are currently working on this. It should also be remembered that the relatively high T_{N-I} values that can now be achieved using alicyclic analogs of very simple aromatic molecules allows much greater scope for the use of structural features that could not possibly be used in fully aromatic, low MW LC systems. An example makes this point clear. Consider the esters shown in Table I.

TABLE I

	$T_{C-N}(°C)$	$T_{N-I}(°C)$
C_5H_{11}—⬡—CO·O—⬡—C_5H_{11}	36	(26)
C_5H_{11}—⬡—CO·O—⬡—C_5H_{11} (**XIX**)	36	48
C_5H_{11}—⬡—CO·O—⬡—C_5H_{11} (**XX**)	31	64.5[a]

[a] From Constant and Raynes (1981).

The disadvantage for all of these esters is their relatively high mps. But lateral substitution can often decrease the mp, particularly if an intermolecular steric effect is involved. This can be achieved by introduction of a 2-fluoro substituent. However, this depresses T_{N-I} for the benzoate esters by 25°C and gives a very low monotropic T_{N-I} value of 1°C. With the cyclohexane esters (Gray et al., 1981), the effect on T_{N-I} is less (average = 11°), and with the BCO ester (Gray and Kelly, 1981d), the fluoro substituent is very effectively shielded and T_{N-I} may even rise for some homologs by as much as 2° (the average decrease is 3°).

The 2-fluoro analogs of the esters **XIX** and **XX** have the constants $T_{C-N} = 17.5°C$, $T_{N-I} = 36.5°C$ and $T_{C-N} = 26°C$, $T_{N-I} = 65°C$, respectively, and many homologs of these two series of esters give room temperature nematic phases and good nematic ranges. Although generally the viscosities of BCO systems are greater than those of cyclohexane systems, the effect is not too pronounced for these esters [$\eta_{20°}$ nematic **XIX** = 23 cP; $\eta_{20°}$ nematic **XX** = 34 cP].

Also, flexible linking units can be tolerated between the rings. Whereas for fully aromatic systems, compounds such as

$$X—⬡—CH_2CH_2—⬡—Y$$

have very low T_{N-I} values, use of alicyclic rings compensates for the linkage and materials such as

$$C_5H_{11}—⬡—CH_2CH_2—⬡—OC_2H_5$$

$T_{C-N} = 18°C$, $T_{t-I} = 47°C$, $\eta_{20°}$ (nematic) = 13 cP have good T_{N-I} values and are also of low viscosity (Carr and Gray, 1982).

Perhaps the lowest viscosity effects are found with the 'compounds (E. Merck, Darmstadt, W. Germany)

(particularly $R = C_3H_7$, $R' = C_2H_5$) for which a viscosity of $\eta_{20°}$ (isotropic) $= 4$–5 cP[†] has been reported; although the T_{N-I} values are strongly monotropic ($< -45°C$), such isotropic liquids are valuable as additives to nematics to lower viscosity.

The curious behavior of the compound

referred to earlier, must now be discussed since it raises the important fact that a group RO is a good terminal group for use with aromatic rings— compare the behavior of the compounds **X** and **XI**—but is a poor terminal group to attach to an alicyclic ring.[‡] Indeed, systems involving the structures

(where A is cyclohexane or BCO) have low mesogenic properties (Gray, 1981; Carr *et al.*, 1981). For the same reason there is an enormous difference in behavior between the following compounds (Carr *et al.*, 1981; see also Carr, 1982): because the latter involves an ether-oxygen to alicyclic-ring function.

$T_{C-I} = 60°C$, $T_{N-I} = [26°C]$[§]   and $T_{C-I} = 63°C$, $T_{N-I} = [-210°C]$[§]

The attachment of RO to an aromatic ring will permit conjugative interactions with the ring. These cannot occur in the alicyclic systems, and it is natural to associate this with the stabilization of the nematic order in the aromatic ethers, i.e., in terms of a higher $\Delta\alpha$ for the aryl ethers. Alternatively, different ring —O—CH$_2$ angles or different dipole distributions may underlie the observed effects. The effect is, however, quite general, and applies also to the movement of the —CH$_2$O— link from next to an aryl

---

[†] Note that η_{nematic} is always smaller than $\eta_{\text{isotropic}}$ (extrapolated to an equivalent temperature).
[‡] See also the effects observed by Dewar for the diesters **XIV** with terminal —OR groups.
[§] Square brackets indicate virtual T_{N-I} values obtained by extrapolation of data from mixtures.

ring to a situation such as we have, for example, in

$$CH_3(CH_2)_nO(CH_2)_m\text{—}\langle\bigcirc\rangle\text{—}$$

It is therefore an aliphatic–O–aliphatic arrangement that depresses T_{N-I} and an alicyclic ring system need not be involved.

2. Other Physical Properties Stemming from the Use of Alicyclic Ring Systems

The commercial interest in novel nematogens has had the consequence that their physical properties have been extensively studied, notably those properties that have a bearing on their use in electro-optical display devices. Thus a growing body of data is becoming available and is making it possible to begin to compare the behavior of related systems.

Perhaps the most fully studied systems are those with the general structure

$$R\text{—}\langle A \rangle\text{—}\langle\bigcirc\rangle\text{—}CN$$

where ring A may be a benzene ring (CB), a cyclohexane ring (PCH), or a bicyclooctane ring (BCO).

Bradshaw *et al.* (1981) recently reviewed the property changes accompanying the change in nature of ring A. We already know that T_{N-I} decreases strongly in the order

$$BCO > PCH > CB$$

but from the work of Bradshaw *et al.* (1981), we can now make the following additional comparisons.

a. Refractive Indices and Birefringence

$$CB \;>\; PCH \approx BCO$$
$$\Delta n \approx 0.25; \qquad \approx 0.14$$

b. Order Parameter (S)

$$BCO > PCH > CB$$

The reduced temperature dependence of S has also been determined and a steady lowering in

$$(-1/S)(dS/dT)$$

occurs from 0.69%/°C for CB, to 0.59%/°C for PCH, to 0.50%/°C for BCO, all at 0.9 T_{N-I}.

c. Viscosities.
The bulk viscosities of the BCO compounds are much greater (≈ 100 cP at 20°C) than those for the PCH and CB compounds (≈ 27 cP at 20°C and 35 cP at 20°C, respectively). The activation energies of CB and PCH materials in both the nematic and isotropic phases are

similar; the activation energy for the BCO system in the nematic phase is, however, much higher.

d. Elastic constants

(i) k_{11} (splay elastic constant). All have similar values at 0.95 T_{N-I}, but the dependence of k_{11} on temperature varies such that $(-1/k_{11})(dk_{11}/dT)$ changes in the order

$$CB \quad > PCH > BCO$$

$$1.49 \quad 1.23 \quad 0.90 \quad \%/°C \text{ at } 0.9 \ T_{N-I}$$

(ii) k_{22}/k_{11} (where k_{22} = twist elastic constant). All the systems have similar values, but the fully aromatic CB compounds do give the lowest values.

(iii) k_{33}/k_{11} (where k_{33} is the bend elastic constant). Significant differences in this ratio arise such that

$$BCO \gg PCH > CB$$

Indeed the ratio for BCO approaches three—a very high value, which is not associated with the formation of the smectic phase.

e. Temperature Dependence of the Threshold Voltage for Twisted Nematic Display Operation at 0.9 T_{N-I}. $(-1/V_c)(dV_c/dT)$ shows

$$BCO < PCH < CB$$

$$0.51 \quad 0.79 \quad 1.21 \quad \%/°C \text{ at } 0.9 \ T_{N-I}$$

This parallels the trend in k_{11} with temperature and suggests that these trends in temperature dependence are explained by $(-1/S)(dS/dT)$.

And so we begin to learn more about the consequences of molecular structural change on important physical parameters other than T_{N-I}, but care must be exercised in extrapolating such results to other systems in which similar changes are made.

For example, consider the esters

Results on these esters (where ring A is varied in the same way from benzoate ester (BE), to cyclohexane ester (CHE), to bicyclooctane ester (BCOE)) are less complete, but data that do exist (Constant and Raynes, 1981) already show two significant differences. Thus, although the temperature dependence of V_c is again in the order BCOE < CHE < BE, the refractive indices and birefringence values are as would be expected, and T_{N-I} and S follow the order BCOE > CHE > BE, it is found that:

(i) Whereas the bulk viscosities of the nematic phases do follow the order BCOE > CHE > BE, the values for BCO esters are now much more comparable with those for the other two ester systems. This may suggest a shielding of the bulk of the BCO ring by the ester function. Thus a dialkyl BCO ester has quite a low viscosity at 20°C in the nematic phase (about 35 cP).

(ii)   The ratio of k_{33}/k_{11} for the dialkyl BCO esters (measured using a mixture of esters doped with 10% of a cyanobiphenyl to give $\Delta\varepsilon$ a positive value) was 1.1 at 20°C, comparable with values for similar mixtures of the other esters. The presence of the BCO ring in the esters does not therefore significantly alter k_{33}/k_{11}, so that such esters give a similar electro-optical steepness of the threshold to that given by the CH esters and B esters. Even these early results must therefore warn us against making sweeping generalizations about effects of structure change on physical behavior. Rather a steady accumulation of data must be awaited and careful comparisons made on the basis of the established facts. It would seem likely that a similar situation would apply to LC polymers.

C. Cholesteric Systems

Since the cholesteric phase is a spontaneously twisted analog of the nematic phase, it is not surprising that molecular structural changes affect the transition temperatures of cholesteric and nematic phases in the same way. Consequently, the above generalizations about structure change in relation to T_{N-I} values apply to T_{Ch-I} values.

Only the pitch of the twist and its sense need therefore be considered additionally. Twist sense is very sensitive to structure change and for cholesterogens derived from sterols such as cholesterol the situation has been discussed by Baessler and Labes (1970). However, for commercially useful cholesterogens we do not usually want to have the presence of a bulky and often UV-sensitive steryl skeleton, and most of the important low MW materials are simply chiral analogs of nematogens, the chirality being introduced by having a chiral centre in a terminally situated alkyl or alkyoxy group. The effect of moving this center up to or away from the neighboring ring (aromatic or alicyclic) of the core structure has already been discussed in terms of the SOL/SED (ROD/REL) rule, and typical examples have been quoted.

Turning therefore to twisting power, this decreases as the chiral center is moved away from the ring system, i.e., as n is increased in a structure such as

$$CH_3CH_2\overset{*}{C}H(CH_3)(CH_2)_n - \hspace{-0.5em} \bigcirc \hspace{-0.5em} - \hspace{-0.5em} \bigcirc \hspace{-0.5em} - \hspace{-0.5em} \bigcirc \hspace{-0.5em} - CN$$

Practically speaking, the best value of n to use is 1. This achieves a good tightness of the pitch without involving the very great reduction in T_{N-I} that arises with $n = 0$.

It is also noted that for an equivalent separation of the chiral center from the ring system longer pitch values are found for aryl ethers than for the aryl–alkyl analogs, i.e., replacement of the ring —CH_2— by ring —O— decreases the twisting power.

Attachment of a chiral alkyl group to a cyclohexane ring is found to give considerably less twisting than in the case of the benzene analog. This is not the case for the more rigid bicyclooctane ring, and suggests that the greater flexibility of the cyclohexane ring may act against the twisting effect of the chiral center.

Results are now beginning to emerge on chiral BCO systems (Gray and Kelly, 1982b), and the following examples are of some interest:

R—[BCO]—CO·O—[ring, F]—$CH_2CH(CH_3)CH_2CH_3$

(a)   (selective reflection in UV at room temp) R = C_8H_{17}, T_{C-Ch} = 22°C, T_{Ch-I} = 36.5°C

R—[BCO]—CO·O—[ring, F]—$CH_2CH_2CH(CH_3)CH_2CH_3$

(b)   (selective reflection of visible light at room temp) R = C_7H_{15}, T_{C-Ch} = 24°C, T_{Ch-I} = 32.5°C

The cholesteric phase of this ester supercools for long periods at room temperature, at which it reflects orange light. The color reflected changes to green near the Ch–I temperature. Note how advantage has been taken of the effect of increasing pitch by moving the chiral center away from the ring to convert the non-color-reflecting phases of the esters (a) into the thermochromic materials (b).

(c)   The esters analogous to those in (a) but without the fluoro substituent also reflect UV light from their cholesteric phases at room temperature. On the other hand, the esters

RO—[ring]—CO·O—[ring]—$CH_2\overset{*}{C}H(CH_3)CH_2CH_3$

reflect visible light from their cholesteric phases. If a comparison of these esters with

R—[BCO]—CO·O—[ring]—$CH_2\overset{*}{C}H(CH_3)CH_2CH_3$

is fair, then the BCO ring encourages a tighter twisting than a benzene ring when the chiral center is on the right of the molecule.

(d)   However, when the chiral group is attached directly to the BCO ring, it is clear that a lower twisting power arises than when the same chiral group is attached to the benzene ring of the phenolic moiety. Thus the 2-methylbutyl

$$CH_3CH_2CH(CH_3)CH_2- \bigcirc\hspace{-0.5em}\bigcirc -CO\cdot O- \bigcirc\hspace{-0.5em}\bigcirc -R$$

derivatives ($R = C_7H_{15}$, $T_{C-Ch} = 3.5°C$, $T_{Ch-I} = 9.5°C$[†]) selectively reflect colored light from their cholesteric phases. Contrast the esters in (a) with reversed terminal substituents, which reflect light in the UV. The above ester with $R = C_7H_{15}$ has a cholesteric phase that appears red at $0°C$, the color shifting toward blue as the temperature rises.

Finally, it is noted that the esters in (b) represent the first pure (i.e., single component) systems of a nonsteryl type to give selective reflection of colored light from their cholesteric phases at room temperature.

D. Smectic Systems

In low MW LC systems of commercial interest one usually wishes to avoid smectic properties. This is not always the case, however, for latent smectic properties can be valuable for increasing the sensitivity of the color response of cholesteric phases to temperature. Also, certain display devices require room temperature smectic A phases, e.g., the laser-writing blackboard (Taylor and Kahn, 1974), and chiral smectic C phases hold out promise of a very fast switching display (Meyer et al., 1975; Clark and Lagerwall, 1981) based on the ferroelectric properties of such phases.

The problem with smectic phases is their extensive polymorphism. Nine types, S_A to S_I, are now well established (Luckhurst and Gray, 1979), and two more probably exist (Gane et al., 1981). Of these, some are uncorrelated 2-dimensional smectics, i.e., the structural arrangement in a given layer is not correlated[‡] with that in neighboring smectic layers. Clearly, this is the case where there is a disordered arrangement in the layer planes (S_A and S_C), but this group also includes S_F and S_I (Gane et al., 1981), and some S_B phases. Long range structural correlation (3-dimensional) over many layers occurs with some S_B phases, and with S_E, S_G, and S_H phases.[§]

[†] Note the greater effect of chain branching on T_{Ch-I} when the chiral group is on the BCO ring; for the reversed system in (a) with $R = C_7H_{15}$, T_{Ch-I} is $34.5°C$.

[‡] However, the tilt direction in well-aligned tilted phases may be the same from one layer to another.

[§] The S_D phase is unique in being optically isotropic and appears to involve a structure with an effective cubic symmetry.

Although correlated smectic phases can be envisaged for LC main chain polymers, these structures would be so akin to solids that it is doubtful if, like correlated low MW smectics, these would be of commercial interest. With LC side chain polymers it is hard to envisage long range correlation of the smectic ordering of the side chain groups that would not be interfered with by the polymer backbone.

It would seem, therefore, that the smectic characteristics most likely to arise, and possibly to be avoided, in LC polymers would be those of the A, the uncorrelated B, and the C, F, and I types. Certain structural points may usefully be summarized at this point (Table II).

TABLE II

Uncorrelated Smectic Phases

Phase Type	Tilt	Structure in Layer	Tilt Direction in Layer
A	None	Disordered	—
B	None	Hexagonal	—
C	Tilted	Disordered	Uniform in aligned samples
F	Tilted	Pseudohexagonal	To sides of hexagonal net (Gane et al., 1981)
I	Tilted	Pseudohexagonal	To apices of hexagonal net (Gane et al., 1981)

We are therefore concerned in this section with those structural features that encourage the molecules to pack not only parallel, but also in a manner such that layers are formed. Factors affecting the tilt situation in the layers are fairly subtle (Van der Meer and Vertogen, 1979).

Surprisingly in some ways, it is not possible to draw up a well defined list of molecular features that promote smectic properties irrespective of type. Even the well established ground rule that increasing the length of terminal alkyl or alkoxy chains will enhance smectic behavior (see Figs. 1 and 2) is not without exception, for in the CCH series [structure **XVI**], the reverse applies; whereas the C_7 member is purely nematic, the earlier homologs form smectic $B^†$ and nematic phases. This may be a unique case, and studies by Leadbetter et al. (1982) suggest that the comparable sizes of the CN and C_3H_7 groups are responsible for the particularly marked tendency of that homolog to align in a smectic manner. However, the occurrence of even one exception to a general rule must necessarily make us cautious about generalizing on the overall subject of structure versus smectic tendencies.

This case does serve to focus attention on why it is difficult to lay down firm guidelines on structural features that determine such a basic matter as

---

† The C_3 and C_5 members also give even more ordered S phases below their S_B phases.

whether a compound will or will not exhibit smectic properties. We are trying to decide those structural features that will determine whether or not parallel molecules will prefer to arrange themselves with their centers lying in a plane, and this is clearly a matter of molecular packing. It is therefore hardly surprising that subtle changes in molecular architecture may bring about very large changes in smectic LC character.

The sensitivity of smectogenic character to structure cannot be more clearly illustrated than by reference (Goodby *et al.*, 1980) to the influence of simply increasing (one CH_2 group at a time) the length of one terminal alkyl chain in a molecule such as

$$C_5H_{11}O-\!\!\bigcirc\!\!-CH\!\!=\!\!N-\!\!\bigcirc\!\!-C_mH_{2m+1}$$

As can be seen from Fig. 4, smectic properties do in this case increase at the expense of nematic properties as *m* is increased, but note the variations

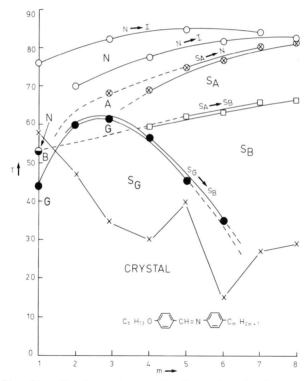

Fig. 4. Plot of transition temperatures against the number of carbon atoms (*m*) in the alkyl chain for the series of *N*-(4′-*n*-hexyloxybenzylidene)-4-*n*-alkylanilines. (From Goodby *et al.*, 1980.)

in smectic behavior over the first four members, which, on cooling, give the following phase sequences:

$m = 1$ Isotropic $\rightarrow$ N $\rightarrow$ S$_B$ $\rightarrow$ S$_G$
    2   Isotropic $\rightarrow$ N $\rightarrow$ S$_G$
    3   Isotropic $\rightarrow$ N $\rightarrow$ S$_A$ $\rightarrow$ S$_G$
    4   Isotropic $\rightarrow$ N $\rightarrow$ S$_A$ $\rightarrow$ S$_B$ $\rightarrow$ S$_G$

From Fig. 4 it can be seen that while the S$_A$ and S$_B$ thermal stabilities increase with m, the S$_G$ properties rise to a maximum at C$_3$ and then decrease. Similarly, in molecules such as (Gray and Harrison, 1971a,b)

merely increasing n has very dramatic effects. When this structure is drawn with due regard to stereochemistry, it is found that when n is 0 or even, the terminal 3-methylphenyl group lies *on*-axis with the rest of the molecule, whereas odd values of n cause the 3-methylphenyl group to project and broaden the molecule greatly, because the group lies *off*-axis. As a result, when n increases, we observe dramatic differences in phase sequence on cooling the isotropic liquids, such that the compounds alternate between being nematic and smectic, or purely smectic. The following phase sequences illustrate the point:

$n = 0$ Isotropic $\rightarrow$ N $\rightarrow$ S$_A$ $\rightarrow$ S$_B$
    1   Isotropic $\rightarrow$ S$_B$ $\rightarrow$ S$_E$
    2   Isotropic $\rightarrow$ N $\rightarrow$ S$_A$ $\rightarrow$ S$_B$ $\rightarrow$ S$_E$
    3   Isotropic $\rightarrow$ S$_A$ $\rightarrow$ S$_B$ $\rightarrow$ S$_E$

Subtle changes not only in configuration, but even in conformation, can therefore have large effects on smectic versus nematic liquid crystal behavior. Drawing general guidelines in this area is therefore not only difficult but dangerous. The above example of ω-phenylalkyl esters does, however, suggest that any deviation from molecular linearity, e.g., molecular broadening, does depress nematic and smectic A characteristics significantly. This appears to be general, and lateral substitution of a mesogen also has this effect, provided that the molecule is broadened (see earlier comments). Consequently, lateral substitution resulting additionally in molecular twisting (a steric effect) always depresses smectic properties very strongly. Thus, in esters of the general structure (Gray et al., 1981)

X = fluoro strongly represses smectic properties relative to those for esters with X = H. For example, when ring A is cyclohexane, X = H, and alkyl = alkyl' = n-pentyl, the ester ($T_{C-N} = 36°C$, $T_{N-I} = 48°C$) shows a monotropic smectic phase at 29°C; the corresponding system with X = F gives $T_{C-N} = 17.5°C$, $T_{N-I} = 36°C$, and no smectic properties. Indeed, in this series of esters the homologs with very long chains, e.g., alkyl = C_7 and alkyl' = C_{11} or alkyl = C_9 and alkyl' = C_7, are still purely nematic.

Likewise, for the chiral esters (Gray and Kelly, 1982b)

$$C_8H_{17} - \langle \rangle - CO\cdot O - \langle \rangle - (CH_2)_2\overset{*}{C}H(CH_3)CH_2CH_3$$

when X = H we find a material that gives only an S_B phase ($T_{S_B-I} = 49°C$), whereas with X = F we have a room temperature cholesteric ($T_{C-Ch} = 21.5°C$, $T_{Ch-I} = 32.5°C$).

In this context it is worth noting that replacement of a benzene ring in a suitable mesogen by a cyclohexane or a bicyclo[2.2.2]octane ring not only enhances nematic properties, but depresses smectic properties—see the earlier comments on systems **XI**, **XII**, and **XV**. However, when smectic properties do occur in cyclohexane and BCO systems they are usually ordered S_B phases and not S_A phases (Carr, 1982). For example, the compound **XXI** melts at 46°C to give an S_B phase that persists until 108°C.

$$C_5H_{11} - \langle \rangle - CH_2CH_2 - \langle \rangle - C_5H_{11}$$

(**XXI**)

When such a compound is mixed with nematic cyanobiphenyls **XI** however, even compositions containing as much as 90% of **XXI** are purely nematic. This is almost certainly because the cyanobiphenyls **XI**, dependent on chain length, have either latent or real S_A (not S_B) properties. Since S_A and S_B phases are always immiscible (Sackmann and Demus, 1966), the smectic characteristics, real or latent, of **XXI** and **XI** are in opposition and strongly depress one another, giving a net nematic behavior. From such a temperature versus composition diagram of state it is possible to extrapolate the N–I transition line to 100% of **XXI** and obtain a latent T_{N-I} value for the smectogen; the value is 74°C.

It will be interesting to see whether, in comb polymers, alternation of side groups having opposing smectic tendencies (e.g., S_A and S_B) give rise to nematic polymers.

In concluding this section, the reader may be reminded of a few other structural features (some of which have already been mentioned in relation to T_{N-I} effects (Gray, 1976)) that influence smectic character:

(i)   Terminal —CN, —NO$_2$, and —OCH$_3$ groups tend to repress smectic properties.

(ii)   Terminal —CO·Oalkyl and —CH=CH·CO·Oalkyl groups promote smectic properties, as do terminal groups such as —CO·NH$_2$ and —OCF$_3$.

(iii)   Ionizing groups such as —CO·O$^-$M$^+$ and —NH$_3^+$X$^-$ usually give smectic behavior.

(iv)   Alkyl chain branching has a more marked depressing effect on nematic properties than on smectic A properties, and may even enhance S$_B$ properties in some cases.

(v)   Central ester groups, unlike terminal ester groups [see (ii)], do not strongly promote smectic properties. Many esters are pure nematogens.

(vi)   Substituents that are dipolar and do not broaden the molecule may enhance smectic properties more than nematic.

Some at least of the above evidence suggests that groups increasing the lateral dipoles in molecules enhance smectic properties, whereas dipoles acting along the long molecular axis enhance nematic properties. Certainly, an alignment of lateral dipoles could lead to a layer arrangement of the parallel molecules. Indeed the McMillan (1973) theory of tilting in the smectic C phase presupposed (as the evidence suggested at that time) that the molecules of compounds forming S$_C$ phases must have two large outboard terminal dipoles and preferably a central lateral dipole.

Thus di-n-alkoxyazoxy compounds are S$_C$/N in nature, whereas the di-n-alkylazoxy analogs are S$_A$/N in nature (de Jeu, 1980):

Similarly, 4-n-alkoxybenzoic acids are S$_C$/N, whereas the 4-n-alkylbenzoic acids are S$_A$/N (Demus et al., 1974).

There is no doubt that such lateral dipoles do tend to promote S$_C$ character, but the presence of terminal outboard dipoles is not essential (Goodby et al., 1977) to the occurrence of S$_C$ properties. For example, the ester

with only the central ester group as a dipolar function still exhibits S$_C$ properties, although the thermal stability of the S$_C$ phase is certainly lower than that of the ester

$$C_8H_{17}O-\langle\ \rangle-CO\cdot O-\langle\ \rangle-OC_8H_{17}$$

E. Diskogens

Due initially to the studies of Chandrasekhar *et al.* (1977), attention has centered recently around compounds consisting of flat, disk-shaped molecules that can pack together to form flexible columns. These columns then constitute a diskotic LC phase, quite distinct from smectic or nematic LC phases, and having a negative sign of the optic axis. The nature and properties of such phases were recently reviewed by Billard (1980), and Destrade *et al.* (1980).

Chandrasekhar first observed diskotic phases in the disklike systems afforded by the hexaacyloxybenzenes such as **XXII**, and similar phases were found for compounds such as **XXIII** and related systems (Billard, 1980).

(XXII)   (XXIII)

Clearly the structural requirements for formation of such phases must be a disk-shaped molecule, although, oddly enough, the earliest example of such a phase (but not recognized as such at the time) was provided by diisobutyl silanediol (Eaborn, 1952; Eaborn and Hartshorne, 1955). This has a most unpromising molecular structure **XXIV** for formation of such a phase until it is realized (Bunning *et al.*, 1980, 1982) that the molecules

(XXIV)   (XXV)

probably form dimers **XXV** that stack one above the other, forming a columnar structure.

The relevance of such diskotic phases to polymer systems, LC main chain, or LC side chain polymers is hard to judge. Presumably a fairly straight polymer backbone with a sufficient density of lateral functions radiating outward (either side chain groups or lateral substituents on a main chain backbone) could constitute a columnar structure, but whether this approximates closely enough to a stacking (fairly flexible) of individual disk-shaped molecules cannot easily be judged.

References

Baessler, H., and Labes, M. M. (1970). *J. Chem. Phys.* **52**, 631.

Billard, J. (1980). *In* "Liquid Crystals of One- and Two-Dimensional Order" (W. Helfrich and G. Heppke, eds.), Springer Series in Chemical Physics, Vol. 11, pp. 383–395. Springer-Verlag, Berlin and New York.

Billard, J. (Personal communication)

Blumstein, A., and Hsu, E. C. (1978). *In* "Liquid Crystalline Order in Polymers" (A. Blumstein, ed.), pp. 105–140. Academic Press, New York.

Bradshaw, M. J., and Raynes, E. P. (1981). *Proc. Annu. Freiburg Liq. Cryst. Conf.*

Bradshaw, M. J., McDonnell, D. G., and Raynes, E. P. (1981). *Mol. Cryst. Liq. Cryst.* **70**, 1567.

Bunning, J. D., Goodby, J. W., Gray, G. W., and Lydon, J. E. (1980). *In* "Liquid Crystals of One- and Two-Dimensional Order" (W. Helfrich and G. Heppke, eds.), Springer Series in Chemical Physics, Vol. 11, pp. 397–402. Springer-Verlag, Berlin and New York.

Bunning, J. D., Lydon, J. E., Eaborn, C., Jackson, P. M., Goodby, J. W., and Gray, G. W. (1982). *J. Chem. Soc., Faraday Trans.* **78**, 713.

Byron, D. J., Gray, G. W., and Worrall, B. M. (1965). *J. Chem. Soc.* p. 3706.

Carr, N. (1982). To be published.

Carr, N., and Gray, G. W. (1982). To be published.

Carr, N., Gray, G. W., and Kelly, S. M. (1981). *Mol. Cryst. Liq. Cryst.* **66**, 267.

Chandrasekhar, S., Sadashiva, B. K., and Suresh, K. A. (1977). *Pramana* **9**, 471.

Clark, N. A., and Lagerwall, S. T. (1981). *Proc. Int. Liq. Cryst. Conf., 8th, Kyoto* Pap. I-29.

Coates, D., and Gray, G. W. (1976). *Microscope* **24**, 117.

Constant, J., and Raynes, E. P. (1981). *Mol. Cryst. Liq. Cryst.* **70**, 1383.

Cser, F., and Nyitrai, K. (1976). *Magy. Kem. Foly.* **82**, 207.

de Jeu, W. H. (1980). "Physical Properties of Liquid Crystalline Materials." Gordon & Breach, New York.

de Jeu, W. H., van der Veen, J., and Goossens, W. J. A. (1973). *Solid State Commun.* **12**, 405.

Demus, D., Demus, H., and Zaschke, H. (1974). "Flüssige Kristalle in Tabellen." VEB Dtsch. Verlag Grundstoffind., Leipzig.

Destrade, C., Bernaud, M. C., Gasparoux, H., Levelut, A.-M., and Tinh, N. H. (1980). *In* "Liquid Crystals" (S. Chandrasekhar, ed.), pp. 29–33. Heyden, London.

Deutscher, H.-J., Kuschel, F., Schubert, H., and Demus, D. (1974). D.D.R. Patent 105,701.

Dewar, M. J. S., and Goldberg, R. S. (1970a). *J. Org. Chem.* **35**, 2711.

Dewar, M. J. S., and Goldberg, R. S. (1970b). *J. Am. Chem. Soc.* **92**, 1582.

Dewar, M. J. S., and Griffin, A. C. (1975). *J. Am. Chem. Soc.* **97**, 6622.

Dewar, M. J. S., and Riddle, R. M. (1975). *J. Am. Chem. Soc.* **97**, 6658.

Dewar, M. J. S., Griffin, A. C., and Riddle, R. M. (1974). *In* "Liquid Crystals and Ordered Fluids" (J. F. Johnson and R. S. Porter, eds.), Vol. 2, pp. 733–741. Plenum, New York.

Dunmur, D. A., and Tomes, A. E. (1980). Private communication of new results.

Eaborn, C. (1952). *J. Chem. Soc.* p. 2840.

Eaborn, C., and Hartshorne, N. H. (1955). *J. Chem. Soc.* p. 549.

Eidenschink, R. (1979). *Kontakte* No. 1, p. 15.

Eidenschink, R. (1980). *Kontakte* No. 3, p. 12.

Eidenschink, R., Erdmann, D., Krause, J., and Pohl, L. (1977). *Angew. Chem.* **89**, 10.

Eidenschink, R., Erdmann, D., Krause, J., and Pohl, L. (1978). *Angew. Chem.* **90**, 13.

Finkelmann, H. (1980). *In* "Liquid Crystals of One- and Two-Dimensional Order" (W. Helfrich and G. Heppke, eds.), Springer Series in Chemical Physics, Vol. 11, pp. 238–251. Springer-Verlag, Berlin and New York.

Finkelmann, H., Ringsdorf, H., and Wendorff, J. H. (1978a). *Makromol. Chem.* **179**, 273.

Finkelmann, H., Happ, M., Portugall, M., and Ringsdorf, H. (1978b). *Makromol. Chem.* **179**, 2541.

Finkelmann, H., Koldehoff, J., and Ringsdorf, H. (1978c). *Angew. Chem.* **90**, 992.

Gane, P. A. C., Leadbetter, A. J., and Wrighton, P. G. (1981). *Mol. Cryst. Liq. Cryst.* **66**, 567.

Goodby, J. W., Gray, G. W., and McDonnell, D. G. (1977). *Mol. Cryst. Liq. Cryst. (Lett.)* **34**, 183.

Goodby, J. W., Gray, G. W., Leadbetter, A. J., and Mazid, M. A. (1980). *In* "Liquid Crystals of One- and Two-Dimensional Order" (W. Helfrich and G. Heppke, eds.), Springer Series in Chemical Physics, Vol. 11, pp. 3–18. Springer-Verlag, Berlin and New York.

Gray, G. W. (1974). *In* "Liquid Crystals and Plastic Crystals" (G. W. Gray and P. A. Winsor, eds.), Vol. 1, pp. 103–152. Horwood, Chichester, England.

Gray, G. W. (1975). *J. Phys. (Orsay, Fr.)* **36**(Cl), 337.

Gray, G. W. (1976). *Adv. Liq. Cryst.* **2**, 1.

Gray, G. W. (1978). "Advances in Liquid Crystals for Material Applications." BDH Chem., Poole, England.

Gray, G. W. (1979). *In* "The Molecular Physics of Liquid Crystals" (G. R. Luckhurst and G. W. Gray, eds.), pp. 1–29, 263–284. Academic Press, New York.

Gray, G. W. (1981). *Mol. Cryst. Liq. Cryst.* **66**, 3.

Gray, G. W., and Harrison, K. J. (1971a). *Mol. Cryst. Liq. Cryst.* **13**, 37.

Gray, G. W., and Harrison, K. J. (1971b). *Symp. Faraday Soc.* No. 5, p. 54.

Gray, G. W., and Kelly, S. M. (1980). *J. Chem. Soc., Chem. Commun.* p. 465.

Gray, G. W., and Kelly, S. M. (1981a). *J. Chem. Soc. Perkin 2* p. 26.

Gray, G. W., and Kelly, S. M. (1981b). *Angew. Chem., Int. Ed. Engl.* **20**, 393.

Gray, G. W., and Kelly, S. M. (1981c). *Mol. Cryst. Liq. Cryst.* **75**, 95.

Gray, G. W., and Kelly, S. M. (1981d). *Mol. Cryst. Liq. Cryst.* **75**, 109.

Gray, G. W., and Kelly, S. M. (1982a). To be published.

Gray, G. W., and Kelly, S. M. (1982b). *Mol. Cryst. Liq. Cryst.* To be published.

Gray, G. W., and McDonnell, D. G. (1975). *Electron. Lett.* **11**, 586.

Gray, G. W., and McDonnell, D. G. (1976). *Mol. Cryst. Liq. Cryst.* **37**, 189.

Gray, G. W., and McDonnell, D. G. (1977). *Mol. Cryst. Liq. Cryst. (Lett.)* **34**, 211.

Gray, G. W., and Mosley, A. (1976). *J. Chem. Soc. Perkin 2* p. 97.

Gray, G. W., and Worrall, B. M. (1959). *J. Chem. Soc.* p. 1545.

Gray, G. W., Hartley, J. B., Ibbotson, A., and Jones, B. (1955). *J. Chem. Soc.* p. 4359.

Gray, G. W., Harrison, K. J., and Nash, J. A. (1973a). *Electron. Lett.* **9**, 130.

Gray, G. W., Harrison, K. J., Nash, J. A., and Raynes, E. P. (1973b). *Electron. Lett.* **9**, 616.

Gray, G. W., Harrison, K. J., Nash, J. A., Constant, J., Hulme, D. S., Kirton, J., and

Raynes, E. P. (1974). *In* "Liquid Crystals and Ordered Fluids" (J. F. Johnson and R. S. Porter, eds.), Vol. 2, pp. 617–643. Plenum, New York.

Gray, G. W., Hogg, C., and Lacey, D. (1981). *Mol. Cryst. Liq. Cryst.* **67**, 657.

Ibrahim, I. H., and Haase, W. (1981). *Mol. Cryst. Liq. Cryst.* **66**, 509.

Leadbetter, A. J. (1982). To be published.

Leadbetter, A. J., and Mehta, A. I. (1981). Private communication of unpublished results.

Leadbetter, A. J., Richardson, R. M., and Collin, C. N. (1975). *J. Phys. (Orsay, Fr.)* **36**(Cl), 37.

Leadbetter, A. J., Durrant, J. L. A., and Rugman, M. (1977). *Mol. Cryst. Liq. Cryst. (Lett.)* **34**, 231.

Luckhurst, G. R., and Gray, G. W., eds. (1979). "Molecular Physics of Liquid Crystals." Academic Press, New York.

McMillan, W. L. (1973). *Phys. Rev. A* **8**, 1921.

Maier, W., and Saupe, A. (1958a). *Z. Naturforsch., A* **13A**, 564.

Maier, W., and Saupe, A. (1958b). *Z. Naturforsch., A* **14A**, 882.

Maier, W., and Saupe, A. (1960). *Z. Naturforsch., A* **15A**, 287.

Meyer, R. B., Liebert, L., Strzelecki, L., and Keller, P. (1975). *J. Phys., Lett. (Orsay, Fr.)* **36**, 69.

Perplies, E., Ringsdorf, H., and Wendorff, J.H. (1976). *J. Polym. Sci., Polym. Lett. Ed.* **13**, 243.

Sackmann, H., and Demus, D. (1966). *Mol. Cryst. Liq. Cryst.* **2**, 81.

Shibaev, V. P., and Platé, N. A. (1977). *Vysokomol. Soedin., Ser. A* **19**(9), 23.

Shibaev, V. P., Freidzon, J. S., and Platé, N. A. (1976). *Dokl. Akad. Nauk SSSR* **227**, 1412.

Taylor, G. N., and Kahn, F. J. (1974). *J. Appl. Phys.* **45**, 4330.

Van der Meer, B. W., and Vertogen, G. (1979). *J. Phys. (Orsay, Fr.)* **40**, 222.

Van Meter, J. P. (1973). *Eastman Org. Chem. Bull.* **45**, No. 1, 1.

Van Meter, J. P., and Klandermann, B. H. (1973). *Mol. Cryst. Liq. Cryst.* **22**, 285.

Young, W. R., and Green, D. C. (1972). *IBM Res. Rep.* RC 4121.

Young, W. R., and Green, D. C. (1974). *Mol. Cryst. Liq. Cryst.* **26**, 7.

Young, W. R., Haller, I., and Green, D. C. (1972a). *IBM Res. Rep.* RC 3827.

Young, W. R., Haller, I., and Green, D. C. (1972b). *J. Org. Chem.* **37**, 3707.

2

Synthesis, Structure, and Properties of Liquid Crystalline Side Chain Polymers

H. Finkelmann

Physikalisch-Chemisches Institut
Technische Universität Clausthal
Clausthal-Zellerfeld
Federal Republic of Germany

I. Introduction

The spontaneous, anisotropic orientation of low molecular weight (MW) mesogenic molecules in the liquid crystalline state results in materials that are of both theoretical and commerical interest. Hence, if one can introduce these liquid crystalline (LC) properties into macromolecules, several new properties are expected. Mesogenic polymers exhibiting liquid crystalline behavior can be synthesized in a variety of different ways, as indicated by the increasing number of publications in the past few years.

To formulate a systematic classification of mesogenic polymers, we shall describe their characteristic molecular structure and phase behavior. If we exclude solid polymers with ordered structures, and only consider homogeneous phases that are in a liquid crystalline state, mesogenic polymers

POLYMER LIQUID CRYSTALS

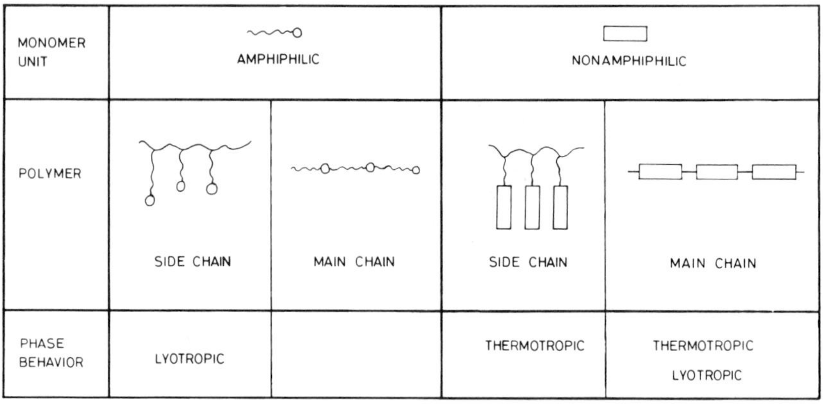

MONOMER UNIT	AMPHIPHILIC		NONAMPHIPHILIC	
POLYMER	SIDE CHAIN	MAIN CHAIN	SIDE CHAIN	MAIN CHAIN
PHASE BEHAVIOR	LYOTROPIC		THERMOTROPIC	THERMOTROPIC LYOTROPIC

Fig. 1.   Classification of mesogenic polymers.

can be subdivided following the scheme given in Fig. 1. It is most convenient first to describe the molecular structure of the monomer unit, following the classification of Gray and Winsor (1974) for conventional liquid crystals. The monomer units can be either amphiphilic or nonamphiphilic. Subsequently, one can distinguish whether the mesogenic moiety forms a part of the polymer main chain or is attached to the side chain. A third distinction involves the phase behavior. The system can be lyotropic or thermotropic, depending upon whether the mesophase is observed by variation of solvent content or by variation of temperature.

At the present time there have been only a few systematic investigations of polymers having lyotropic repeating units. Some papers are summarized by Kelker and Hatz (1980) and Elias (1977). Siol (1979) has investigated polymerization of amphiphilic monomers. The phase behavior of lyotropic liquid crystalline side chain polymers in aqueous solution has recently been compared with the phase behavior of the corresponding monomers (Finkelmann *et al.*, 1982).

Nonamphiphilic monomer units are characterized by their rigid rodlike molecular structure typical of low MW liquid crystals. If these monomer units are joined to form the main chain, the polymer backbone itself becomes rigid and rodlike. The rodlike structure causes anisotropic packing of the macromolecules, as predicted theoretically by Flory (Chapter 4, this volume). For mesogenic polymers, however, it is not necessary that the monomer units be rigid. Chain rigidity can be caused by the secondary structure of the polymer, as exemplified by cellulose derivatives and polyglutamates. The liquid crystalline state can be observed for these "mesogenic main chain polymers" in solution as well as in the melt.

The nonamphiphilic mesogenic moiety can also be attached to the side chain. In contrast to the previously described mesogenic main chain polymers, in the liquid crystalline state only the original mesogenic side chains are responsible for the liquid crystalline order. This order is more or less independent of the conformation of the polymer main chain. Therefore it would be assumed that these "mesogenic side chain polymers" are very similar to the conventional low MW mesogens. In this chapter the synthesis of mesogenic side chain polymers will be described, and the influence of chemical constitution on the mesophase will be compared to low MW nonamphiphilic mesogens (see also Gray, Chapter 1, this volume). Cross-linking of these polymers to form liquid crystalline polymer networks will also be considered.

II. Synthesis of Polymers Having Mesogenic Side Chains

The mesogenic side chain polymers consist of two components, the mesogenic moieties and the polymer main chain to which they are attached. A large number of known mesogenic molecules (Demus *et al.*, 1976) can be used as components. Moreover, a variety of different polymer main chains are available. The combination of these components allows a manifold variation of different types of mesogenic side chain polymers. These polymers can be prepared by three different types of polymerization specified in Table I. The most convenient method is to introduce into a mesogenic

TABLE I

Synthesis of Mesogenic Side

PRINCIPLES		EXAMPLES
ADDITION POLYMERIZATION	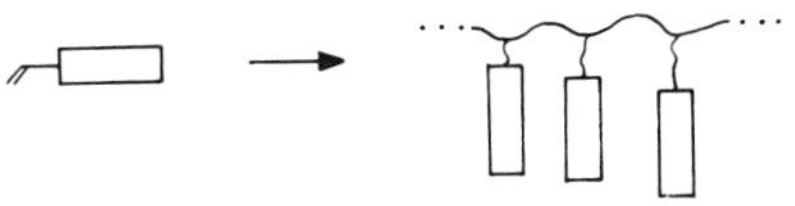	– POLYACRYLATES – POLYMETHACRYLATES – POLYSTYRENE DERIVATIVES
CONDENSATION POLYMERIZATION	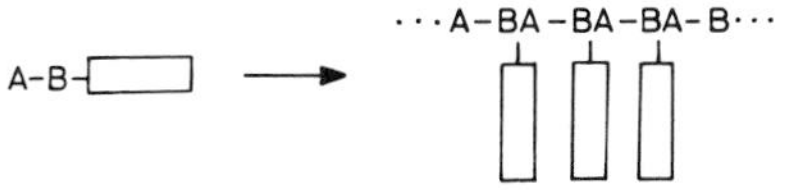	
MODIFICATION OF POLYMERS	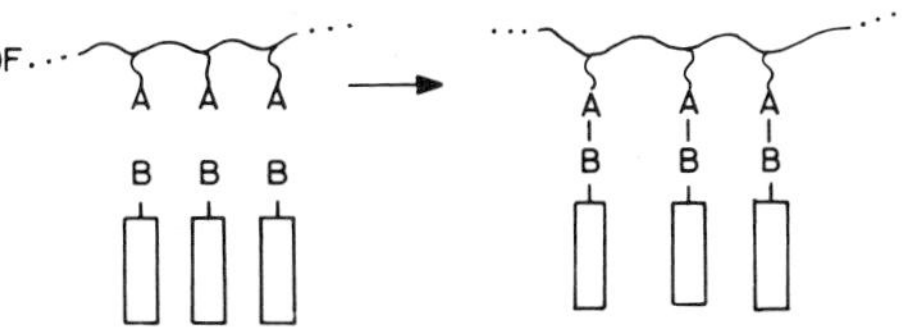	– POLYSILOXANES

molecule a reactive group capable of undergoing addition polymerization. A number of polymerizable monomers have been synthesized. They are comprehensively summarized by Blumstein and Hsu (1978) and by Shibaev and Plate (1978). In most cases the polymerizable group is a methacrylate or an acrylate, which forms a flexible vinyl backbone.

The second possibility is to introduce into the low MW mesogen a reactive group capable of undergoing a polycondensation reaction. In this way polymers containing heteroatoms in the backbone can be synthesized. However, at present, no mesogenic side chain polymers have been prepared via polycondensation.

The third synthesis route starts with reactive polymers. They can be modified to mesogenic side chain polymers by using suitable reactive mesogenic monomers (Paleos, et al., 1981; Finkelmann and Rehage, 1980a,b). An example of this type is the smooth addition of vinyl-substituted mesogenic monomers (Finkelmann and Rehage, 1980a) to poly[oxy(methylsilylene)]:

$$
\cdots-\underset{\underset{H}{\overset{\overset{CH_3}{|}}{|}}{Si}-O-\cdots + CH_2{=}CH-\boxed{} \xrightarrow{[Pt]} \cdots-\underset{\underset{CH_2}{\overset{\overset{CH_3}{|}}{|}}{Si}-O-\cdots
$$

$\boxed{}$ = mesogenic molecule

In this case a quantitative addition reaction can be easily detected by the disappearance of the Si–H vibrational absorption. In most cases, however, analogous reactions involving polymers are difficult to perform.

The large number of variations mentioned above suggests that it should be easy to prepare polymers with mesogenic side chains. Several experiments, however, have demonstrated that a direct linkage of rigid mesogenic groups to a polymer backbone is generally not sufficient to obtain a mesogenic polymer. In most cases only isotropic polymer melts are observed or solid polymers with a liquid crystalline structure. The latter structure vanishes irreversibly if the polymer is heated above its glass transition temperature T_g. When these polymers are subsequently cooled below T_g, isotropic glasses are obtained. Thus direct linkage of mesogenic groups to the polymer main chain results in a different system than the low MW LC. In fact, because of the covalent linkage to the main chain, motion of the mesogenic groups is drastically reduced. Translational diffusion is no longer independent of the neighboring molecules. As space filling models show, large mesogenic groups are prevented from attaining parallel alignment because of strong steric hindrance. This also causes a stiffening of the polymer chain, which is often indicated by an increase of T_g compared to the same polymer without a mesogenic side chain. This problem can be solved by simple model considerations (Finkelmann *et al.*, 1978a), which will be described next.

A. Model Considerations

Two extreme cases are considered. In the first case a polymer is dissolved in a LC matrix. Apart from local anisotropic conformations of the polymer chain due to the anisotropic matrix (Dubault *et al.*, 1978), motions of the polymer segments and the mesogenic molecules are not coupled, as represented schematically in Fig. 2a. Second, if rigid mesogenic side chains are directly attached to the polymer backbone (Fig. 2c) motions of the polymer segments and mesogenic groups are directly coupled. In the liquid state above T_g the polymer tends to adopt a statistical chain conformation that hinders anisotropic orientation of the mesogenic side chains. Also, steric hindrance prevents mesogenic order. In addition to these extreme conditions intermediate states could be realized if the mesogenic groups were fixed to the polymer backbone via a flexible spacer as indicated in Fig. 2b. The flexible spacer decouples motions of the main chain and side chain and alleviates steric hindrance. Under these conditions the mesogenic side chains can be anisotropically ordered in the liquid state even though the polymer main chain tends to adopt a statistical chain conformation. Consequently, variation of the spacer length should clearly influence the LC order of the side chain. This will be discussed in Section III.C.

These model considerations can be easily understood in terms of nematic or cholesteric order. The statistical distribution of the centers of gravity of the mesogenic molecules are compatible with a more or less statistical chain

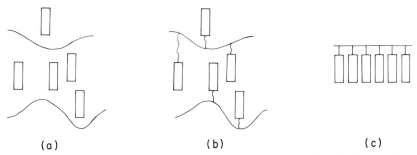

Fig. 2. Liquid crystalline molecules: (a) dissolved in a polymer, (b) linked to a polymer via a flexible spacer, and (c) directly linked.

conformation. However, in case of smectic phases a three-dimensional coil is no longer consistent with the layered structure; instead, the backbone is restricted to lie in a two-dimensional plane, as deduced from the X-ray investigations of Wendorff *et al.* (1978).

These considerations can be easily verified: a flexible spacer, preferably an alkyl chain, must be inserted between the mesogenic group and the main chain. A simple synthesis route for a typical mesogenic side chain polymer with spacers is illustrated for a polymethacrylate (Finkelmann *et al.*, 1978b):

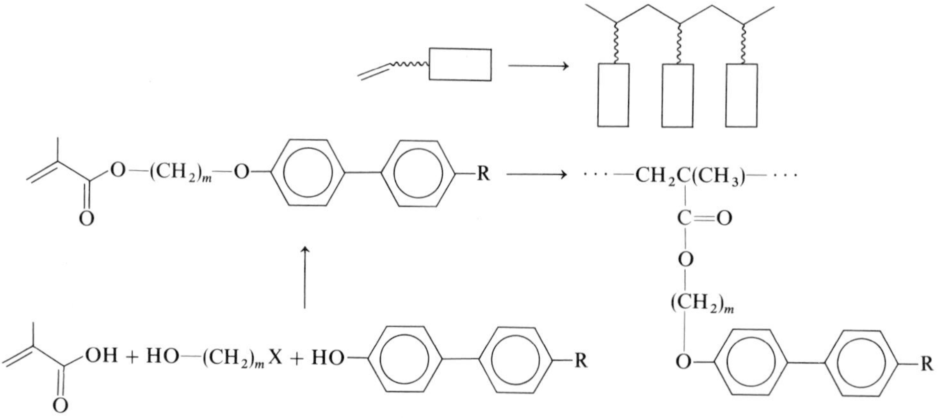

The mesogenic diphenyl derivative is linked via a flexible alkyl chain consisting of m methylene groups to the polymethacrylate main chain.

In the last few years several syntheses indicated that mesogenic polymers can be prepared following these model considerations. Mesophases can be observed reversibly above the glass transition temperature and, upon heating, the polymers become isotropic at the clearing temperature T_{cl}.

B. Phase Behavior

The phase behavior of conventional low MW mesogens has been investigated by both theoretical and experimental methods (Kelker and Hatz, 1980). One can ask how this phase behavior will be influenced by linkage of the mesogenic molecules to the polymer main chain, and whether there are still analogies to the low MW mesogens. Furthermore, the polymeric behavior is of interest.

The essential features of the phase behavior of mesogenic polymers can be established by differential scanning calorimetry (DSC), which turns out to be an appropriate and convenient method. First, the characteristic DSC trace of a low MW LC, a phenyl benzoate exhibiting one liquid crystalline phase, is illustrated in Fig. 3a. The cooling curve is characterized by the high temperature exotherm, indicating a first-order transformation from the isotropic phase to a mesophase, and the low-temperature exotherm indicating a first-order transformation of the mesophase to crystal. On the other hand, Fig. 3b shows the thermodynamic behavior of the noncrystallizing polymer, poly[oxy(methylsilylene)], which exhibits a transition from the isotropic liquid to a glass at low temperatures. This transition is typically indicated by a "jump" in the DSC curve. If the mesogenic phenyl benzoate is linked to a siloxane main chain as shown in Fig. 3c, a change in the DSC curve is evident. Compared to Fig. 3a, a glass transition is observed instead of low temperature crystallization. If Fig. 3c is compared to 3b, the phase transformation from mesophase to isotropic is seen to occur at a higher temperature. This example indicates the characteristic combination of properties for this material: (i) a glass transition characteristic of the polymer backbone, and (ii) a phase transformation from mesophase to the isotropic phase due to the mesogenic side chains. Complementary studies with a polarizing microscope reveal a very important property of the polymers:

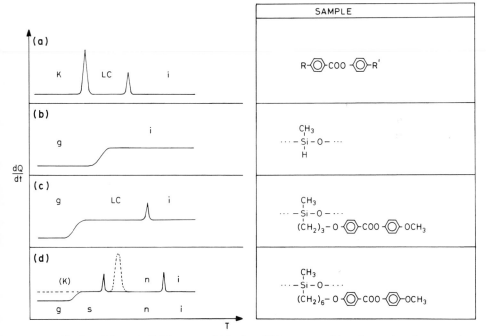

Fig. 3. Schematic DSC curves of (a) a low MW mesogen, (b) an isotropic polymer, and (c), (d) mesogenic polymers.

the texture observed in the LC state of the polymer can be frozen without change in the glassy state. This procedure may have practical applications, because it allows the preservation of information given by the texture in the glassy state.

Figure 3d shows schematically a more complicated system, which was recently investigated by Frenzel (1981). The polymer exhibits two mesophases and a low temperature glass transition. The lower temperature mesophase is monotropic. This phase can be observed if the sample is annealed some degrees above the glass transition temperature. Very slow crystallization sets in, as indicated by the appearance of a new peak drawn with a dashed line. The polymer becomes partially crystalline and, simultaneously, the change of specific heat at the glass transition decreases.

So far, these DSC measurements have indicated combinations of properties of polymers and mesogens. A typical property of conventional semi-crystalline polymers is their biphasic character due to coexisting crystalline and amorphous domains. This raises the question of whether an anisotropic mesophase is thermodynamically homogeneous or should be defined in a way similar to the partially crystalline polymer. Pressure, volume, temperature (PVT) measurements give a more detailed insight into the phase behavior of the mesogenic polymers. In Fig. 4 the specific volume of a mesogenic polymethacrylate, which shows the same phase diagram as the polymer in Fig. 3c, is measured as a function of the temperature for different pressures. Upon lowering the temperature, the phase transformation from the isotropic melt to the polymeric mesophase is indicated by the "jump" in the $\bar{V}(T)$ curve. The pressure dependence of this transformation can be analyzed by the Clausius–Clapeyron equation

$$(dp/dT)_{\mathrm{tr}} = \Delta H/T\Delta V \qquad (1)$$

which describes a first-order transition. The enthalpies for the transformation determined by DSC at 1 bar agree within experimental error with those calculated from Eq. (1). Since this transformation can be described by Eq. (1), it must be first order, by analogy to low MW mesogens. Optical measurements confirm these results (see also Section III.C.2).

A second important conclusion from these measurements is that the change in specific volume is independent of path. For a given specific volume, defined at a particular p and T, this volume can be obtained by varying either the pressure or temperature, and is independent of path and time. This indicates that $\bar{V}$ is a function of only p and T. Hence the polymeric mesophase is a thermodynamically homogeneous phase similar to those phases of low MW mesogens. This is in clear contrast to the biphasic partially crystalline polymers, where $\bar{V}$ also depends on thermal prehistory.

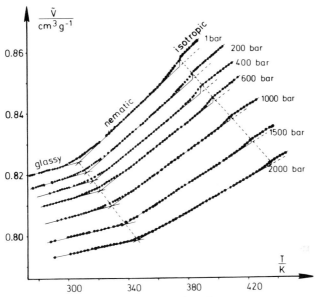

Fig. 4. PVT behavior of the mesogenic polymethacrylate

(Reproduced from Fenzel and Rehage, 1980.)

III. Influence of the Chemical Constitution on the Type of Mesophase Formed

For low MW mesogens extensive theoretical and experimental studies have been performed to relate the type of mesophase and the chemical constitution (Kelker and Hatz, 1980). Linkage of mesogenic side chains to a polymer backbone also results in thermodynamically homogeneous thermotropic mesophases. However, because of the polymer main chain, additional variations of the chemical constitution are possible. These are the chemical composition of the backbone, degree of polymerization, tacticity, and the chemical constitution of the flexible spacer. Further on, analogies between the mesogenic monomers and the corresponding mesogenic side chain polymers regarding variations of the chemical constitution will be examined.

A. Variation of the Mesogenic Group

1. Nematic and Smectic Polymers

If a low MW rodlike mesogenic molecule is substituted at the ends of the rods, in most cases a systematic change in the phase behavior is observed (see also Gray, Chapter 1, this volume). This is schematically represented in Fig. 5. If the length m of the substituent is short (e.g., short alkyl chains), nematic phases are observed. With increasing length of the substituent, a low-temperature smectic phase becomes stable and the nematic phase is suppressed.

According to model considerations for polymers (Section II.A), the flexible spacer partially decouples the main chain and side chain. Consequently, one should expect for the polymers that variation of the length of the substituents on the mesogenic side chains will also change the character of the mesophase. Table II lists polymers in which substituted phenyl benzoate (PB) moieties are linked to the backbone. For the low MW PBs, the phase behavior indicated schematically in Fig. 5 is well established (Demus *et al.*, 1976). A similar progression of mesophase type is also observed for polymers when the substituent of the PB is changed from methoxy to hexyloxy. Even though the mesogenic moiety is attached by different flexible spacers (examples 1–4 in Table II), or to different polymer chains (examples 5–6), this behavior is preserved. The change from nematic to smectic with increasing length of the substituents on the mesogenic side chain also applies to the length of the flexible spacer. This is demonstrated in Table III for various mesogenic moieties attached to the main chain by different spacers. These examples indicate that the influence of the substituents upon the type of mesophase formed is essentially identical for low MW and polymeric

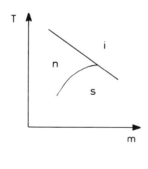

Fig. 5. Schematic phase behavior of a mesogen depending on the chain length m of the spacer.

TABLE II

Phase Behavior of Mesogenic Polymers with Varying Length of Substituents in the Mesogenic Moiety[a]

No	POLYMER		TRANSITIONS (K)				$\Delta H_{LC \to i}$ (J/g)
1	$\cdots -CH_2-\overset{\displaystyle CH_3}{\underset{\displaystyle COO-(CH_2)_2-O-\bigcirc-COO-\bigcirc-OCH_3}{C}}-\cdots$	g	369	n	394	i	2.3
2	$- OC_6H_{13}$	g	410	s	451	i	11.5
3	$\cdots -CH_2-\overset{\displaystyle CH_3}{\underset{\displaystyle COO-(CH_2)_6-O-\bigcirc-COO-\bigcirc-OCH_3}{C}}-\cdots$	g	309	n	374	i	23
4	$- OC_6H_{13}$	g ~ 300		s	374	i	14.0
5	$\cdots -\overset{\displaystyle CH_3}{\underset{\displaystyle (CH_2)_3-O-\bigcirc-COO-\bigcirc-OCH_3}{Si}}-O-\cdots$	g	288	n	334	i	2.2
6	$- OC_6H_{13}$	g	288	s	385	i	11.6

[a] g = glassy, n = nematic, s = smectic

TABLE III

Phase Behavior of Mesogenic Polymers with Varying Flexible Spacer Length[a]

No	POLYMER		TRANSITIONS (K)				
1	$\cdots -CH_2-\overset{\displaystyle CH_3}{\underset{\displaystyle COO-(CH_2)_2-O-\bigcirc-COO-\bigcirc-CH=N-\bigcirc-CN}{C}}-\cdots$	g	361	n	580	i	
2	$- (CH_2)_6 -$	g	324	s	607	i	
3	$\cdots -\overset{\displaystyle CH_3}{\underset{\displaystyle (CH_2)_3-O-\bigcirc-COO-\bigcirc-OCH_3}{Si}}-O-\cdots$	g	288	n	334	i	
4	$- (CH_2)_6 -$	g	278	s	319	n	381	i
5	$\cdots -CH_2-\overset{\displaystyle CH_3}{\underset{\displaystyle COO-(CH_2)_2-O-\bigcirc-COO-\bigcirc-\bigcirc-OCH_3}{C}}-\cdots$	g	—	n	551	i	
6	$- (CH_2)_6-O -$	g	333	s	398	n	535	i

[a] g = glassy, n = nematic, s = smectic; polymers 1 and 2 from Zentel (1980).

mesogens, so that polymeric nematic and smectic mesophases can be achieved by varying the length of the substituents.

2. Cholesteric Polymers

We have seen that by varying the length of substituents on the mesogenic groups nematic and smectic phases can be prepared. If, in case of a low MW mesogen, a chiral substituent is added, a nematic phase can be changed to a cholesteric or chiral nematic phase. The cholesteric phase is characterized by the statistical distribution of the centers of gravity of the molecules, as in the case of the nematic phase, and an orientational order of the long molecular axes of the molecules characterized by the director. Due to chirality of the molecules, the director is spontaneously twisted in distance r_{ab} perpendicular to the director through an angle θ (Fig. 6). The resulting helicoidal structure is characterized by the pitch P,

$$P = 2\pi r_{ab}/\theta \tag{2}$$

which gives the cholesteric phase interesting optical properties. These are

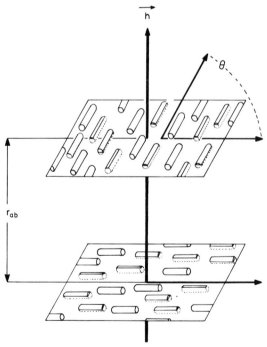

Fig. 6. Structure of the cholesteric phase: θ = twist angle, r_{ab} = distance of molecules perpendicular to the director, h = pitch axis.

a high optical rotation and a circular dichroism, due to a selective reflection of circularly polarized light. The wavelength of reflected light λ_R depends directly upon the helical pitch

$$\lambda_R = \bar{n}P \tag{3}$$

where $\bar{n}$ is the average refractive index of the phase.

In view of possible commercial applications as filters or reflectors, one might try to prepare cholesteric polymers by linking a chiral mesogenic group to the polymer backbone. However, all attempts involving homopolymers failed because, until now, chiral monomers yielded only smectic mesophases. Nevertheless, cholesteric polymers can be prepared by copolymerization. The first cholesteric polymer was obtained by distortion of the layered smectic structure using monomers of different lengths m of the flexible spacer (Finkelmann *et al.*, 1978c). This is indicated schematically in Fig. 7. Polymerization of the cholesteryl derivatives with $m = 2, 6,$ or 12 results in smectic homopolymers as shown in Fig. 7a. For copolymers having spacers differing sufficiently in length, for example, $m = 2$ and $m = 12$, cholesteric copolymers are produced as indicated in Fig. 7b. However, the cholesteric phase is only observed when the mole ratio of the monomers is 1:1.

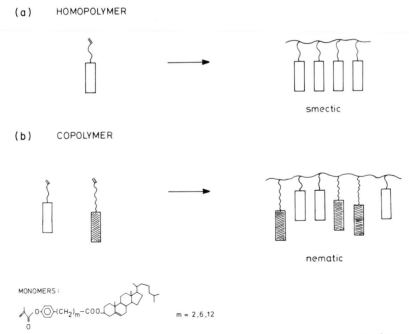

Fig. 7.   Scheme of (a) smectic order in homopolymers, (b) nematic order in copolymers.

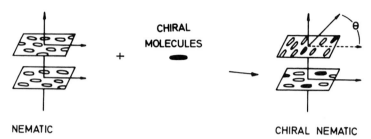

Fig. 8.   Change of a nematic phase to a cholestoric phase by addition of chiral molecules.

This can be demonstrated by small-angle X-ray diffraction because the sharp small-angle peaks become smeared out when the smectic phase is transformed to cholesteric.

A second but more practicable method of observing chiral nematic phases is well known for mesogens: the preparation of "induced cholesteric phases" (Fig. 8). A low MW nematic phase can be converted to a cholesteric phase by addition of chiral molecules, which need not be mesogenic or covalently attached to the nematogen. The induced cholesteric twist angle θ is directly proportional to the concentration and is also a function of the molecular structure of the chiral molecule (Finkelmann and Stegemeyer, 1978). Copolymerization of a monomer capable of producing a nematic polymer with a chiral comonomer also results in a cholesteric phase. This was accomplished for the first time using polymethacrylates (Finkelmann *et al.*, 1978d), and

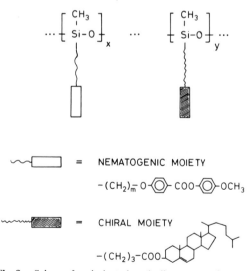

Fig. 9.   Scheme for cholesteric polysiloxane copolymers.

was studied in more detail for polysiloxanes as illustrated in Fig. 9 (Finkelmann and Rehage, 1980b). The cholesteric polymers exhibit the same optical properties as low MW cholesteric phases. If prepared between thin glass slides, the cholesteric phase spontaneously forms a Grandjean texture that reflects circularly polarized light of wavelength λ_R. This light-reflecting texture can be frozen in the glassy state, provided no smectic phase exists at low temperatures. Cholesteric films that may be commerically useful can be prepared in this way. Figure 10 illustrates the optical behavior of cholesteric polysiloxanes having spacer length $m = 6$ (see Fig. 9) as a

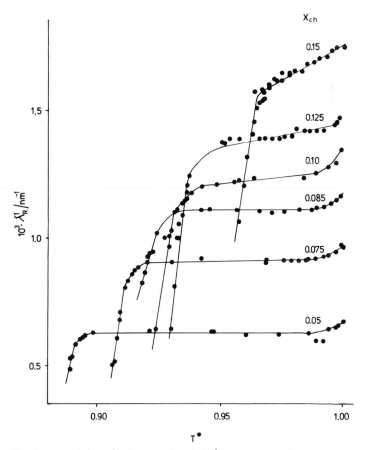

Fig. 10. Inverse of the reflection wavelength λ_R^{-1}, versus the reduced temperature, T^*, for cholesteric copolymers having different mole fractions, x_{ch}, of the chiral comonomer; ($T^* = T_m/T_{cl}$; T_m = measuring temperature, T_{cl} = clearing temperature). (Reproduced from Finkelmann and Rehage, 1980b with permission of Hüthig and Wepf Verlag, Basel.)

function of the temperature and the mole fraction x_{ch} of the chiral comonomer. These copolymers also exhibit a smectic phase at low temperature. In direct analogy to the low MW mesophases the cholesteric twist (and therefore λ_R^{-1}) becomes larger with increasing mole fraction of the chiral comonomer. The temperature dependence of λ_R^{-1} is nearly linear and is strongly dependent upon temperature near the cholesteric-to-smectic phase transition. This can be attributed to pretransitional effects. It is notable that, in contrast to low MW cholesteric phases, λ_R^{-1} very slowly equilibrates to the corresponding temperature in the temperature interval of divergence. This indicates that the decrease of λ_R^{-1} with decreasing temperature involves rearrangement of the mesogenic groups into smectic clusters near the transition to the smectic phase. The required change in conformation of the polymer backbone is the cause of the time dependence.

B. Variation of the Polymer Main Chain

The structure of low MW mesogens can be varied in numerous ways to influence the type of mesophase and physical properties. This may be done by changing the rigid mesogenic moieties or their substituents. By linking these moieties to a polymer backbone, the possibilities are multiplied because a large variation is also possible for the chemical structure, tacticity, and molecular weight of the polymer backbone. The influence of anisotropic orientation on the structure of the polymer backbone is also of interest. However, at present only very few of these aspects have been investigated. Some examples are discussed below.

1. Variation of Chemical Structure

Recalling the model considerations in Section II.A, a flexible spacer decouples motions of the main chain and side chain. In reality there will be a more or less strong coupling of motions, resulting in varying interactions of the main chain and side chains. Two aspects should be considered. On the one hand, the linkage of the mesogenic moieties to the polymer backbone restricts translational and rotational motions. Motions of the mesogenic side chains are now correlated to motions of the neighboring molecules via the backbone, which influences interactions of the mesogenic side chains. Consequently the flexibility of the backbone, which can be characterized by the statistical chain length, directly affects the behavior of mesophase. On the other hand, mobility of chain segments of the polymer backbone is strongly influenced by the bulky mesogenic groups. This directly affects the glass transition of the polymer.

TABLE IV

Phase Behavior of Polymers with Constant Mesogenic Moieties
Linked to Different Polymer Main Chains[a]

POLYMER	TRANSITIONS (K)					ΔT

POLYMER	TRANSITIONS (K)	ΔT
![structure: methacrylate] CH_2-C with CH_3 and $COO-R$	g 369 n 394 i	25
![structure: acrylate] CH_2-C with H and $COO-R$	g 320 n 350 i	30
![structure: siloxane] $O-Si$ with CH_3 and CH_2-R	g 288 n 334 i	46

[a] $R = (CH_2)_2-O-\hexagon-COO-\hexagon-OCH_3$

In Table IV mesogenic phenyl benzoates of the same chemical structure
are attached via flexible spacers of approximately the same length to different
polymer backbones. If the main chain and side chain are decoupled, essentially no influence on the mesophase should be observed, so long as the
chemical constitutions of the main chain are very similar. However, the
nematic-to-isotropic transition is strongly shifted toward lower temperature
with increasing flexibility of the polymer backbone. Comparison of polymethacrylates and polyacrylates shows that the shift of the phase transformation can only be associated with the change in flexibility of the backbone.
If the mesogenic moiety is attached to a very flexible polysiloxane chain,
a further decrease is observed. According to these results, restriction of
motion of the mesogenic moieties by the backbone moves the range of
stability of the mesophase toward higher temperature (see also Section
III.B.2). Influence of the bulky mesogenic moieties on the glass transition
can be seen most clearly for polysiloxanes. In contrast to the mesogenic
polymer, the initial poly[oxy(methylsilylene)] has a T_g of 140 K.

An increasing flexibility of the backbone can also be achieved by copolymerization if the bulky mesogenic groups are replaced with small groups
(Blumstein et al., 1978; Shibaev et al., 1978). By analogy to the previous
results, lowering the glass transition temperature by increasing the flexibility
of the main chain reduces the mesophase-to-isotropic transition temperature.

2. Effect of Tacticity

The initial investigations of mesogenic polymers attempted to determine whether the ordered state of low MW mesogenic monomers influences the kinetics of the polymerization reaction and the tacticity of the resulting polymers, as compared to a polymerization carried out in the isotropic phase. Careful measurements by different authors (Cser *et al.*, 1978; Tanaka *et al.*, 1976, 1978) indicate that the kinetics as well as the tacticity are not strongly influenced by order in the monomer mesophase. However, all these investigations are difficult to interpret because in every case the monomer exhibits a mesophase of different type from that of the polymer. Hence only heterogeneous reactions were investigated, where, for example, the kinetics is not influenced only by the anisotropic order of the monomers.

Concerning the interactions of the main chain and mesogenic side chains, the influence of tacticity on the type of mesophase is of interest. As demonstrated in the previous section, a shift of the glass transition due to a change of chemical constitution of the main chain causes a shift of anisotropic–isotropic transition in the same direction. For conventional poly(methyl methacrylate) (PMMA) a change in the tacticity strongly influences the glass transition (Krause and Roman, 1965; Kusy, 1976; Shetter, 1963): isotactic PMMA, 318–330 K; syndiotactic PMMA: 388–393 K. Therefore, judging from these results, the tacticity of the polymer backbone must shift the temperature range of stability of the mesophase corresponding to the shift of T_g. The only results consistent with this view are those of Newman *et al.* (1978) and Hahn *et al.* (1981).

Dependence of the type of mesophase upon tacticity is to be expected only if the polymer backbone directly influences the packing of the mesogenic side chains because of steric hindrance. Even in the case of direct linkage of the mesogenic side chains, Blumstein *et al.* (1978) and Newman *et al.* (1978) did not find an influence of tacticity on the packing of mesogenic side chains.

3. Effect of Molecular Weight

The average molecular weight of the polymer and the molecular weight distribution are polymer specific properties that have to be considered in connection with the properties of the mesogenic polymer. In Table V three characteristic examples are given in which the phase behavior of a mesogenic monomer is compared with that of the corresponding polymer. In all three cases polymerization stabilizes the mesophase. For system 1 the monomer is not mesomorphous, whereas the polymer is nematic. In system 2 the nematic monomer yields a highly ordered smectic polymer. The transition to the isotropic phase is strongly shifted to higher temperature. For system 3 both monomer and polymer exhibit smectic and nematic phases. Again, the phase transition temperatures of the polymer are shifted to higher temperatures.

TABLE V

Comparison of the Phase Behavior of Monomers with the Corresponding
Mesogenic Polymers

	TRANSITIONS (K)	
	MONOMER	POLYMER
(1) $\quad$ O-(CH$_2$)$_2$-O-⬡-COO-⬡-OCH$_3$	K $\quad$ 370 $\quad$ i	g 374 n 394 i
(2) $\quad$ O-(CH$_2$)$_2$-O-⬡-COO-⬡-OC$_6$H$_{13}$	K 320n 326 i	g 333 s 388 i
(3) $\quad$ O-(CH$_2$)$_6$-O-⬡-COO-⬡-⬡	K 337 s 341 n 365 i	g 405 s 437 n 457 i

Stabilization of the mesophase is observed for all monomer–polymer systems investigated. In principle, upon changing from monomer to polymer the chemical constitution of the system, polarizibility, and form anisotropy should remain constant. Therefore, the strong shift of phase behavior must be attributed to restriction of translational and rotational motions of the mesogenic side chains. This suggests that a strong shift of phase behavior is expected in changing from the monomer to the dimer, trimer, etc. This shift should vanish when a chain length is reached for which no correlation exists between the motions of an additional monomer and the initial monomer. Investigation of the variation with molecular weight indicates no shift of phase behavior at higher degrees of polymerization.

C. Variation of the Flexible Spacer

1. Effect of Spacer Length

The effect of lengthening the flexible spacer has already been discussed in Section III.A. Increasing the spacer length stabilizes smectic phases and, according to the model considerations given in Section II.A, the flexibility increases.

We have so far considered the spacer as the flexible linkage of the mesogenic groups to the polymer backbone. Flexibility of the system, however can also be changed, if not every monomer is linked to a mesogenic group. In this way, main chain segments can be considered to form a part of the spacer. An example is shown in Table VI. In VIa the mesogenic moieties are attached to each monomer unit via spacers of increasing length, which

TABLE VI

Phase Behavior of Mesogenic Polysiloxanes[a]

$$CH_3-\overset{\underset{|}{O}}{\underset{|}{Si}}-(CH_2)_m-O-\bigcirc-COO-\bigcirc-OCH_3$$

m	TRANSITIONS (K)
(a) 3	g 288 n 334 i
4	g 288 n 368 i
6	(g 278 s 319) k 330 n 381 i

$$CH_3-\overset{\underset{|}{O}}{\underset{|}{Si}}-(CH_2)_m-O-\bigcirc-COO-\bigcirc-OCH_3$$

$$CH_3-\overset{\underset{|}{O}}{\underset{|}{Si}}-CH_3$$

m	TRANSITIONS (K)
(b) 3	g 276 n 294 i
4	g 267 s 323 i
6	g 263 s 350 i

[a] (a) Mesogenic moiety attached to each monomer unit, (b) mesogenic moiety attached to every second monomer unit.

causes a low temperature smectic phase when $m = 6$. If, on the average, only every second monomer unit is linked to a mesogenic moiety, as in VIb, an additional dimethylsiloxane monomer unit elongates the spacer. In this case, only the polymer with $m = 3$ still exhibits a nematic phase. This implies that segments of the backbone act to lengthen the mesogenic moiety. The increasing flexibility of the system is also indicated by the decrease of T_g as compared to that of the homopolymer in Table VIa.

2. Effect on the Order of Nematic Polymers

For low MW LC, molecular motions are restricted only by anisotropic interactions with neighboring molecules. In addition, the mesogenic moieties in polymers are restricted by linkage via the spacer to the main chain. The length of the flexible spacer influences the motions of side chain and main chain, and should therefore directly affect the state of order exhibited by the anisotropically ordered side chains. To get a more quantitative insight into these interactions, one may investigate the state of order in cholesteric polymers (see Section III.A.2).

The order of a nematic mesophase can be described by the order parameter introduced by Maier and Saupe (1958):

$$S = \langle P_2(\theta) \rangle \tag{4}$$

where θ describes the angle between the principal long axis of the molecule and the ordering axis of the liquid crystal (director), P_2 is the second Legendre polynomial, and the brackets describe an ensemble average. For a cholesteric phase in which the molecules lack an axis of rotational symmetry (chiral molecules do not possess a plane or center of symmetry), the order must be described by the set of four molecular parameters (Goossens, 1979; Straley, 1974)

$$F_1 = \tfrac{3}{2}(\cos^2\theta - \tfrac{1}{3}), \qquad F_2 = \cos 2\psi \sin^2\theta$$
$$F_3 = \cos 2\varphi \sin^2\theta, \qquad F_4 = \tfrac{1}{2}\cos 2\varphi \cos 2\psi\,(1 + \cos^2\theta) \tag{5}$$

Here the ensemble average of F_1 denotes the Maier–Saupe basic nematic order parameter S, which describes the uniaxial ordering of the long molecular z axis with respect to the Z axis, φ describes rotation about the Z axis, and ψ denotes rotation of the molecule around its long z axis (Fig. 11).

To determine experimentally the influence of a flexible spacer on the order, only the length of the spacer is allowed to vary, whereas the polymer backbone and the mesogenic moiety remain unchanged. Under these conditions it can be assumed that the molecular properties such as polarizability and form anisotropy of the mesogenic group remain constant. The order parameters are then dependent only upon the flexible spacer. A suitable system that fulfills these conditions is shown in Fig. 12b (Finkelmann and Rehage, 1980b). For the cholesteric copolymers, only the number of methylene units in the flexible spacer is changed, whereas all other molecular properties remain the same. These polymers will be compared with the chemically analogous monomeric system in Fig. 12a.

The basic nematic order parameter $S = \langle F_1 \rangle$ is determined directly from birefringence measurements (Maier and Saupe, 1961). The mean refractive indices of the cholesteric phase, $n_{e,n*}$ and $n_{o,n*}$, can be converted to the mean refractive indices of the corresponding untwisted nematic phases the relation given by Müller and Stegemeyer (1973):

$$n_{o,n} = n_{e,n*}, \qquad n_{e,n} = (2n_{o,n*}^2 - n_{e,n*}^2)^{1/2} \tag{6}$$

Using the extrapolation method suggested by Haller et al. (1973), calculation of the order parameter S is straightforward.

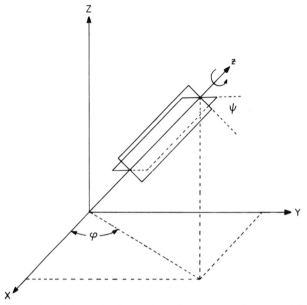

Fig. 11. Orientation of the molecular z axis; φ denotes a rotation around the Z axis and φ denotes a rotation of the molecule around its long z axis.

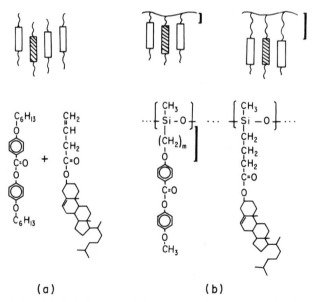

(a)   (b)

Fig. 12. Induced cholesteric phases: (a) monomer mixture, and (b) chemically analogous polymer with varying length of spacer: $m = 3, 4, 6$.

In Fig. 13b the temperature dependence of the order parameter S is shown for polymers with spacer length $m = 3$ (Benthack *et al.*, 1981). Qualitatively it should be noted that the slope of S with temperature corresponds to the data for the low MW system in Fig. 13a. Furthermore, at the clearing temperature $T^* = 1$, S vanishes discontinously, indicating a first-order cholesteric-to-isotropic transition, in agreement with the conclusions from

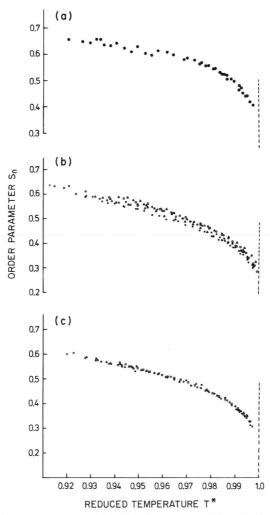

Fig. 13.   Order parameter S versus the reduced temperature, $T^* = T_m/T_{cl}$ (T_m = measuring temperature, T_{cl} = clearing temperature) for: (a) monomer (see Fig. 11a), (b) polymers with spacer length $m = 3$ (see Fig. 11b), (c) polymers with spacer length $m = 6$.

PVT measurements (see Section II.B). For a definite reduced temperature, say $T^* = 0.95$, the polymer with $m = 3$ shown in Fig. 13b exhibits an order parameter of 0.55. In Fig. 13c the polymer with $m = 6$ also has order parameter 0.55 at $T^* = 0.95$. Within the accuracy of these measurements, it must be concluded that, although the distance of the mesogenic group to the main chain varies, there is no difference of orientation of the principal axis of the mesogenic groups with respect to the director. The very similar low MW mixture in Fig. 13a has $S = 0.6$ at $T = 0.95$, hence there is a slight decrease of S on changing from the monomer to the polymer.

Further characterization of the degree of order of cholesteric polymers can be obtained experimentally by measuring the "helical twisting power" (htp) (Finkelmann and Stegemeyer, 1978). The htp is the slope of the reciprocal of the pitch P versus the mole fraction x_{ch} of added chiral molecules at a defined reduced temperature:

$$\text{htp} \equiv \left(\frac{dP^{-1}}{dx_{ch}}\right)_{T^*} = \bar{n}\left(\frac{d\lambda_R^{-1}}{dx_{ch}}\right)_{T^*} \tag{7}$$

P can be determined by measuring the wavelength of reflection (Eq. (3)) and the mean refractive index, $\bar{n}$; furthermore P is a function of molecular properties and the order parameters of Eq. (5) (Goossens, 1979). We assumed that the molecular properties of our polymers remain constant, although the spacer length is varied. Consequently, upon varying the spacer length, any change in htp is related to a change of the order parameters. According to model calculations (Straley, 1974) and the theory of Goossens (1979), the pitch is mainly influenced by the ratio of F_1 to F_4. F_1 is not changed within experimental error by varying the spacer length as seen above. We can therefore conclude that under these conditions a variation of the htp is mainly related to a change in F_4, which directly reflects rotation of the mesogenic groups around their long molecular axes. The experimental results are shown in Fig. 14 (Finkelmann and Rehage, 1980b). Curve I corresponds to the low MW mixture of Fig. 11a. The hatched plane indicates the area of curves obtained for low MW cholesteric phases of various cholesteryl esters in different nematic host phases. The variation of the slope is due to different order parameters of the nematogenic host phases and to changes of the chemical structure. In contrast to these monomers, curve II shows that a much stronger induced twist is observed for the polymers of Fig. 12b having $m = 3$. Because the system is chemically related to the systems represented in curve I, the strong shift of htp must be attributed to the mode of attachment of the mesogenic side chains to the polymer backbone. When the spacer is lengthened to $m = 4$ (curve III) and to $m = 6$ (curve IV), the slope decreases and approaches that of the low MW mixture (curve I). In comparison to these results, curve V shows the highest slope ever observed for a

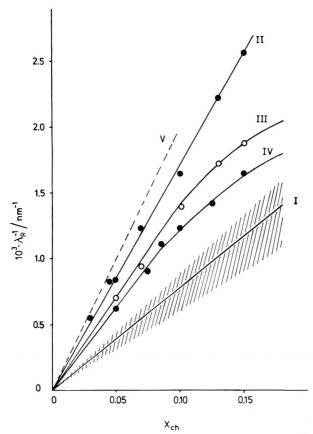

Fig. 14.   Inverse wavelength of reflection, λ_R^{-1}, versus mole fraction, x_{ch}, of the chiral comonomer for induced cholesteric phases (see Fig. 12) at $T^* = 0.95$ (curvg III, $T^* = 0.98$); curve I: monomer (see Fig. 12a); curve II: polymers with spacer length $m = 3$ (see Fig. 12); curve III: polymers with spacer length $m = 4$; curve IV: polymers with spacer length $m = 6$; curve V: monomer mixture DDCO/MBBA. (Reproduced from Finkelmann and Rehage, 1980, with permission of Hüthig and Wepf Verlag, Basel.)

low MW system (DDCO/MBBA; DDCO = 4.5 diphenyldinaphtol[2,l-e:1,2-9][1,4]diazocine, MBBA = 4-butyl-N-(4-methoxybenzylidene)aniline).

  These observations indicate a strong influence on the order of the mesogenic groups arising from the spacer length. With increasing spacer length htp decreases, indicating a decrease of F_4. Consequently, an increase of the spacer length is associated with less hindered rotation of the mesogenic groups around their long molecular axes. This confirms the principle that an increasing spacer length decouples motions of mesogenic side chains and the polymer backbone.

IV. Liquid Crystalline Elastomers

Linkage of conventional low MW mesogenic molecules to linear polymers has resulted in a new class of polymeric substances that combine the properties characteristic of the mesogen and the polymer. Above T_g these polymers exhibit an almost free motion of the chain segments. Because of this mobility the main chains are able to diffuse past each other, a property of the liquid state. On the other hand, in the liquid state the mesogenic moieties fixed via spacers to the backbone create an ordered mesophase more or less

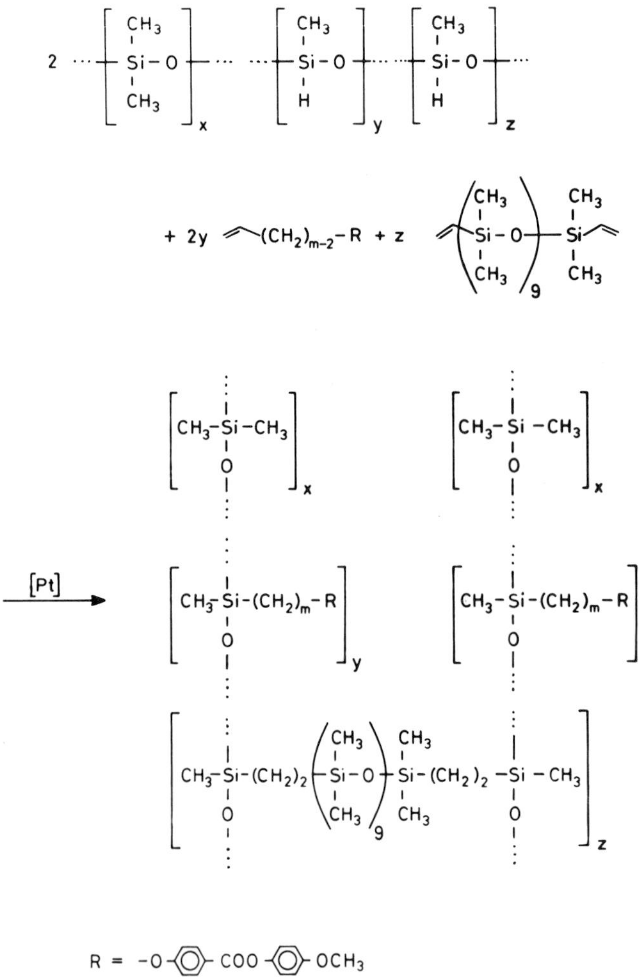

$$R = -O-\bigcirc-COO-\bigcirc-OCH_3$$

Fig. 15. Synthesis of LC polysiloxane networks.

independently of the main chain. If linear polymers are cross-linked by suitable agents, the chains can no longer diffuse past each other, although the mobility of the chain segments is still nearly free. In this way a fluid linear polymer is converted to a material that retains its shape and that can be viewed (Rehage, 1977) as an elastic liquid. If mesogenic groups are attached to the chain segments, elastic, form retaining liquid crystalline networks can be obtained. This concept has been realized with polysiloxanes (Finkelmann *et al.*, 1981) that can be converted to rubbers via the scheme presented in Fig. 15. By analogy to linear liquid crystalline polymers, variation of the substituents of the mesogenic groups allowed synthesis of nematic, cholesteric, and smectic elastomers. These liquid crystalline polymer networks have elastic properties typical of cross-linked polymers. The mesogenic side chains can be macroscopically oriented by mechanical deformation (see de Gennes, Chapter 5, this volume), which results in interesting opto-elastic properties.

References

Benthack, H., Finkelmann, H., and Rehage, G. (1981). *Int. Symp. Macromol. 27th, Abstr. Commun.* **2**, 961.

Blumstein, A., and Hsu, E. C. (1978). *In* "Liquid Crystalline Order in Polymers" (A. Blumstein, ed.), p. 150. Academic Press, New York.

Blumstein, A., Osada, Y., Clough, S. B., Hsu, E. C., and Blumstein, R. B. (1978). *In* "Mesomorphic Order in Polymers" (A. Blumstein, ed.), ACS Symposium Series, No. 74, p. 56. American Chemical Society, Washington, D.C.

Cser, F., Nyitrai, K., and Hardy, G. (1978). *In* "Mesomorphic Order in Polymers" (A. Blumstein, ed.), ACS Symposium Series, No. 74, p. 95. American Chemical Society, Washington, D.C.

Demus, D., Demus, H., and Zaschke, H. (1976). "Flüssige Kristalle in Tabellen." VEB Dtsch. Verlag Grundstoffind., Leipzig.

Dubault, A., Casagrande, C., and Veyssie, M. (1978). *Mol. Cryst. Liq. Cryst.* **41**, 239.

Elias, H. G., ed. (1977). "Polymerization of Organized Systems," Midland Macromolecular Monographs, Vol. 3. Gordon & Breach, New York.

Finkelmann, H., and Rehage, G. (1980a). *Makromol. Chem. Rapid Commun.* **1**, 31.

Finkelmann, H., and Rehage, G. (1980b). *Makromol. Chem., Rapid Commun.* **1**, 733.

Finkelmann, H., and Stegemeyer, H. (1978). *Ber. Bunsenges. Phys. Chem.* **82**, 1302.

Finkelmann, H., Ringsdorf, H., and Wendorff, J. H. (1978a). *Makromol. Chem.* **179**, 273; (Hüthig and Wepf Verlag, Basel).

Finkelmann, H., Happ, M., Portugall, M., and Ringsdorf, H. (1978b). *Makromol, Chem.* **179**, 2541; (Hüthig and Wepf Verlag, Basel).

Finkelmann, H., Ringsdorf, H., Siol, W., and Wendorff, H. J. (1978c). *Makromol. Chem.* **179**, 829; (Hüthig and Wepf Verlag, Basel).

Finkelmann, H., Koldehoff, J., and Ringsdorf, H. (1978d). *Angew. Chem., Int. Ed. Engl.* **17**, 935.

Finkelmann, H., Kock, H. J., and Rehage, G. (1981). *Makromol. Chem., Rapid Commun.* **2**, 317

Finkelmann, H., Lühmann, B., and Rehage, G. (1982). *Colloid Polymer Sci.* **260**, 56.

Frenzel, J. (1981). Ph.D. Thesis. Tech. Univ. Clausthal, Clausthal-Zellerfeld.

Frenzel, J., and Rehage, G. (1980). *Makromol. Chem., Rapid Commun.* **1**, 129.

Goossens, W. J. A. (1979). *J. Phys., Colloq. (Orsay, Fr.)* **40**, 158.

Gray, G. W., and Winsor, P. A., eds. (1974). "Liquid Crystals & Plastic Crystals," Vol. 1. Wiley, New York.

Hahn, B., Wendorff, J. H., Portugall, M., and Ringsdorf, H. (1981). *Colloid Polymer Sci.* **259**, 875.

Haller, I., Huggins, H. A., Lilienthal, H. R., and McGuire, T. R. (1973). *J. Phys. Chem.* **77**, 950.

Kelker, H., and Hatz, H. (1980). "Handbook of Liquid Crystals." Verlag Chemie, Weinheim.

Krause, S., and Roman, N. (1965). *J. Polym. Sci.* **173**, 1631.

Kusy, R. P. (1976). *J. Polym. Sci., Polym. Chem. Ed.* **14**, 1527.

Maier, W., and Saupe, A. (1958). *Z. Naturforsch., A* **13A**, 564.

Maier, W., and Saupe, A. (1961). *Z. Naturforsch., A* **16A**, 816.

Müller, W. U., and Stegemeyer, H. (1973). *Ber. Bunsenges. Phys. Chem.* **77**, 20.

Newman, B. A., Frosini, V., and Maganini, P. L. (1978). *In* "Mesomorphic Order in Polymers" (A. Blumstein, ed.), ACS Symposium Series, No. 74, p. 71. American Chemical Society, Washington, D.C.

Paleos, C. M., Filippakis, S. E., and Magomenou, G. (1981). *J. Polym. Sci., Polym. Chem. Ed.* **19**, 1427.

Perplies, E., Ringsdorf, H., and Wendorff, H. J. (1974). *Ber. Bunsenges. Phys. Chem.* **78**, 921.

Rehage, G. (1977). *Ber. Bunsenges. Phys. Chem.* **81**, 969.

Shetter, J. A. (1963). *J. Polym. Sci., Part B* **1**, 209.

Shibaev, V. P., and Plate, N. A. (1978). *Polym. Sci. USSR (Engl. Transl.)* **19**, 1065.

Shibaev, V. P., Platé, N. A., and Freidzon, Y. S. (1978). *In* "Mesomorphic Order in Polymers" (A. Blumstein, ed.), ACS Symposium Series, No. 74, p. 33. American Chemical Society, Washington, D.C.

Siol, W. (1979). Ph.D. Thesis. Univ. Mainz.

Straley, J. P. (1974). *Phys. Rev. A* **10**, 1881.

Tanaka, Y., Hototsuyanagi, M., Shimura, Y., Okada, A., Sakuraba, H., and Sakata, T. (1976). *Makromol. Chem.* **177**, 3035.

Tanaka, Y., Yamaguchi, F., Shiraki, M., and Okada, A. (1978). *J. Polym. Sci., Polym. Chem. Ed.* **16**, 1027.

Wendorff, H. J., Finkelmann, H., and Ringsdorf, H. (1978). *J. Polym. Sci., Polym. Symp.* No. 63, p. 245.

Zentel, R. (1980). Diplomarbeit. Univ. Mainz.

3

Rigid and Semirigid Chain Polymeric Mesogens

A. Ciferri

Istituto di Chimica Industriale
University of Genoa
Genoa, Italy

I. Generalities

Polymer liquid crystals may be formed from flexible polymers having mesogenic side chains organized into nematic, cholesteric, and smectic mesophases. This area of polymer liquid crystals was reviewed by Finkelmann (Chapter 2, this volume) and is not considered in this chapter. We shall concern ourselves with the alternative approach for obtaining polymer liquid crystals, that in which the liquid crystalline character is a property of the chain backbone. This goal can be achieved by building up asymmetric

63

particles based on a linear sequence of units that may, but often do not, possess intrinsic mesogenic character. This class of polymeric liquid crystalline materials is particularly suited for applications involving mechanical properties (Ciferri and Ward, 1979). This chapter describes the chain structure and the equilibrium behavior of these systems.

A. Theory

The theory of polymer liquid crystals has been greatly expanded in recent years. This theory attaches great importance to the geometrical asymmetry of the particle. For instance, according to Flory and Ronca (1979a), for an undiluted system, a critical value of the axial (length to diameter) ratio is given by

$$x_{crit} = L/d = 6.417 \tag{1}$$

When the axial ratio is much greater than 6.4, we are properly in the field of rigid and semirigid polymers. Then the simple geometrical asymmetry of the particle is sufficient to stabilize mesophase, even for a rather diluted system. In fact, the following (approximate) relationship (Flory, 1956a) illustrates how the minimum polymer volume fraction v_2 at which a nematic phase is stable in solution is decreased by an increase of the axial ratio:

$$v_2^* \simeq 8/x(1 - 2/x) \tag{2}$$

On the other hand, for axial ratios smaller than the critical value, the role of hard rod repulsion is still necessary but is not sufficient to assure the stability of the mesophase. According to a recent Flory and Ronca (1979b) theory asymmetric attractive interactions must be added to asymmetric repulsion. These asymmetric, soft interactions are of the same type as those originally considered by Maier and Saupe (1958a,b, 1960) to be solely responsible for the thermotropic behavior of undiluted systems. In this case we are more properly in the field of low molecular weight (MW) liquid crystals (LC). The interplay between the geometrical asymmetry, represented by the axial ratio, and the intensity of the orientation-dependent attraction, represented by a characteristic temperature T^*, is illustrated by the equation (Flory and Ronca, 1979b)

$$\tilde{T}_{NI} = T_{NI}/xT^* \tag{3}$$

Here $\tilde{T}_{NI}$ is a reduced nematic → isotropic transition temperature. This theory requires that T_{NI} increase with increasing axial ratio below the 6.4 limit. The value of T_{NI} should actually reach infinity when the axial ratio is greater than 6.4 and the anisotropic phase is sufficiently stabilized by hard rod repulsion. However, when the axial ratio is smaller than 6.4, T^* falls into the measurable range, with both hard and soft interactions contributing to the stabilization of the nematic phase of undiluted, thermotropic systems.

The inclusion of repulsive interaction in the classical treatment of the thermotropic behavior of low MW LC (Flory and Ronca, 1979b) was followed (Warner and Flory, 1980) by the inclusion of attractive interaction in the classical treatment of the lyotropic high MW LC. The result is that a consistent set of theories describes the behavior of both low and high MW rigid mesogens in their diluted or undiluted state. Treatments of mixtures or polydisperse systems having a distribution of axial ratios are also available (Flory and Abe, 1978; Flory and Frost, 1978; Frost and Flory, 1978; Abe and Flory, 1978; Flory, 1978a). Partial rigidity has been also theoretically considered (Flory, 1978; Marrucci and Sarti, 1979; Matheson and Flory, 1981), although not yet in the case of thermotropic systems. The experimental behavior of undiluted and diluted systems to be described in Sections II and III, respectively, will be compared with the predictions of the above theories.

B. Asymmetric Molecules

With this theoretical framework we consider now the types of chemical bonds that must be used to build asymmetric particles. Typical polymers consist of a linear sequence of monomer residues in which main chain bonds undergo extensive internal rotation. The resulting molecular shape is that of the familiar rather symmetric random coil (Flory, 1969). However, an asymmetric shape may be obtained by using a flow field, or by using special solvents, or particular types of bonds. The role of flow and other external fields is described in Krigbaum (Chapter 10, this volume). Solvents may induce asymmetry by promoting aggregation, by altering the potentials of internal rotation, or by favoring intramolecular hydrogen bonding. For instance, poly(γ-benzyl-L-glutamate) (PBLG) contains a peptide bond fixed in its planar trans configuration by its double bond character (Walton and Blackwell, 1973) and two adjacent rotating bonds

$$-\overset{|}{\underset{|}{C}} \overset{\psi}{\underset{}{\rightarrow}} \overset{\overset{O}{\|}}{C} \cdots N \overset{\phi}{\underset{\underset{H}{|}}{\leftarrow}} \overset{|}{\underset{|}{C}}-$$

which generally assure random coil behavior to the chain. However, in helicogenic solvents, all rotatable bonds are forced into equivalent positions by hydrogen bonding. A helix develops, and liquid crystalline behavior may be observed. Hydrogen bonding, along with other intermolecular interactions, also contributes to the stabilization of extended conformations in mesogenic cellulose by restricting rotation about the glycoside linkage (Walton and Blackwell, 1973). Numerous examples of alteration of chain rigidity induced by solvents can be given. Cellulose derivatives have been extensively studied in this respect (Buntjakov and Averyanova, 1972). The driving force for a more rigid conformation is not the eventual formation of the mesophase at sufficiently high polymer concentration (Pincus and de Gennes, 1978), but rather the particular affinity of a given solvent for some groups of the macromolecule, causing them to become exposed, with a consequent alteration of conformation and rigidity (Ciferri, 1975; Puett and Ciferri, 1971). Noticeable also is the case of poly(terephthaloyl-*p*-aminobenzhydrazide), which, in the absence of aggregation or coil → helix transition, forms anisotropic phases in sulfuric acid but not in organic solvents (Bianchi *et al.*, 1981).

Asymmetric conformations may alternatively be obtained using monomers that, irrespective of the solvent, give rise to main chain bonds with severely restricted internal rotations. Some poly(alkyl isocyanates) forming liquid crystals (Aharoni, 1979) may be a relevant example. The molecule contains a sequence of amide groups,

$$-\overset{\overset{\displaystyle R}{|}}{\underset{\displaystyle O}{C}}-N-\overset{\overset{\displaystyle R}{|}}{\underset{\displaystyle O}{C}}-N-\overset{\overset{\displaystyle R}{|}}{\underset{\displaystyle O}{C}}-N-$$

none of which is free to rotate because of the partial double bond character and to steric interference between the *n*-alkyl side chain and the carbonyl group (Burr and Fetters, 1976). These amide groups form a sequence of alternating cis and trans planar configurations giving rise to a rather rigid structure (Tsvetkov, 1976). Asymmetric structures may also be obtained using bonds for which internal rotation occurs, but has no effect on flexibility. The best known examples of polymer liquid crystals, the para-linked aromatic polyamides and polyesters, fall into this category. The amide bond is again fixed in the planar trans configuration because of its partial double bond character. Rotation can occur about the adjacent Ph–C and N–Ph bonds. If the angles α and β of the amide group are equal, these bonds are parallel. Moreover, if the benzene rings are connected to the chain through the para position, the direction of these bonds in all monomer units is the same. For this type of chain internal rotation cannot cause a bending of the chain on itself. The chain is naturally extended like a crankshaft with all axes of rotation parallel to each other and distributed along a cylinder parallel to the chain itself (Tsvetkov, 1976). The situation for the para-linked aromatic polyesters is entirely similar, provided NH is replaced by O (Hummel and Flory, 1980). Tsvetkov (1976) pointed out an analogy with low MW LC, where diene acids form a stable mesophase, whereas the same is not true for normal aliphatic acids. A regular alternation of rotating and nonrotating bonds is the factor that assures chain extension in these cases.

It is to be noted that the monomeric unit of a mesogenic polymer should possess adequate rigidity so that a large axial ratio results from polymerization. However, the axial ratios of the units may be below the value necessary to assure mesogenic behavior to the corresponding monomer, e.g., poly(p-hydrobenzoic acid), poly(p-benzamide). Cases in which the monomers are extended enough to be themselves mesogens are, however, also known. This is, for instance, the case in rigid low MW mesogens polymerized in a regular alternation with flexible methylene units (cf. Section II.B). Thus not all bonds along the chains are required to contribute to the asymmetry; localized asymmetry may be enough to ensure mesogenic behavior.

C. Controlling the Melting Temperature

In the search for polymer liquid crystals intended for both scientific investigation or industrial application, one may be inclined to select the monomers that give rise to the most rigid, extended structures. Unfortunately, the requirement of a completely rigid structure must be mitigated with the requirements of polymers with accessible melting temperatures or sufficient solubility. Rigid structures such as poly(p-hydroxybenzoic acid), cellulose, and poly(p-phenyleneterephthalamide) are untractable as undiluted systems, since their melting temperature is close to the decomposition temperature. The latter polymers are only tractable as a diluted system, but in rather exotic solvents. The high melting behavior of extended structures is related to low melting entropy. Considerable effort is presently devoted to modifying the molecular architecture of rigid systems in order to reach more practical conditions for industrial processing and scientific investigation. In a sense this effort parallels that which took place with low MW mesogens when compounds exhibiting mesophases at room temperature were needed for display applications (Gray, Chapter 1, this volume). However, as long as the primary application of rigid and semirigid mesogens remains that of polymers with superior mechanical properties, convenient crystal → nematic transition temperatures (T_{CN}) should be of the order of 250–350°C.

The general methods of aiming at an increase of the melting entropy are illustrated in Table I. Copolymerization and the use of asymmetrical units depress the diffusion of the crystalline solid by reducing the probability of finding a run of crystallizable units. Kinks or bends based on essentially rigid bonds (e.g., ortho or meta bisphenols, isophthalic or phthalic acid) act in a similar manner. The use of flexible bonds and sequences tends to increase the entropy of the melt. Generally these alterations also tend to decrease the axial ratio of the molecule and the mesogenic behavior may eventually disappear. For practical applications it may be desirable to lower T_{CN} (to $\simeq 250°C$) but not T_{NI}. Appropriate study may suggest which strategies to use toward this goal. For instance, two kinks along the polymer repeat may not destroy the overall rodlike nature (Preston, 1975). Also, as in the typical cases of poly(n-hexyl isocyanate) and PBLG, a long and flexible chain occurring as a side chain increases the melting entropy without compromising the rigidity of the backbone.

TABLE I

Methods Used for Depressing T_{CN}

Copolymerization	
Asymmetric units	
Kinks or bends	
Flexible bonds	$-O-$, $-CH_2-$, $-NH-NH-$
Flexible sequences	$\bullet\bullet\bullet-CH_2-CH_2-CH_2-\bullet\bullet\bullet$
Flexible side chains	
Keep MW low	
Use of a diluent	

Some control over the melting temperature can also be achieved by using rather low MW samples. It will be shown in Section II.B that both T_{CN} and T_{NI} can be reduced by decreasing MW (provided the inherent viscosity of samples is smaller than ≈ 0.4 dl g^{-1}). For fiber application one is generally interested in high MW samples, since these give greater values of strength. However, for the polyesters, an increase of MW can be obtained by post-polymerization in the fiber form, so that processing of low MW samples becomes attractive (Schaefgen *et al.*, 1975; Prevorsek, Chapter 12, this volume).

D. Determining Chain Rigidity

The above considerations indicate some of the strategies that are followed for the synthesis and the processing of rigid and semirigid chains exhibiting mesogenic behavior. The next problem of a general nature is the quantitative characterization of the degree of chain extension. The simplest case would be that in which internal rotation is totally inhibited giving rise to rigid, rodlike molecules (Fig. 1a) in which the rod length is equal to the contour length L and much greater than the diameter d. Semirigid chains may instead be modeled as a continuously bending wormlike chain (Fig. 1b) or as a collection of rigid rods connected by flexible joints (Fig. 1c). The flexible joint may sometimes involve a sequence of flexible bonds as in Fig. 1d. In most cases the so-called persistence length q furnishes a measure of chain extension. Suppose a freely rotating chain is located at the center of a Cartesian coordinate system with its first link along y. Then q is the average sum of the projections of the remaining bonds along y, and is a measure of the memory of the direction taken by the first segment (Flory, 1969). This parameter can be calculated theoretically from conformational analysis, or can be determined experimentally using techniques such as light scattering, viscosity, and flow birefringence, all of which are sensitive to chain dimensions. From the experimentally determined radius of gyration R_G the persistence length is derived using statistical theories for wormlike chains (Kratky and Porod, 1949; Benoit and Doty, 1953) giving

$$\overline{R_G^2} = q^2\{\tfrac{1}{3}(L/q) - 1 + (2q/L) - (2q^2/L^2)[1 - \exp(-L/q)]\} \tag{4}$$

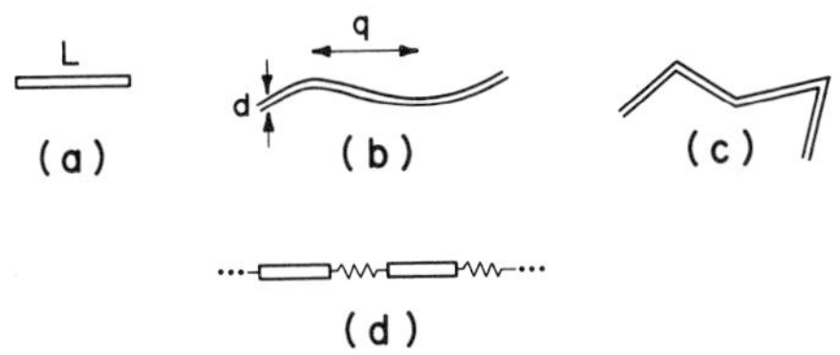

Fig. 1. Models for rigid ($d \ll L < q$) and semirigid ($d \ll L \gg q$) chains.

where L is the contour length. The latter treatment is appropriate for chain behavior intermediate between that of a rigid rod (R_G proportional to the number N of bonds) and a flexible coil (R_G proportional to $N^{1/2}$). The persistence length can also be determined using statistical theories for flexible chains if the chain is long enough to exhibit Gaussian behavior. The latter theories yield the value of the Kuhn segment length A related to q by (Flory, 1969)

$$A = \overline{r_0^2}/L = q/2 \tag{5}$$

where r_0 is the unperturbed end-to-end distance of the chain. A quantitative comparison of the q values determined for various polymers will be used here as a measure of their relative chain extension. It is somewhat more arbitrary to define a chain as a rigid or as a semirigid one. In fact, while q is independent of molecular weight, the dependence of the radius of gyration upon N is typical of rodlike molecules only at relatively low N, and is always typical of flexible Gaussian coils at rather large N. Thus, we shall define as rigid a chain for which the contour length is smaller or comparable with the persistence length. A semirigid chain is then a chain for which the contour length is greater than the persistence length (but L/q still below the Gaussian limit).

It should be noted that the persistence length, determined from dilute solution studies, will be used to assess the axial ratio, which controls mesophase formation in concentrated solutions and melts according to Eqs. (2) and (3). The question has arisen whether the persistence length may be affected by polymer concentration, or actually have a different value in isotropic and anisotropic phases. Generally no experimental or theoretical data indicate an effect of polymer concentration, although, as discussed in Section I.B, the persistence length may vary from one solvent to another. However, in mixed solvents with one strongly interacting component, an alteration of rigidity due to changes of composition or polymer concentration may indeed occur (Ciferri, 1975). Concerning the possibility that a semirigid polymer may assume an overall more extended conformation in the anisotropic than in the isotropic phase, present treatments for semirigid polymers (cf. Section I.A) suggest that mesophase formation is controlled by the axial ratio of the rigid, persistent segment at the transition point and not by the overall conformation.

E. Recognizing the Mesophase

For both diluted and undiluted systems it is often difficult to assess the occurrence of the mesophase and contradictory assignments are sometimes reported. Often the microscopic texture characteristic of low MW mesogens

(Demus and Richter, 1978) is not seen in polymers. What appears to be an anisotropic, birifrengent, liquid phase may be associated with gelation (crystallization) in concentrated solutions (Balbi *et al.*, 1980), or with a dispersion of higher melting crystallites in a partially molten polymer (Lader and Krigbaum, 1979). Moreover, time effects are often present both in solution (Balbi *et al.*, 1980) and in the melt. Especially when the T_{CN} and T_{NI} temperatures are close to each other, the formation of the texture may be observed on cooling, but not on heating (e.g., Asrar *et al.*, 1982a).

The preferred technique for studying liquid crystalline mesophases, optical polarized microscopy, is therefore only really useful when a texture develops. In favorable circumstances textures typical of low MW LC have been observed for polymers. Batonnets of smectic A structure growing into a fan-shaped texture upon cooling have been observed (Fig. 2a) for homopolyesters of *p,p'*-bibenzoic acid with aliphatic diols (Krigbaum *et al.*, 1982; Meurisse *et al.*, 1981). Cholesteric textures of the planar type have been

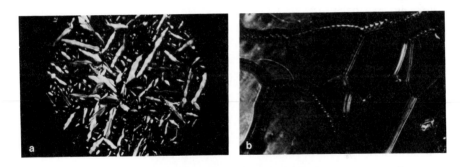

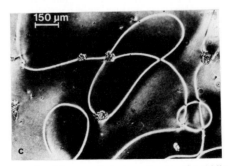

Fig. 2. Typical textures for: (a) a smectic homopolyester of *p,p'*-bibenzoic acid with hexamethylene glycol at 205°C (Krigbaum *et al.*, 1982) (40 ×); (b) a cholesteric homopolyester of 4,4'-dihydroxy-α-methylstilbene and adipic acid in a mechanical mixture with 80% of the corresponding nematogen containing adipic acid $T = 200$°C (Krigbaum *et al.*, 1981) (40 ×); (c) nematic PBA in solution of DMAc/3% LiCl. (Reprinted with permission from Conio *et al.*, *Macromolecules* **14**, 1084 (1981). Copyright 1981, American Chemical Society.)

observed, for instance, for solutions of cellulose derivatives (Werbowyi and Gray, 1980) and for plasticized homopolyesters or copolyesters (Fig. 2b) (Krigbaum et al., 1981). Cholesteric fingerprint patterns are easily observed with PBLG solutions (see, e.g., DuPré, Chapter 7, this volume). The typical threadlike pattern of the nematic phase (Fig. 2c) is often observed for solutions of rigid polymers (Panar and Beste, 1977; Aharoni, 1979; Millaud et al., 1978a, 1979a; Conio et al., 1981), but not always for thermotropic polyesters (e.g., Asrar et al., 1982a). Antoun et al. (1981) report schlieren textures for a polyester melt with disinclination points with four or two dark brushes. The number of threads decreased as the temperature was increased between T_{CN} and T_{NI}.

The suggestion that the high viscosity of the melt may hinder the formation of the texture (Krigbaum et al., 1981; Griffin and Havens, 1981) has not been substantiated. Asrar et al. (1982a) observed that some polyesters develop the schlieren texture irrespective of MW. Panar and Beste (1977) noticed for poly(p-benzamide) (PBA) solutions a tendency toward a uniform ordering, both spontaneously and in the magnetic field, manifested by the gradual disapperance of the threadlike pattern. Millaud et al. (1978a, 1979a) observed that for solutions of aromatic polyamides and polyazomethines disinclination lines with strength $+1$ (Chandrasekhar, 1977) were formed by a mechanical disturbance of a thin layer of solution between glass plates, but tended to disappear as the relaxation proceeded. The evolution of the threadlike pattern occurred more quickly when the molecular weight decreased or the polymer concentration increased, but was not necessarily correlated with the viscosity of the medium. The results were interpreted in terms of high values of the elastic constant k_{33} for polymers. Additional considerations on domain formation in polymers are presented in this book (Meyer, Chapter 6; Asada, Chapter 9; Krigbaum, Chapter 10). Much study is needed in this area.

Differential scanning calorimetry (DSC) is also employed, not only for melts (e.g., Griffin and Havens, 1981), but also for solutions (e.g., Navard et al., 1981). However, its interpretation is often complicated by multiple peaks due to recrystallization or to several crystalline forms (Lader and Krigbaum, 1979; Griffin and Havens, 1981). Annealing and comparison of heating and cooling curves with regard to the evolution and supercooling of peaks may sometimes help in the assignment of the transitions.

When the nematic texture is not observed and DSC not informative the occurrence of the mesophase can be established by developing electro-hydrodynamic instabilities in the polymer sample (Krigbaum et al., 1980). This technique and other techniques based on flow and magnetic fields are described by Krigbaum (Chapter 10, this volume). Small angle light and x-ray scattering techniques (e.g., Antoun et al., 1981), and IR and NMR spectroscopy have also been employed.

Finally, an extremely useful and reliable tool for assessing the type of mesophase is the determination of the phase diagram of the polymer with a reference mesogen. The compatibility of similar mesophases allows the determination of the unknown mesophase from the extrapolation of mutual solubility lines (Fayolle *et al.*, 1979; Griffin and Havens, 1980). Use of the contact method (Kofler and Kofler, 1954) allows great rapidity in the assessment of the phase diagram.

II. Undiluted Systems

We can now begin a more specific analysis of the various polymer mesogens that have been reported, starting with undiluted systems. An impressive amount of work is being performed in the field of thermotropic polyesters, which is almost impossible to discuss properly in a short review. The simple processing of these systems allows one to predict industrial developments in this field, although no development has yet been announced. The polyesters are a large group, which it is convenient to subdivide into three subgroups: the rigid homopolyesters; the semirigid homopolyesters having rigid groups and flexible spacers; and the copolyesters. (Undiluted, nonpolyester systems have also been reported, some of which are included in Table II.)

A. Rigid Homopolyesters

Completely rigid polymers of high symmetry such as poly(p-hydroxy-benzoic acid) and poly(p-phenyleneterephthalate) (Cottis *et al.*, 1973)

$$\cdots - O - \bigcirc - CO - O - \bigcirc - CO - \cdots$$

$$\cdots - CO - \bigcirc - CO - O - \bigcirc - O - \cdots$$

(I)

exhibit melting temperatures too high (e.g., 467°C, Frosini *et al.*, 1977) to be conveniently handled. The rigidity of these chains has been evaluated theoretically from conformational analysis by Erman *et al.* (1980). For the two above structures they obtained $q = 740$ and 751 Å. Taking as chain diameter a value of ~ 5 Å, the axial ratio of a chain containing one persistence length (corresponding to a degree of polymerization of ~ 110 and MW $\sim 13,500$) is ~ 150. A chain of this length can certainly be polymerized, but even if decomposition did not occur, one could not expect it to show thermotropic behavior since, according to the theory embodied in Eq. (3), its T_{NI} would be infinite. This type of polymer could exhibit lyotropic behavior (Eq. (2)) only if a suitable solvent existed.

Less symmetric structures allow a reduction of melting temperature below the decomposition temperature. Examples taken from the patent literature (Jackson *et al.*, 1979; Payet, 1978; Schaefgen *et al.*, 1975) are

The latter contains as unsymmetrical monomer and an ether bond. A T_{CN} temperature of the order of 340°C was reported for these polymers. No detailed study of the mesophase was reported. The high modulus of the fibers spun from the melt was taken as evidence of a nematic mesophase in the melt. An important characteristic of the patented process is the post-polymerization step. An oriented fiber was first obtained by spinning a low MW nematic melt. The low MW avoids high melt viscosity and high spinning temperatures. Later, postpolymerization in the solid fiber allowed an increase of MW and of mechanical properties (see also Prevorsek, Chapter 12, this volume). We note that these systems are thermotropic only in the sense that they show a crystal → nematic transition, but are not demonstrably thermotropic with respect to the nematic → isotropic transition, in line with the above considerations pertaining to extremely high axial ratios. From the point of view of fiber application, polymers unable to attain the isotropic melt should be desirable since the orientation can not be dissipated during processing.

B. Semirigid Homopolyesters

1.   Nematic Systems

A quite different situation is observed with the homopolyesters in which the repeating unit has rigid segments, which are often mesogenic even before polymerization, and flexible spacers. Roviello and Sirigu (1977) prepared aliphatic esters of 4,4′-dihydroxy-α-methylstilbene (**III**). Members with n

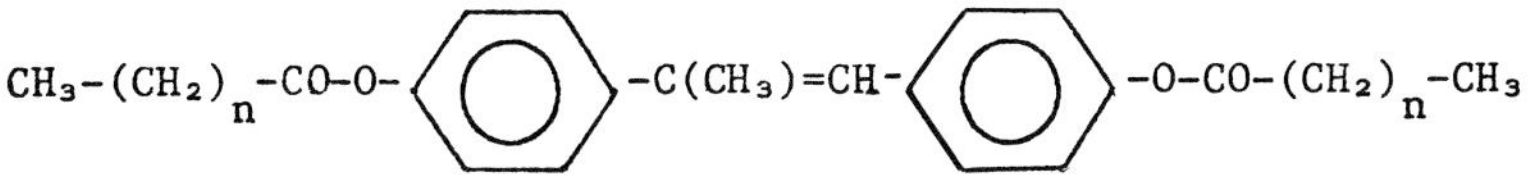

$$CH_3-(CH_2)_n-CO-O-\langle\bigcirc\rangle-C(CH_3)=CH-\langle\bigcirc\rangle-O-CO-(CH_2)_n-CH_3$$

(**III**)

between one and four exhibited nematic mesophases with T_{CN} between 107 and 53°C and T_{NI} between 113 and 97°C, both transition temperatures decreasing with n, as often observed with low MW LC (Gray, Chapter 1, this volume). Even the compound without the ester bond (**IV**) is a nematogen,

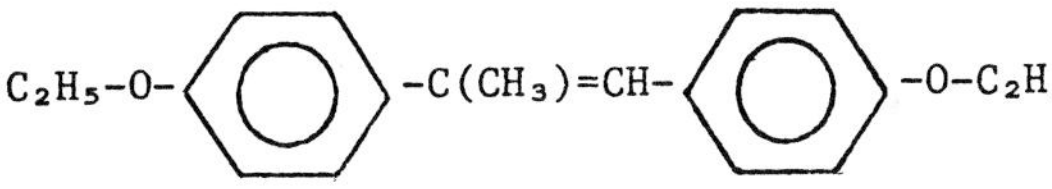

$$C_2H_5-O-\langle\bigcirc\rangle-C(CH_3)=CH-\langle\bigcirc\rangle-O-C_2H_5$$

(**IV**)

with $T_{CN} = 106°C$ and $T_{NI} = 121°C$ (Cox, 1972). Roviello and Sirigu (1982) then used these mesogenic units to prepare polyesters according to

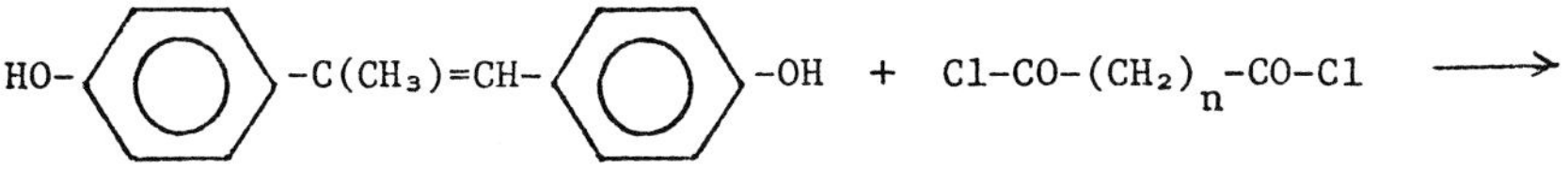

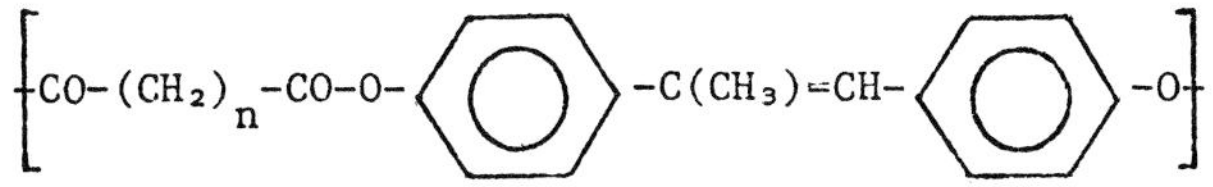

$$HO-\langle\bigcirc\rangle-C(CH_3)=CH-\langle\bigcirc\rangle-OH \ + \ Cl-CO-(CH_2)_n-CO-Cl \ \longrightarrow$$

$$\left[CO-(CH_2)_n-CO-O-\langle\bigcirc\rangle-C(CH_3)=CH-\langle\bigcirc\rangle-O\right]$$

(**V**)

The transition behavior, as a function of n, is illustrated in Fig. 3. For these polymers $n \geq 6$ since polymers with $n \leq 4$ were not fusible. The mesophase

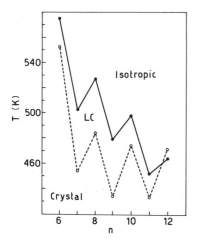

Fig. 3. Variation of T_{CN} and T_{NI} temperatures with the number of methylene for homopolyesters of 4,4'-dihydroxy-α-methylstilbene and aliphatic acids. (From Roviello and Sirigu, 1982, with permission of Hüthig and Wepf Verlag, Basel.)

was indicated to be nematic, although no schlieren patterns were reported. Note that these polymers are truly thermotropic, in the sense that the nematic → isotropic transition temperature is now within the measurable range. These results illustrate a number of features that were also observed in work by Griffin and Havens (1979, 1981), Strzelecki and Van Luyen (1980), Asrar *et al.* (1982a), and Antoun *et al.* (1981) on other polyesters formed by rigid units and flexible spacers. The first feature is the effect of an increase of the number of methylene units, which reduces both the crystal → nematic and the nematic → isotropic transition temperatures, with an evident even–odd effect on both transitions. This effect is similar to that observed for low MW LC (Gray, Chapter 1, this volume) and side chain LC polymers (Finkelmann, Chapter 2, this volume). However, some differences also exist. Upon increasing n the smectic mesophase is seldom observed (see, however, Van Luyen and Strzelecki, 1980) and, due to the different ways in which n affects T_{CN} and T_{NI}, there is only a discrete range of n in which the mesophase is stable (Asrar *et al.*, 1982a). The second feature is the effect of the degree of polymerization. By comparing the absolute values of the transition temperatures of the low MW analogs and of the polymers we note that polymerization increases both transition temperatures. The latter effect is better shown in Fig. 4 for homopolyesters based on 4,4'-dihydroxybiphenyl and dibasic acids with seven or eight methylene units (Asrar *et al.*, 1982a).

$$\left[CO-(CH_2)_n-CO-O- \bigcirc\!\!\bigcirc -O \right]$$

(VI)

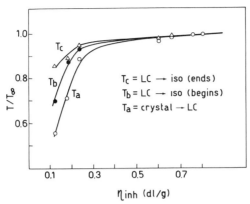

Fig. 4. Variation of the ratio of transition temperatures T_a ($\bigcirc$), T_b ($\bullet$), T_c ($\triangle$) to the corresponding value measured at the highest inherent viscosity, with inherent viscosity. The polymer is a homopolyester based on dihydroxybiphenyl and aliphatic acids with 7 or 8 methylene units. (From Asrar *et al.*, 1982a.)

Transition temperatures were determined by hot-stage microscopy. Temperature T_a refers to the crystal → liquid crystal transition and temperatures T_b and T_c refer to the onset and to the end of the liquid crystal → isotropic transition, respectively. The ratio of these transition temperatures to the corresponding values found for the sample with the highest inherent viscosity is plotted against inherent viscosity. We observe that the dependence of transition temperatures upon MW is large when the inherent viscosity <0.4. Moreover, for the low MW samples the range between crystal melting and complete isotropization is broader and includes a biphasic region where anisotropic and isotropic phases coexist. The coexistence of two phases in the melt is attributed to the MW dispersion of the sample (Krigbaum *et al.*, 1981; Flory and Abe, 1978). The possibility that other factors may also be responsible for the occurrence of a "degree of nematicity" is currently under consideration.

Polymers of type **VI** were also reported by Van Luyen and Strzelecki (1980) and by Blumstein *et al.* (1979). The latter also investigated polymers derived from sebacyl chloride and bisphenols

$$\text{HO}-\!\!\left\langle\!\!\bigcirc\!\!\right\rangle\!\!-Y-\!\!\left\langle\!\!\bigcirc\!\!\right\rangle\!\!-\text{OH}$$

where Y is —CH=CH— or —CH=N—. Similar polymers were also described by Iimura *et al.* (1980).

The systems investigated by Griffin and Havens (1979, 1981) are

$$Cl-\overset{O}{\underset{\|}{C}}-\bigcirc\!\!\!-O-(CH_2)_{\overline{x}}-O-\bigcirc\!\!\!-\overset{O}{\underset{\|}{C}}-Cl + HO-\bigcirc\!\!\!-O-(CH_2)_{\overline{y}}-O-\bigcirc\!\!\!-OH \longrightarrow$$

$$\left\{\overset{O}{\underset{\|}{C}}-\bigcirc\!\!\!-O-(CH_2)_{\overline{x}}-O-\bigcirc\!\!\!-\overset{O}{\underset{\|}{C}}-O-\bigcirc\!\!\!-O-(CH_2)_{\overline{y}}-O-\bigcirc\!\!\!-O\right\}$$

y = 6 , or 8 , or 10; x = 2 ÷ 10 , 12

(VII)

Three series of polyesters were prepared corresponding to the three diphenols having the indicated value of y. Within each series the effect on the transition temperature of increasing x from 2 to 12 was investigated. The mesophase was demonstrated to be nematic from phase diagram studies (Griffin and Havens, 1980). Both T_{CN} and T_{NI} decreased with the length of the aliphatic chain, with an odd–even effect. The clearing entropies and enthalpies, ΔS_{NI} and ΔH_{NI}, did not exhibit a corresponding regular trend with increasing n (as observed with low MW LC, Marzotko and Demus, 1975), but the even members had larger ΔH_{NI} than the odd ones. Absolute values of ΔH_{NI} were in the range of 0.45 to 2.20 kcal mol^{-1}, somewhat greater than those for low MW LC (range for the latter is 0.02–2.30 kcal mol^{-1}, Marzotko and Demus, 1975).

The ratio of transition entropies $\Delta S_{NI}/\Delta S_{CI}$, which could afford a measure of the relative order in the phases involved, is difficult to evaluate because of the unknown degree of crystallinity of the solid polymer. With a debatable approximation Griffin and Havens concluded that this ratio is in the range 0.05–0.23, significantly greater than the corresponding ratio for low MW LC (0.02–0.03, Marzotko and Demus, 1975). When a reliable determination of $\Delta S_{NI}/\Delta S_{CI}$ is available it will be possible to use it to make useful suggestions concerning the structure of the nematic phase of a semirigid polymer (Krigbaum and Salaris, 1978).

Thermodynamic data were also obtained by Antoun *et al.* (1981) for polyesters from 4,4′-dihydroxy-α,ω-diphenoxyalkanes and terephthalic acid

$$\left[O-\bigcirc\!\!\!\!\bigcirc\!\!\!\!-O-(CH_2)_n-O-\bigcirc\!\!\!\!\bigcirc\!\!\!\!-O-CO-\bigcirc\!\!\!\!\bigcirc\!\!\!\!-CO\right]$$

(VIII)

for which the effect of n was investigated, and for polyesters from 1,10-

bis(p-carboxyphenoxy)decane and substituted hydroquinones

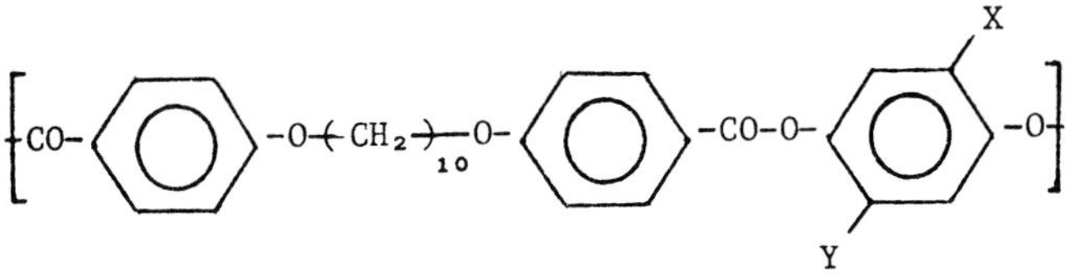

(IX)

for which the effect of substitution was investigated. Increasing n in series **VIII** from five to ten again produced a depression of transition temperatures and an even–odd effect. For this series, ΔS_{NI} increased regularly with n, from 0.37 to 1.2 cal mol^{-1} K^{-1}, with a corresponding trend for ΔH_{NI} from 0.22 to 0.67 kcal mol^{-1}. Polyesters of series **IX** exhibited greater ΔS_{NI} and ΔH_{NI} than those of series **VIII** (e.g., $\Delta S_{NI} = 1.7$ cal mol^{-1} K^{-1}; $\Delta H_{NI} = 0.97$ kcal mol^{-1} when X $=$ Y $=$ H). Still greater values resulted from unsymmetrical substitution on the hydroquinone ring, i.e., for X $=$ H and Y $=$ CH$_3$, $\Delta S_{NI} = 4.6$ cal mol^{-1} K^{-1} and $\Delta H_{NI} = 2.5$ kcal mol^{-1}, accompanied by a reduction of T_{NI}. Interpretation of these thermodynamic data in terms of order of the mesophase was attemped. However, the interpretation is complicated by the fact that ΔS_{NI} also reflects changes in the isotropic state.

An additional parameter that affects this class of polyesters is the length of the rigid block. Van Luyen and Strzelecki (1980) investigated first the effect of the number of methylene units on the transitions of the polymer

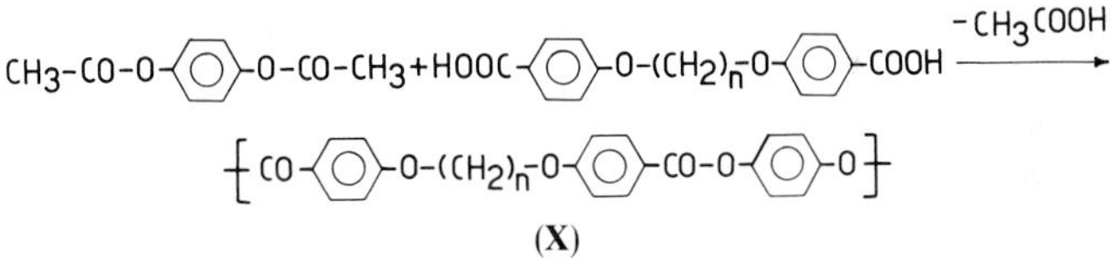

(X)

The depression of the transition temperatures was of the same form as that we have already discussed, although they also reported the occurrence of a smectic mesophase when $n > 6$. The novelty of their study is, however, the attempt to alter the size of the rigid block while keeping constant the number of methylene units. They considered several polyesters, among which are those having the rigid blocks

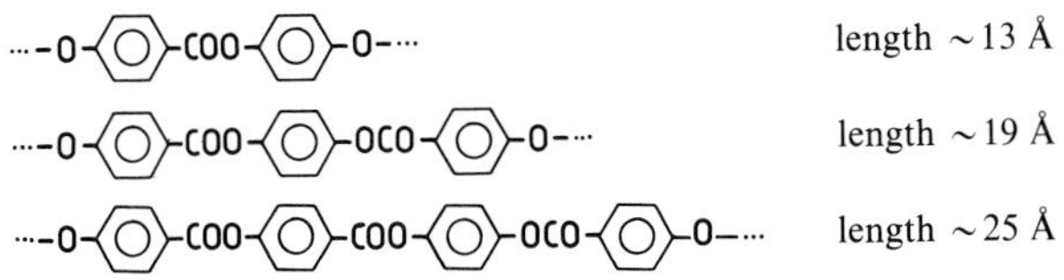

(XI)

for a series of polyesters, all having $n = 9$. The values of T_{NI} temperatures, which increased with the length of the rigid block, were, respectively, 200, 305, and 350°C. We can see that the structure of the short rigid block in **XI** is the same as for the rigid homopolyesters in **I**. The latter had a persistence length of ~ 750 Å and were not expected to exhibit a measurable T_{NI}. The fact that T_{NI} for the present polymers has decreased to a measurable value suggests a simple interpretation of the behavior. For this class of polymers the axial ratio corresponding to the length of the rigid block is never greater than six. Thus we are in the range of validity of Eq. (3), which suggests that the liquid crystalline behavior may be simply controlled by the axial ratio of the rigid block aided by attractive dispersion interactions. However, when the axial ratios of the blocks in **XI** and the corresponding T_{NI} temperatures are used in conjunction with Eq. (3), using the $\tilde{T}_{NI}^{-1}$ reported by Flory and Ronca (1979b), the calculated T^* is not a constant. Thus the theory valid for independent particles with axial ratios <6.4 cannot be applied to segmented polymers having $x < 6.4$. The result is not unexpected since the data illustrated above indicate a dependence of T_{NI} upon the number of methylene units and upon MW, which is not included in the theory. If one treats the flexible segment as a diluent of the rigid blocks and follows Warner and Flory (1980) in replacing T^* with $T^* v_r$ where v_r is the volume fraction of rigid blocks, Eq. (3) still fails to represent the data of Van Luyen and Strzelecki (Asrar et al., 1982a). A satisfactory theory for the behavior of this class of nematogens is still not available.

We also do not have enough information concerning the organization of the flexible spacers within the nematic phase, and few fiber spinning data have been reported. A report by Liebert et al. (1980) indicates that orientation may be obtained by crystallization from a mesophase subjected to a magnetic field. Bernstein (1981) has performed theoretical calculations suggesting that the flexible spacers extending from a surface may adopt the extended trans conformation. Volino et al. (1981) report some evidence suggesting that the flexible unit is partially aligned along the director in a nematic phase. If this turns out to be the real situation, a larger persistence length than that corresponding to the rigid segment may have to be considered in future elaboration of the theory. Moreover, high orientation might be easily achieved by fiber spinning.

2. Smectic and Cholesteric Systems

Occurrence of smectic H mesophases, which would transform to nematic phases on increasing temperature, was reported by Van Luyen and Strzelecki (1980) for polymers **X** having $n > 6$. Blumstein et al. (1979) reported from X-ray evidence that polymer **VI** with $n = 8$ exhibited a smectic mesophase above 200°C. This assignment is, however, in conflict with that made by

Van Luyen and Strzelecki (1980), who reported the same polymer to be nematic between 210 and 275°C. A very clear-cut, smectic mesophase was exhibited (Krigbaum et al., 1982; Meurisse et al., 1981) by some homopolyesters of p,p'-bibenzoic acid with aliphatic diols.

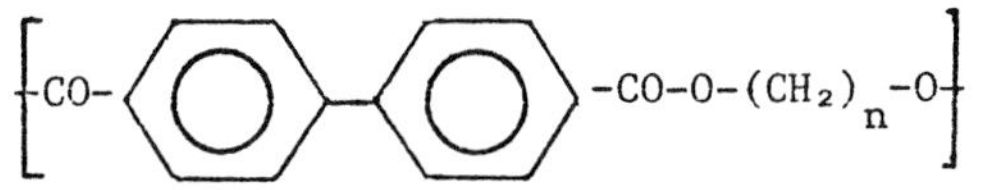

(XII)

The polymer with $n = 6$ exhibited (Fig. 2a) batonnets of the smectic A mesophase growing into a fan shaped texture upon cooling at 205°C. It is interesting that while polymers **VI** with the sequence

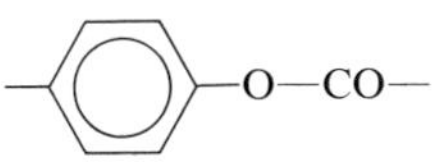

are often nematic, polymers **XII** with sequence

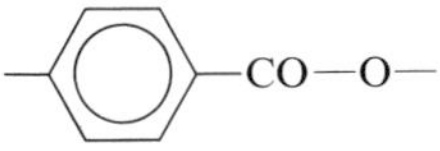

are instead smectic. The role of the number of methylene units and their even or odd nature should, however, always be considered in making this type of comparison. In fact, for polymers **XII** an increase of n caused the familiar depression of transition temperatures with an even–odd effect, but the mesophase was more evident for the even members while for polymers **VI**, odd members were nematic and even members were smectic (Asrar et al., 1982a).

Meurisse et al. (1981) also investigated the behavior of terphenyl derivatives

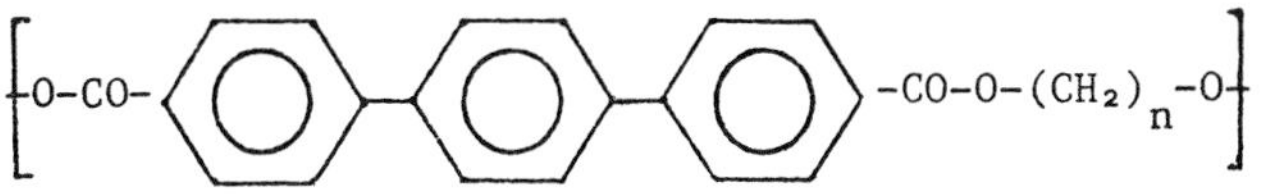

(XIII)

prepared because some low MW dialkyl-p-terphenyl 4-4″-carboxylates exhibit smectic S_E, S_C, and S_A mesophases (Demus and Zaschke, 1974). The polyesters with $n = 2$ and 4 were prevalently nematic, but there was an indication of a smectic C (or A) → nematic transition. Since the nematic range of the polymer with $n = 2$ is only 8°C, it was only upon quenching from the isotropic phase that batonnets evolving into a fan-shaped texture were observed. Polymers with $n = 5$, 6, and 10 showed more definite smectic phases.

Cholesteric phases were observed when chiral (+)-3-methyl adipate was inserted into the flexible spacer of nematogenic polymers (Krigbaum et al.,

1981; Asrar et al., 1982a; Van Luyen et al., 1980; Vilasagar and Blumstein, 1980; Blumstein and Vilasagar, 1981). Interest in cholesteric rather than nematic mesophases is related to the possibility of producing solid films having an ultrahigh modulus along any arbitrary direction in the plane of the film (Krigbaum et al., 1979). It is known that with nematic polymers one can only produce fibers because of the uniaxial alignment characteristic of the mesophase. Films have, however, even greater practical interest than fibers. Biaxially oriented films can be produced starting from high modulus fibers through lamination of several sheets, each sheet being a composite containing uniaxially oriented fibers (Halpin, 1975). Random composites are also being considered (Prevorsek, Chapter 12, this volume). A simpler approach would be the solidification of a cholesteric mesophase with Grandjean texture, having the helical axes perpendicular to the plane and the molecular axes oriented in all directions in the plane of the film. Krigbaum et al. (1981) investigated homopolyesters containing the α-methylstilbene unit (**XIV**) and copolymers containing both the adipic acid and the $(+)$-3-methyladipic acid residue. The mesophase of the homopolyester ($T_{CCh} = 110°C$, $T_{ChI} = 250°C$) exhibited fanlike structures, suggesting a pitch of the

(**XIV**)

cholesteric helix smaller than the wavelength of visible light. In fact, upon mixing the polymer with p-azoxyanisole or with the corresponding polymeric nematogen (containing adipic acid residues), bright colors typical of a cholesteric planar texture were observed (cf. Fig. 2b). Copolymers also exhibited a wavelength of maximum reflectivity within the visible region. The planar cholesteric texture was spontaneously formed between slide and cover glass and, very interestingly, the texture and the colors could be preserved in the solid film by quenching. Control of the helical pitch by copolymerization was also performed in the systems investigated by Van Luyen et al. (1980), Asrar et al. (1982a), Vilasagar and Blumstein (1980), and Blumstein and Vilasagar (1981).

It is interesting to observe that while the cholesteric texture is preserved in the solid by quenching, application of a flow field to cholesteric solutions produces high modulus fibers (Ballard et al., 1964; Franks and Varga, 1979). The flow field is expected to cause a transformation to a nematic mesophase. However, it is at least conceivable that, with proper control of the orientation of the cholesteric mesophase and of the flow profile, morphologies typical of natural fibers (e.g., cotton: nematic core, cholesteric surface) could be obtained in the solid fiber.

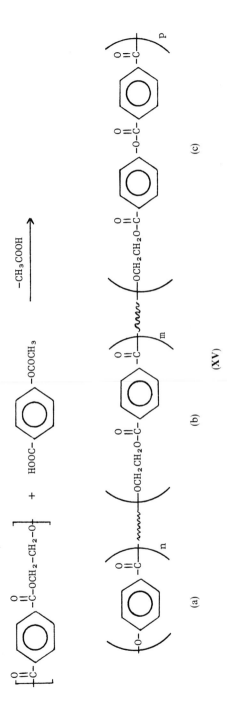

C. Copolymers

The third type of liquid crystalline polyester is the random copolyester. These systems are suitable for high modulus applications. However, they may exhibit wide biphasic regions, a feature that might not be desirable in polymer processing (the isotropic phase may not achieve the optimal orientation during flow). The copolymers prepared by Tennessee Eastman investigators (Jackson and Kuhfuss, 1976; Kuhfuss and Jackson, 1974; Wooten et al., 1979) are formed by transesterification of poly(ethylene terephthalate) with p-acetoxybenzoic acid (**XV**, p. 83). The final copolymer contains segments of PET (b) in addition to segments of types a and c. The net effect is to enrich the PET chain with oxybenzoyl units, reducing the content of flexible methylene units. More than 30% of oxybenzoyl units must be introduced to observe stable mesophases. For example, the polymer with 60% oxybenzoyl units has $T_{CN} \sim 260°C$ and $T_{NI} > 300°C$ (the polymer decomposes before reaching the isotropic state). The behavior of these copolymers may be related to the behavior of the model compound **XVI**, which contains three

$$CH_3-CO-O-\langle O \rangle-CO-O-\langle O \rangle-CO-O-\langle O \rangle-CO-O-C_2H_5$$

(**XVI**)

p-oxybenzoyl units and is a thermotropic nematic (Vorlander, 1923). However, the random nature of the sequence (established mainly from NMR data) does not facilitate the evaluation of axial ratio or persistence length and interpretations of T_{NI} in terms of the theory. Nevertheless, T_{NI} increases, as expected, with the content of oxybenzoyl units (Wooten et al., 1979).

Identification of the mesophase was a rather difficult task since a schlieren pattern does not readily form (Mackley et al., 1981). However, under an electric field (Krigbaum et al., 1980) instabilities typical of the nematic phase were observed. Under a magnetic field (Mackely et al., 1981) disinclinations with strength ± 1 were also detected. Recently (Jackson and Kuhfuss, 1980) a somewhat similar synthesis was reported using p-acetamidobenzoic acid instead of p-acetoxybenzoic acid.

$$\left\{ \overset{O}{\overset{\|}{C}}-\langle O \rangle-\overset{O}{\overset{\|}{C}}-O-CH_2-CH_2-O \right\} + CH_3-\overset{O}{\overset{\|}{C}}-NH-\langle O \rangle-\overset{O}{\overset{\|}{C}}-OH \xrightarrow{-CH_3 COOH}$$

$$\left\{ \overset{O}{\overset{\|}{C}}-\langle O \rangle-\overset{O}{\overset{\|}{C}}-NH-\langle O \rangle-\overset{O}{\overset{\|}{C}}-O-CH_2-CH_2-O \right\}$$

(**XVII**)

The resulting polymer is a copoly(ester–amide), which contains segments of the type indicated, segments of PET, and segments resulting from the self-condensation of *p*-acetamidobenzoic acid.

Additional copolyesters are described in the patent literature (Kleinschuster *et al.*, 1975; Pletcher, 1976; Schaefgen, 1978; Elliot, 1978; Irwin, 1979, 1980; Jackson and Kuhfuss, 1979; Jackson *et al.*, 1979; Jackson and Morris, 1980; Calundann, 1978, 1979, 1980; Cottis *et al.*, 1976). For example, du Pont patents describe random copolyesters where Ar_1 and Ar_2 are one or two of the following:

$$-O-Ar_1-O-\overset{\overset{O}{\|}}{C}-Ar_2-\overset{\overset{O}{\|}}{C}-$$

(XVIII)

$$Ar_1 = \text{—}\langle O \rangle\text{—} , \text{—}\langle O \rangle\text{—O—}\langle O \rangle\text{—} , \text{—}\langle O \rangle\text{—}$$
$$\hspace{3cm}| \hspace{6cm} |$$
$$\hspace{3cm}Cl \hspace{5.5cm}CH_3$$

$$Ar_2 = \text{—}\langle O \rangle\text{—} , \text{—}\langle O \rangle\text{—}\langle O \rangle\text{—} , \text{—}\langle O \rangle\text{—O—}\langle O \rangle\text{—} , \text{—}\langle O \rangle\text{—O—}CH_2\text{—}CH_2\text{—O}\text{—}\langle O \rangle\text{—} ,$$

$$\text{—}\langle \hexagon \rangle\text{—} , \text{—}\langle O \rangle\overset{\overset{CH_3}{|}}{\underset{\underset{CH_3}{|}}{C}}\langle O \rangle\text{—} , \text{—}\langle O \rangle\text{—} , \langle O O \rangle$$

The use of one or more diacetates of hydroquinone ($-O-Ar_1-O-$) (some of them asymmetrically substituted) and/or more than one aromatic or cycloaliphatic dicarboxylic acid ($-CO-Ar_2-CO-$) reduces the melting temperature, but sufficient rigidity can be maintained to assure formation of a nematic melt. Random copolyesters were prepared by transesterification in the melt. These polymers yielded high modulus fibers, but no detailed study of the mesophase was reported.

Volksen *et al.* (1979) considered copolymers based on blocks of biphenyl terephthalate (a) and *p*-hydroxybenzoic acid (b). These copolymers melted

$$\{O\text{—}\langle O \rangle\text{—}\langle O \rangle\text{—}O\text{—}CO\text{—}\langle O \rangle\text{—}CO\} \hspace{1cm} \{O\text{—}\langle O \rangle\text{—}CO\text{—}O\text{—}\langle O \rangle\text{—}CO\}$$

(a)   (b)

(XIX)

at ~400°C, but again no detailed study of the mesophase was reported. Still other polyesters are being reported in the literature (Jin *et al.*, 1980). Among the most recent reports are a series of polyesters with varying composition of linear (para) and nonlinear (meta) links (Griffin and Cox,

1980). One system investigated was chloro-1,4-phenylene 4,4'-oxydi-benzoate–terephthalate with varying amounts of isophthalate link. Quite unexpectedly, the copolymer remains liquid crystalline even with about 80% meta units. More details on this system are reported by Prevorsek (Chapter 12, this volume).

III. Diluted Systems

We begin the analysis of diluted or lyotropic systems starting with the phase diagram reported by Miller *et al.* (1974) for PBLG in dimethyl-formamide. In Fig. 5 we compare it with the theoretical diagram resulting from Flory's (1956a) original treatment of rigid rods having an axial ratio of 100. In the theoretical diagram the polymer volume fraction v_2 is plotted against the polymer–solvent interaction parameter χ_1 defined by the equation (Flory, 1953)

$$\chi_1 = z \Delta W_{12}/kT \tag{6}$$

where $\Delta W_{12} = W_{12} - \frac{1}{2}(W_{11} + W_{22})$ is the change in energy for the formation of unlike contact pairs in a lattice of coordination number z. In the first approximation, χ_1 is inversely proportional to temperature and can be correlated with the abscissa of the experimental diagram, which directly represents the temperature. The shape of the two diagrams is quite similar. We recognize the wide biphasic region at low T, corresponding to high χ_1 or poor solvents, and the narrow biphasic region at high T, corresponding to good solvents. In these regions isotropic and anisotropic phases coexist. Outside these regions only pure phases are stable. We shall now analyze in detail the role of the various parameters affecting these regions.

A. The Role of Temperature

According to the theoretical diagram in Fig. 5 the narrow region runs parallel to the ordinate axis. Experimentally, however, we observe a curvature of the narrow region toward high concentration as the temperature is increased. A similar trend was also observed for cellulose acetate in trifluoro-acetic acid (Navard *et al.*, 1981). This indicates that even in lyotropic systems thermotropic behavior occurs. The theoretical diagram, however, essentially denies thermotropic behavior in the narrow region. How can we interpret this deviation between theory and experiment? Miller *et al.* (1974) considered the effect of temperature on the rigidity of the chain. The persistence length is known to decrease often with temperature, and this effect is equivalent

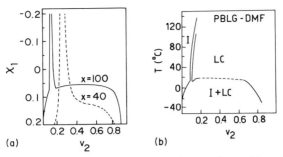

Fig. 5. (a) Theoretical phase diagrams for rigid rods according to Flory (1956a) for axial ratios of 100 and 40. (b) Experimental phase diagrams for PBLG in DMF reported by Miller *et al.* (1974).

to a decrease of axial ratio. According to Eq. (2) this would correspond to a bending of the narrow region toward higher concentration as T is increased. In order to fit the data a reduction of x from ~ 100 to ~ 60 over a $\sim 100°C$ range was required. [Using the earlier treatment of a semirigid chain (Flory, 1956b), the flexibility parameter f changes from 0.06 at low T to 0.11 at high T. However, this treatment does not offer a satisfactory description of semirigid chains (Flory, 1981).]

Are there other contributions to this effect? One must consider other types of energy aside from conformational energy. In fact, for low MW LC it is known that thermotropic behavior occurs even in the presence of a diluent [e.g., *p*-methoxybenzylidene-*p'*-*n*-butylaniline in several nonmesogenic solvents in the composition range $0.9 < v_2 < 1$ (Martire *et al.*, 1976; Kronberg and Patterson, 1976)]. Martire *et al.* (1976) and Kronenberg and Patterson (1976) attempted a description of these effects by introducing different activity coefficients in isotropic and anisotropic phases. The latter investigators assumed the same Flory–Huggins expression for the chemical potential of the components in the two pases, but allowed the χ_1 parameter to assume different values in the isotropic and anisotropic phases. By treating the depression of the nematic isotropic transition temperature by a solvent in a way similar to that employed for the depression of the melting temperature, they obtained expressions that described the experimental behavior. This approach is complicated by the choice of the standard state for the pure nematogen (which requires specification of its order parameter S), and by the fact that the classical Flory–Huggins expression does not describe the chemical potentials in the anisotropic phases (Flory, 1956a; Krigbaum and Ciferri, 1980).

To describe the behavior of polymers, Toriumi (1981) used the chemical potentials for the two phases that appear in Flory's (1956a) original treatment of rigid rod equilibria, but assumed the interaction parameter for the anisotropic phase (χ_1') to differ from that for the isotropic one (χ_1). He assumed different (arbitrary) functional relationships between χ_1' and χ_1 and concluded that different χ_1 values, particularly when their difference increases with increasing temperature, cause a curvature of the narrow region toward high concentration. Some of his phase diagrams, calculated assuming $\chi_1' = a\chi_1 + b$, are reproduced Fig. 6. These calculations are empirical in nature, but suggest that the W_{22} term entering the expression of ΔW_{12} (Eq. 6) could assume different values in the isotropic and anisotropic phases.

A molecular theory by Warner and Flory (1980) offers a justification for the thermotropic behavior observed in diluted systems that is consistent with the phenomenological considerations presented above. These authors extended the hard-rod theory with anisotropic dispersion forces—embodied in Eq. (3)—to the case in which a diluent is present. The T^* term representing the intensity of the asymmetric polymer–polymer attraction in the neat liquid is reduced to $T^* v_2$ in the presence of a diluent, but still contributes to the stabilization of the mesophase. Direct polymer–solvent interaction (of the usual symmetric or asymmetric type) was neglected by assuming $\chi_1 = 0$. In the expression of the chemical potential of the solvent in the anisotropic phase (Flory, 1956a; Flory and Ronca, 1979a) χ_1 was replaced by a term $S^2 T^*/T$ containing the order parameter. A phase diagram calculated according to this theory is included in Fig. 6. In the limit $T^*/T \to 0$ (no asymmetric attraction) the result of the earlier theory with $\chi_1 = 0$ is obtained.

The phase diagrams in Figs. 5 and 6 also allow some observations concerning the wide region. In some instances the wide region includes two nematic phases and a critical point. Similar features were observed by Khokhlov (1979), who has recently reexamined some of the approximations concerning the interaction parameter. The occurrence of the two nematic phases in the wide region is primarily related to a relatively high axial ratio (>50), as observed also by Wee and Miller (1978). Due to occurrence

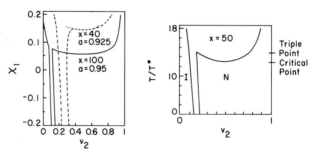

Fig. 6. Left: Phase diagram calculated according to Flory's (1956a) theory plus the assumption $\chi_1' = a\chi_1$ for axial ratios of 100 and 40. Right: Phase diagram according to the Warner–Flory theory (Warner and Flory, 1980).

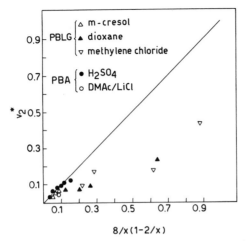

Fig. 7. Critical polymer volume fraction plotted versus inverse axial ratio according to Eq. (2) for PBLG and PBA in several solvents.

of crystallization (cf. Section III.D), it is unlikely that these theoretical predictions can be reliably verified (see, however Miller *et al.*, 1974).

B. The Role of MW and Rigidity

1. Rigid Systems

We now consider the effect of MW and rigidity on the narrow region. This allows a test of the predictions of Eq. (2). Figure 7 is a plot of the critical polymer volume fraction versus the inverse axial ratio according to Eq (2). Data included are those of Hermans (1962) for PBLG in *m*-cresol, of Robinson *et al.* (1958) for PBLG in dioxane and methylene chloride, of Kwolek *et al.* (1977) for poly(*p*-benzamide) (PBA) in H_2SO_4, and of Balbi *et al.* (1980) for PBA in *N,N*-dimethylacetamide (DMAc)/3% LiCl. PBA, along with the related poly(*p*-phenyleneterephthalamide) (PPD-T) (**XX**) is described in the

$$\left[-NH-\langle\bigcirc\rangle-CO- \right]\qquad\qquad PBA$$

$$\left[-NH-\langle\bigcirc\rangle-NH-CO-\langle\bigcirc\rangle-CO- \right]\qquad PPD-T$$

(**XX**)

original Kwolek (1972) patent dealing with the preparation of ultrahigh modulus fibers from lyotropic mesophases. Data included in Fig. 7 were analyzed for their conformity with the model of rigid molecules as defined in Section I.D. This required an analysis of corresponding persistence lengths. For PBA, Arpin and Strazielle (1977) measured q in 96% H_2SO_4 (1977), which was found to be of the order of 400 Å ($q = 600–400$, 400 and 320 Å, using light scattering, viscosity, and flow birefringence techniques, respectively; corresponding figures for PPD-T were 200, 150, and 200 Å). Theoretical estimates by Erman et al. (1980) gave $q = 410$ Å for both PBA and PPD-T. For PBLG persistence length data are scattered between 400 and 900 Å in dimethylformamide, a value of ~ 800 Å being probably a reliable one (Moha et al., 1964). Data included in Fig. 7 satisfy the condition that the contour length is smaller or comparable with q (at least in the solvents in which the latter was determined). The axial ratio was determined from the molecular weight of the polymers on the basis of the geometry of an α helix for PBLG (Walton and Blackwell, 1973), and of an extended trans conformation for PBA (Ciferri and Valenti, 1979). The data in Fig. 7 support the predicted decrease of v_2^* with axial ratio. The experimental v_2^* tend to be slightly smaller than expected from Eq. (2). The results are not greatly affected by solvent type. Also, no difference is exhibited between nematic (PBA) and cholesteric (PBLG) mesophases.

2. Semirigid Systems

Rather different results were obtained for hydroxypropylcellulose (HPC) (**XXI**) in water by Werbowyi and Gray (1980). The critical concentration was

(**XXI**)

unaffected ($v_2^* = 0.37 \pm 0.02$) by an extremely large variation (from 6×10^4 to 1×10^6) of MW. To explain this behavior, we consider the rigidity of cellulose and its derivatives, which is known to be considerably affected by the solvent and by the type and degree of substitution (Cleland, 1971;

Zippler *et al.*, 1969; Henley, 1961). For HPC in ethanol the persistence length was evaluated to be 85 ($\pm$ 20) Å by Werbowyi and Gray (1980) from data by Wirik and Waldam (1970). Assuming that a q value in this range also applies to the HPC/H_2O system, and considering the high values of MW indicated above, it appears that for this system the persistence length was much smaller than the contour length. Thus, at variance with PBA and PBLG, we can classify this polymer as a semirigid one. Theoretical treatments (Flory, 1978b; Marrucci and Sarti, 1979) for semirigid polymers, based on the model of rigid segments connected by flexible joints, indicate that the relevant axial ratio is in this case the length of the rigid segment. The whole length of the molecule is irrelevant for semirigid chains. The independence of v_2^* of MW is thus in line with this theory. Again, the actual value of v_2^* is somewhat smaller than expected. In fact, if we take an axial ratio of ~ 10, corresponding to the above value of q, v_2^* is expected to be ~ 0.7. Thus the experimental v_2^* is about a factor of two smaller than expected from Eq. (2) (a somewhat similar discrepancy was noticed for the rigid polymers, cf. above). Use of the Kuhn segment A [Eq. (5)] rather than q would remove this discrepancy (Flory, 1981). However, the justification for the choice of A rather than q is not on a definite basis at the present time.

Not all cellulosic systems show an invariance of v_2^* with MW. For cellulose in N-methylmorpholine N-oxide/water, Chanzy and Peguy (1980) noticed about a twofold increase in v_2^* when the MW was increased by a factor of about six. A similar trend was noticed by Navard *et al.* (1981) for cellulose acetate in trifluoroacetic acid. Values of persistence lengths for these systems have not been measured, but it is likely that more extended conformations prevail than in the case of HPC/H_2O. The role of solvent type on the critical concentrations is well documented by data of Aharoni (1980a) and of White *et al.* (1980), suggesting a correlation between v_2^* and chain extension for cellulose acetate.

The cholesteric texture of the HPC/H_2O system was vividly demonstrated by Werbowyi and Gray (1980) by a shift of the wavelength of maximum reflectivity from red to violet, reflecting a decrease of the helical pitch as polymer concentration was increased.

C. The Role of MW Distribution and Ternary Systems

The effect of the MW distribution on the narrow region is illustrated in Fig. 8 for PBA in DMAc/3% LiCl. The theoretical treatments considered in Sections III.A and B refer to monodisperse systems. For systems having a molecular weight distribution of the most probable type (which is the one expected for PBA), the Flory and Frost (1978) theory predicts a fractionation

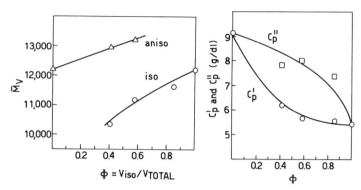

Fig. 8. Left: Variation of molecular weight partitioned in isotropic and nematic phases with the volume function of isotropic phase. Right: Corresponding variation of the composition of isotropic (C_p') and nematic (C_p'') phases. The system is PBA/DMAc + 3% LiCl. (Reprinted with permission from Conio et al., *Macromolecules* **14**, 1084 (1981). Copyright 1981, American Chemical Society.)

effect and a widening of the narrow region. The results on the left diagram indicate the MW of the polymer present in the isotropic and in the anisotropic phases. The phases were separated by ultracentrifugation after equilibrium had been attained. The abscissa represents the volume fraction of the isotropic phase. When only one phase is stable, the isotropic one at $\phi = 1$ or the anisotropic one at $\phi = 0$, all the polymer must be in only one phase and the MW is obviously the same in the two phases. However, by increasing the overall composition within the narrow region the amount of isotropic phase can be decreased. We observe that as this decrease occurs the molecular weight partitioned in the isotropic phase decreases. Simultaneously, the MW of the anisotropic phase increases. It is evident that the maximum MW in the anisotropic phase and the minimum in the isotropic one are reached when the corresponding phase volumes are very small. In other words, there is a fractionation process and its efficiency depends upon the relative volume of the coexisting phases. The data on the diagram on the right reveal a corresponding effect of the ratio of phase volume on the individual concentrations C_p' and C_p'' of isotropic and anisotropic phases. This result means that the width of the narrow region is not fixed, as in the case of monodisperse rods, but may be affected by the overall composition that determines ϕ. The effects observed are in line with the above theory for polydisperse systems. However, there is also a discrepancy. In fact, the extent of fractionation and the maximum width of the narrow region are by no means as large as predicted. The ratio of the limiting values of C_p' and C_p'' should not be greater than 1.6 for a monodisperse system, but should easily reach ten for the most probable distribution (Flory and Frost, 1978). In all cases of polydisperse systems so far investigated, the ratio is close to the value expected for a monodisperse system. This discrepancy is under investigation.

The narrowing of the MW distribution had a remarkable effect on the clarity with which the liquid crystalline pattern can be detected under the polarizing microscope. Figure 2c is a micrograph of a pure anisotropic phase separated by centrifugation from a coexisting isotropic one. The typical threadlike pattern of the nematic phase is seen with great clarity in fractionated samples such as this one.

The large biphasic gap expected for a most probable distribution should not be expected for a narrow distribution. For example, the results obtained for the Poisson distribution, which has a much narrower breadth than the most probable one, are rather similar to those for monodisperse systems (Frost and Flory, 1978).

Ternary diagrams involving two polymers and one solvent have also been considered. Theoretically, two rodlike polymers have been shown to be mutually compatible, at variance with the behavior of two flexible polymers (Abe and Flory, 1978). Two rodlike polymers with different axial ratios may form a single anisotropic phase at a sufficiently high concentration. When the concentration is decreased a biphasic region appears. The width of this region is greater than for monodisperse systems and the anisotropic phase is not only more concentrated but also richer in the high MW polymer than the conjugated isotropic one. If a solution involving a rodlike and a flexible polymer having the same MW is considered, the flexible polymer is predicted (Flory, 1978a) to go into the isotropic phase. Some of the rigid polymer may enter the isotropic phase if its axial ratio is relatively small.

Only the general features of these theories have been verified (Aharoni, 1980b; Hwang *et al.*, 1981; Asrar *et al.*, 1982b). Much work remains to be done. Mixtures of polymers differing in degree of rigidity may be of technological interest (Prevorsek, Chapter 12, this volume).

D. The Wide Region: Crystallization

We now consider the wide region in more detail. The phase diagram for PBA in DMAc–LiCl determined by Balbi *et al.* (1980) is illustrated in Fig. 9. This is an isothermal system in which the interaction parameter was varied by altering the LiCl content. In place of χ_1, the LiCl concentration is reported directly as the state variable. From independent studies we know that both low and high salt concentrations correspond to poor solvents (χ_1 positive) (Ciferri, 1978). Only at intermediate salt concentration is the solvent a good one (χ_1 negative). In fact, it is in this intermediate salt concentration range that we observe a narrow biphasic region. At low salt concentration the

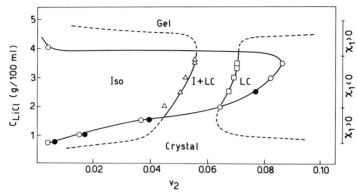

Fig. 9. Phase diagram for PBA in DMAc with varying LiCl content. $T = 20°C$. Broken lines indicate hypothetical wide regions. (From Balbi *et al.*, 1980.)

polymer does not dissolve at all and there is little doubt that the insolubility is due to crystallization. At high salt concentration one observes gelation and it is not clear whether the underlying phenomenon is crystallization or the occurrence of the wide region (dotted line). To complicate matters some slow gelation and aggregation effects also occur in the narrow region and in the pure anisotropic phase. For PBLG Miller *et al.* (1978; see also Tohyama and Miller, 1981) attribute the formation of the gel to spinodal decomposition when the wide region is entered. To clarify which regions of the phase diagram involving mesophases are indeed accessible to stable equilibrium Krigbaum and Ciferri (1980) considered the relationships for the melting temperature depression for both isotropic and anisotropic phases. These are given by the equations

$$
\frac{1}{T_M} - \frac{1}{T_M^\circ} = \frac{Rx}{\Delta H_F}\left[\left(1 - \frac{1}{x}\right)v_1 + \frac{1}{x}\ln v_2 - \chi_1 v_1^2\right]
$$

$$
\frac{1}{T_M} - \frac{1}{T_M^{\circ\prime}} = \frac{Rx}{\Delta H_F'}\left[-\frac{1}{x}\left(\ln\frac{v_2}{x} + (y - 1)v_2 - \ln y^2\right) - \chi_1 v_1^2\right]
$$

(7)

where T_M° and $T_M^{\circ\prime}$ are the melting temperatures of the crystal into undiluted isotropic and anisotropic phases and ΔH_f and $\Delta H_f'$ the corresponding enthalpies, y the disorientation parameter that enters Flory's (1956a) theory, and v_1 the solvent volume function.

The melting temperature depression for the isotropic solution is the classical Flory–Huggins result. The depression for the liquid crystalline phase was obtained by using Flory's (or Flory and Ronca's, 1979a) expression for the chemical potential of the polymer in the anisotropic phase. Solubility

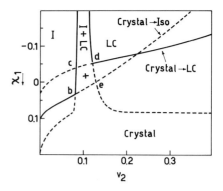

Fig. 10. Superimposition of the phase diagram predicted by Flory (1956b) and of the solubility equilibria calculated according to Eqs. (7). $x = 100$, $T_M = 800$ K, $T'_M = 400$ K, $\Delta H_f/x = 900$ cal/mol, $\Delta H'_f/x = 400$ cal/mol. (From Ciferri and Krigbaum, 1981.)

curves were calculated from Eqs. (7) for plausible values of the parameters involved. These curves were superimposed (Balbi *et al.*, 1980; Ciferri and Krigbaum, 1981) on the phase diagram describing the equilibrium between isotropic and anisotropic phases as calculated by Flory (1956a) or by Flory and Ronca (1979a). A typical example is reproduced in Fig. 10. Some portions of the individual diagrams will be inaccessible when the composite equilibria are considered. These inaccessible regions are indicated by dashed lines, and can exist only under metastable conditions. For example, the whole wide region lies below the crystallization curves and therefore should not be observed if equilibrium prevails. This conclusion should apply to most systems. In fact, only by making the parameters $T_M^{\circ\prime}$ and $\Delta H'_f$ extremely small could the crystallization line of the anisotropic phase be shifted considerably downward. Moreover, peculiar situations within the narrow region may be explained. Consider, for example, the area bcde between the saturation lines of isotropic and anisotropic solutions. Suppose that an anisotropic phase nucleates from the solution having the overall composition represented by the cross. This phase should selectively crystallize since its composition is below the solubility curve of the anisotropic solution. However, since the isotropic solution is unsaturated, the polymer should redissolve. Thus the system may reach a stationary state but cannot exist in thermodynamic equilibrium. This situation suggests an interesting crystallization mechanism that may also be relevant to the fibrogenesis of some biopolymers. If fresh polymer is added so as to keep the overall composition constant at the value of the cross, one can continuously extract a crystalline material having the oriented morphology typical of the mesophase. This is done without the need of ever attaining the high concentration characteristic of the pure mesophase.

Occurrence of crystallization has also been shown often to limit the possibility of investigating the ternary phase diagrams described in the preceding section (Asrar *et al.*, 1982b).

TABLE II

Summary of Results for Some LC Systems

Polymer	Reference	Mesophase	q (Å)
$R-N=C=O \longrightarrow \left\{ \begin{matrix} O \\ \parallel \\ C-N \\ \mid \\ R \end{matrix} \right\}$	Aharoni (1979, 1980c)	Nematic (TCE; Therm: $T_{CN} \sim 177°C$)	425 (PHIC/THF)
$NH_2-\bigcirc(CH_3)-NH_2 + OHC-\bigcirc-CHO \longrightarrow$ $\{N-\bigcirc(CH_3)-N=CH-\bigcirc-CH\}$	Morgan (1976); Millaud et al. (1978b)	Nematic (H_2SO_4; Therm: $T_{CN} \sim 260°C$)	150 (H_2SO_4)
$NH_2-\bigcirc-CO-NH-NH_2 + Cl-CO-\bigcirc-CO-Cl \longrightarrow$ $\{\!\{NH-\bigcirc-CO-NH-NH\}CO-\bigcirc-CO\}$ (X500)	Preston (1975); Morgan (1978)	Isotropic (organic solvent) Nematic (H_2SO_4)	50 (X-500/DMSO)
$NH_2-NH-CO-\bigcirc-CO-NH-NH_2 + Cl-CO-\bigcirc-CO-Cl$ $\longrightarrow \{NH-NH-CO-\bigcirc-CO\}$	Hartzler and Morgan (1977)	Nematic (Organic bases, H_2SO_4)	—
$\{$ (PBT) $\}$	Wolfe et al. (1981)	Nematic (MSA, CSA)	—

E. Additional Systems

Finally, in Table II, we summarize some results for additional systems that have been reported (see also Ciferri, 1977; Preston, 1979). Several polyisocyanates or nylon 1 have been described by Aharoni (1979). These polymers are not technologically important, since they begin to decompose at $\sim 150°$C. The simplest structures are homopolymers of alkyl isocyanates. The side group R must be relatively long, from hexyl to decyl, to assure solubility and hence the observation of nematic mesophases. Lyotropic behavior was observed in solvents such as tetrachloroethane and bromoform. Thermotropic behavior for the undiluted system was also reported, but was complicated by decomposition (Aharoni, 1980c). The rigidity of the poly(n-hexyl isocyanate) (PHIC) chain was investigated. In tetrahydrofuran (THF) q was found to be 425 Å (Berger and Tidswell, 1973), indicating a rather rigid chain. The rigidity is, however, affected by solvent type. For example, more rigid postures were observed in THF than in CHCl$_3$ (Burr and Fetters, 1976).

The polyazomethines formed by reaction of diamines and dialdehydes (Morgan, 1976; Millaud et al., 1978b) form nematic mesophases in solvents such as H$_2$SO$_4$. In this solvent $q = 150$ Å (Millaud and Strazielle, 1979). Use of unsymmetrical monomers (as indicated in Table II) or introduction of flexible spacers (Millaud et al., 1979b) reduces the melting temperature so that thermotropic behavior for the undiluted system could be observed. Fibers spun from the melt were reported (Morgan, 1976).

The polyamide hydrazides and the polyhydrazides (Preston, 1975; Morgan, 1978; Hartzler and Morgan, 1977) contain the hydrazide bond —NH—NH— in addition to the bonds that, as we have seen, cause rigidity in aromatic polyamides. Conformational energy calculations (Allegra, 1975; Laupretre et al., 1980) indicate that while the amide group is frozen in the planar conformation, the hydrazide bond has a substantial degree of internal rotation. Tsvetkov and Shtennikova (1978) suggest that this rotation should not cause flexibility since the N–N axis is parallel to the axes of adjacent phenylene units, a situation similar to that discussed for the aromatic polyamides. Perhaps, they suggest, few amide groups may assume the cis conformation since the hydrazide bond increases their distance along the chain. Whatever the reason, polyamide hydrazides are certainly less extended than polyamides. This is shown by a persistence length not greater than 50 Å for the polyterephthalamide of p-aminobenzhydrazide (X-500) in dimethyl sulfoxide (DMSO) (Bianchi et al., 1979, 1981). Moreover, X-500 does not yield anisotropic solutions in the organic solvents up to $v_2 \simeq 0.19$ (Bianchi et al., 1981), when crystallization occurs. However, the same polymer gives a nematic phase in H$_2$SO$_4$ (Morgan, 1978) at $v_2 \simeq 0.13$. This is again a very dramatic example of the role of the solvent in controlling the formation of the mesophase. Since aggregation phenomena were not observed, the solvent seems to act directly on the rigidity of the chain. This particular polymer, X-500, is also noteworthy since the isotropic → anisotropic transition may be induced by a flow field (Ciferri and Valenti, 1979; Valenti *et al.*, 1981). The polyhydrazides also yielded anisotropic solutions in H_2SO_4 (Morgan, 1978), as well as in aqueous solutions of organic bases (Hartzler and Morgan, 1977). All these polymers may be of technological relevance; the spinning of these polymers is described in the patent literature. The preparation of several compositions, besides those shown in Table II, and of copolymers was reported by Preston (1978, 1979), by Morgan (1978), and by Hartzler and Morgan (1977).

Totally aromatic polymers such as poly(*p*-phenylenebenzobisthiazole) (PBT) are of interest for aerospace applications because of improved thermo-oxidative stability compared with polymers having amide bonds. They also exhibit exceptional mechanical properties and are soluble in solvents such as methanesulfonic (MSA) and chlorosulfuric acid (CSA) in which they form nematic mesophase at concentrations of $\sim 10\%$ (MW 13,000–16,000). Their synthesis and properties were described by Wolfe *et al.* (1981), by Allen *et al.* (1981), and by Hwang *et al.* (1981). The latter study includes the formation of composite films (see also Prevorsek, Chapter 12, this volume).

References

Abe, A., and Flory, P. J. (1978). *Macromolecules* **11**, 1122.

Aharoni, S. M. (1979). *Macromolecules* **12**, 94.

Aharoni, S. M. (1980a). *Mol. Cryst. Liq. Cryst.* (*Lett.*) **56**, 237.

Aharoni, S. M. (1980b). *Polymer* **81**, 26.

Aharoni, S. M. (1980c). *J. Polymer Sci. Polym. Phys. Ed.* **18**, 1303.

Allegra, G. (1975). *Polym. Eng. Sci.* **15**, 207.

Allen, S. R., Filippov, A. G., Farris, R. J., Thomas, E. L., Wong C.-P., Berry, G. C., and Chenevey, E. C. (1981). *Macromolecules* **14**, 1135.

Antoun, S., Lenz, R. W., and Jin, J.-I. (1981). *J. Polym. Sci., Polym. Chem. Ed.* **19**, 1901.

Arpin, M., and Strazielle, C. (1977). *Polymer* **18**, 597.

Asrar, J., *et al.*, (1982a). To be published.

Asrar, J., Preston, J., Ciferri, A., and Krigbaum, W. R. (1982b). *J. Polym. Sci., Polym. Chem. Ed.* **20**, 373.

Balbi, C., Bianchi, E., Ciferri, A., Tealdi, A., and Krigbaum, W. R. (1980). *J. Polym. Sci., Polym. Phys. Ed.* **18**, 2037.

Ballard, D. G. H., Griffiths, J. D. and Watson, J. (1964). U.S. Patent 3,121,766 (to Courtaulds, Ltd.)

Benoit, H., and Doty, P. (1953). *J. Phys. Chem.* **57**, 958.

Berger, M. N., and Tidswell, B. M. (1973). *J. Polym. Sci. Part C* **42**, 1063.

Bernstein, H. J. (1981). Personal communication.

Bianchi, E., Ciferri, A., Tealdi, A., and Krigbaum, W. R. (1979). *J. Polym. Sci. Polym. Phys. Ed.* **17**, 2091.

Bianchi, E., Ciferi, A., Preston, J., and Krigbaum, W. R. (1981). *J. Polym. Sci., Polym. Phys. Ed.* **19**, 863.

Blumstein, A., and Vilasagar, S. (1981). *Mol. Cryst. Liq. Cryst. (Lett.)* **72**, 1.

Blumstein, A., Sivaramakrishnan, K. N., Clough, S. B., and Blumstein, R. (1979). *Mol. Cryst. Liq. Cryst. (Lett).* **49**, 255.

Buntjakov, A., and Averyanova, V. (1972). *J. Polym. Sci. Part C* **38**, 109.

Burr, A. J., and Fetters, L. J. (1976). *Chem. Rev.* **76**, 727.

Calundann, G. W. (1978). U.S. Patents 4,067,852 and 4,130,545 (to Celanese).

Calundann, G. W. (1979). U.S. Patent 4,161,470 (to Celanese).

Calundann, G. W. (1980). U.S. Patent 4,184,996 (to Celanese).

Chandrasekhar, S. (1977). "Liquid Crystals." Cambridge Univ. Press, London and New York.

Chanzy, H., and Peguy, A. (1980). *J. Polym. Sci., Polym. Phys. Ed.* **18**, 1137.

Ciferri, A. (1975). *Polym. Eng. Sci.* **15**, 191.

Ciferri, A. (1977). *Ann. Chim. (Rome)* **67**, 269.

Ciferri, A. (1978). *Int. J. Polym. Mater.* **6**, 137.

Ciferri, A., and Krigbaum, W. R. (1981). *Mol. Cryst. Liq. Cryst.* **69**, 273.

Ciferri, A., and Valenti, B. (1979). *In* "Ultra-High Modulus Polymers" (A. Ciferri and I. M. Ward, eds.), p. 203. Appl. Sci., London.

Ciferri, A., and Ward, I. M., eds. (1979). "Ultra-High Modulus Polymers." Appl. Sci., London.

Cleland, P. L. (1971). *Biopolymers* **10**, 1925.

Conio, G., Bianchi, E., Ciferri, A., and Tealdi, A. (1981). *Macromolecules* **14**, 1084.

Cottis, S. G., Economy, J., and Wohrer, L. C. (1973). Ger. Patent 2,248,127 (to Carborundum).

Cottis, S. G., Economy, J., and Wohrer, L. C. (1976). Ger. Patent 2,507,066 (to Carborundum).

Cox, R. J. (1972). *Mol. Cryst. Liq. Cryst.* **19**, 111.

Demus, D., and Richter, L. (1978). "Textures of Liquid Crystals." Verlag Chemie, Weinheim.

Demus, H., and Zaschke, H. (1974). "Flüssige Kristalle in Tabellen." VEB Dtsch. Verlag Grundstoffind., Leipzig.

Elliot, S. P. (1978). Ger. Patent 2,751,585 (to Du Pont Co.).

Erman, B., Flory, P. J., and Hummel, J. P. (1980). *Macromolecules* **13**, 484.

Fayolle, B., Noel, C., and Billard, J. (1979). *J. Phys. (Orsay, Fr.)* **40**(C3), 485.

Flory, P. J. (1953). "Principles of Polymer Chemistry." Cornell Univ. Press, Ithaca, New York.

Flory, P. J. (1956a). *Proc. R. Soc. London, Ser. A* **234**, 73.

Flory, P. J. (1956b). *Proc. R. Soc. London, Ser. A* **234**, 60.

Flory, P. J. (1969). "Statistical Mechanics of Chain Molecules." Wiley (Interscience), New York.

Flory, P. J. (1978a). *Macromolecules* **11**, 1138.

Flory, P. J. (1978b). *Macromolecules* **11**, 1141.

Flory, P. J. (1981). Discussion at the SML Seminar.

Flory, P. J., and Abe, A. (1978). *Macromolecules* **11**, 1119.

Flory, P. J., and Frost, R. S. (1978). *Macromolecules* **11**, 1126.

Flory, P. J., and Ronca, G. (1979a). *Mol. Cryst. Liq. Cryst.* **54**, 289.

Flory, P. J., and Ronca, G. (1979b). *Mol. Cryst. Liq. Cryst.* **54**, 311.

Franks, N., and Varga, J. (1979). U.S. Patent 4,145,532 (to Akzona Co.)

Frosini, V., Levita, G., Landis, J., and Woodward, E. A. (1977). *J. Polym. Sci., Polym. Phys. Ed.* **15**, 239.

Frost, R. S., and Flory, P. J. (1978). *Macromolecules* **11**, 1134.

Griffin, B. P., and Cox, M. K. (1980). *Br. Polym. J.* **12**, 147.

Griffin, A. C., and Havens, S. J. (1979). *Mol. Cryst. Liq. Cryst. (Lett.)* **49**, 239.

Griffin, A. C., and Havens, S. J. (1980). *J. Polym. Sci., Polym. Lett. Ed.* **18**, 295.

Griffin, A. C., and Havens, S. J. (1981). *J. Polym. Sci., Polym. Phys. Ed.* **19**, 951.

Halpin, J. (1975). *Polym. Eng. Sci.* **15**, 132.

Hartzler, J. D., and Morgan, P. W. (1977). *Contemp. Top. Polym. Sci.* **2**, 19.

Henley, D. (1961). *Ark. Kemi.* **18**, 327.

Hermans, J., Jr. (1962). *J. Colloid Sci.* **17**, 638.

Hummel, J. P., and Flory, P. J. (1980). *Macromolecules* **13**, 479.

Hwang, H. F., Wiff, D. R., and Helminiak, T. (1981). *Org. Coat. Plast. Chem.* **44**, 32.

Iimura, K., Koide, M., Tanabe, H., Ohta, R., and Takeda, M. (1980). Presented at the VIIIth Intl. Liq. Cryst. Conf. Kyoto, 1980.

Irwin, R. S. (1979). U.S. Patent 4,176,223 (to Du Pont Co.).

Irwin, R. S. (1980). U.S. Patent 4,188,476 (to Du Pont Co.).

Jackson, W. J., Jr., and Kuhfuss, H. F. (1976). *J. Polym. Sci., Polym. Chem. Ed.* **14**, 2043.

Jackson, W. J., Jr., and Kuhfuss, H. F. (1979). U.S. Patent 4,140,846 (to Tennessee Eastman Co.)

Jackson, W. J., Jr., and Kuhfuss, H. F. (1980). *J. Appl. Polym. Sci.* **25**, 1685.

Jackson, W. J., Jr., and Morris, J. C. (1980). U.S. Patent 4,181,792.

Jackson, W. J., Jr., Gebeau, G. G., and Kuhfuss, H. F. (1979). U.S. Patent 4,153,779 (to Tennessee Eastman Co.).

Jin, J.-I., Antoun, S., Ober, C., and Lenz, R. W. (1980). *Br. Polym. J.* **12**, 132.

Khokhlov, A. R. (1979). *Vysokomol. Soedin., Ser. A* **21**, 1981.

Kleinschuster, J. J., Pletcher, T. C., Schaefgen, J. R., and Luise, R. R. (1975). Ger. Patents 2,520,819 and 2,520,820 (to Du Pont Co.).

Kofler, L., and Kofler, A. (1954). "Thermomikromethoden." Verlag Chemie, Weinheim.

Kratky, O., and Porod, G. (1949). *Rec. Trav. Chim. Pays. Bas* **68**, 1106.

Krigbaum, W. R., and Ciferri, A. (1980). *J. Polym. Sci., Polym. Lett. Ed.* **18**, 253.

Krigbaum, W. R., and Salaris, F. (1978). *J. Polym. Sci., Polym. Phys. Ed.* **16**, 883.

Krigbaum, W. R., Salaris, F., Ciferri, A., and Preston, J. (1979). *J. Polym. Sci., Polym. Lett. Ed.* **17**, 601.

Krigbaum, W. R., Lader, J., and Ciferri, A. (1980). *Macromolecules* **13**, 554.

Krigbaum, W. R., Ciferri, A., Asrar, J., Toriumi, H., and Preston, J. (1981). *Mol. Cryst. Liq. Cryst.* **76**, 142.

Krigbaum, W. R., Asrar, J., Toriumi, H., Ciferri, A., and Preston, J. (1982). *J. Polym. Sci., Polym. Lett. Ed.* **20**, 109.

Kronberg, B., and Patterson, D. (1976). *J. Chem. Soc. Faraday 2* **72**, 1686.

Kuhfuss, H. F., and Jackson, W. J., Jr. (1974). U.S. Patent 3,804,805.

Kwolek, S. L. (1972). U.S. Patent 3,671,542 (to Du Pont Co.).

Kwolek, S. L., Morgan, P. W., Schaefgen, J. R., and Gulrich, L. W. (1977). *Macromolecules* **10**, 1390.

Lader, H., and Krigbaum, W. R. (1979). *J. Polym. Sci., Polym. Phys. Ed.* **17**, 1661.

Laupretre, F., Monnerie, L., and Fayolle, B. (1980). *J. Polym. Sci., Polym. Phys. Ed.* **18**, 2243.

Liebert, L., Strzelecki, L., Van Luyen, D., and Levelut, A. M. (1980). *Eur. Polym. J.* **16**, 71.

Mackley, M. R., Pinaud, F., and Siekmann, G. (1981). *Polymer* **22**, 437.

Maier, W., and Saupe, A. (1958a). *Z. Naturforsch., A* **13A**, 564.

Maier, W., and Saupe, A. (1958b). *Z. Naturforsch., A* **14A**, 882.

Maier, W., and Saupe, A. (1960). *Z. Naturforsch., A* **15A**, 287.

Marrucci, G., and Sarti, G. C. (1979). *In* "Ultra High Modulus Polymers" (A. Ciferri and I. M. Ward, eds.), p. 137. Appl. Sci., London.

Martire, D. E., Oweimreen, G. A., Agren, G. J., Ryan, S. G., and Peterson, H. T. (1976). *J. Chem. Phys.* **64**, 1456.

Marzotko, D., and Demus, D. (1975). *Liq. Cryst., Proc., Int. Conf., Bangalore, 1973* p. 189.

Matheson, R. R., and Flory, P. J. (1981). *Macromolecules* **14**, 954.

Meurisse, P., Noel, C., Monnerie, L., and Fayolle, B. (1981). *Br. Polym. J.* **13**, 55.

Millaud, B., and Strazielle, C. (1979). *Polymer* **20**, 563.

Millaud, B., Thierry, A., and Skoulius, A. (1978a). *J. Phys. (Orsay, Fr.)* **39**, 1109.

Millaud, B., Thierry, A., and Skoulios, A. (1978b). *Mol. Cryst. Liq. Cryst.* **41**, 263.

Millaud, B., Thierry, A., and Skoulios, A. (1979a). *J. Phys. Lett. (Orsay, Fr.)* **40**, 607.

Millaud, B., Thierry, A., Strazielle, C., and Skoulios, A. (1979b). *Mol. Cryst. Liq. Cryst.* **49**, 299.

Miller, W. G., Jr., Rai, J. H., and Wee, E. L. (1974). *In* "Liquid Crystals and Ordered Fluids" (R. Porter and J. Johnson, eds.), Vol. 2, p. 243. Plenum, New York.

Miller, W. G., Kou, L., Tohyama, K., and Voltaggio, V. (1978). *J. Polym. Sci., Polym. Symp.* **65**, 91.

Moha, P., Weill, G., and Benoit, H. (1964). *J. Chim. Phys.* **61**, 1240.

Morgan, P. W. (1976). D. T. 2620–351 (to Du Pont Co.).

Morgan, P. W. (1978). *J. Polym. Sci., Polym. Symp.* **65**, 1.

Navard, P., Haudin, J. M., Dayan, S., and Sixou, P. (1981). *J. Polym. Sci., Polym. Lett. Ed.* **19**, 379.

Panar, M., and Beste, F. (1977). *Macromolecules* **10**, 1401.

Payet, C. R. (1978). Ger. Patent 2,751,653 (to Du Pont Co.).

Pincus, P., and de Gennes, P. G. (1978). *J. Polym. Sci., Polym. Symp.* **65**, 85.

Pletcher, T. C. (1976). U.S. Patents 3,991,013 and 3,991,014 (to Du Pont Co.).

Preston, J. (1975). *Polym. Eng. Sci.* **15**, 199.

Preston, J. (1978). *In* "Liquid Crystalline Order in Polymers" (A. Blumstein, ed.), p. 141. Academic Press, New York.

Preston, J. (1979). *In* "Ultra-High Modulus Polymers" (A. Ciferri and I. M. Wards, eds.), p. 155. Appl. Sci. London.

Puett, D., and Ciferri, A. (1971). *Biopolymers* **10**, 547.

Robinson, C., Ward, J. C., and Bevers, R. B. (1958). *Discuss. Faraday Soc.* **25**, 29.

Roviello, A., and Sirigu, A. (1977). *Gazz. Chim. Ital.* **107**, 333.

Roviello, A., and Sirigu, A. (1982). *Makromol. Chem.* **183**, 895.

Schaefgen, J. R. (1978). U.S. Patents 4,075,262 and 4,118,372 (to Du Pont Co.).

Schaefgen, J. R., Pletcher, T. C., and Kleinschuster, J. J. (1975). Belg. Patent No. 828,935 (to Du Pont Co.).

Schultz, G. V., and Penzel, E. (1968). *Makromol. Chem.* **112**, 260.

Strzelecki, L., and Van Luyen, D. (1980). *Eur. Polym. J.* **16**, 299.

Tohyama, K., and Miller, W. G. (1981). *Nature (London)* **289**, 813.

Toriumi, H. (1981). Unpublished results.

Tsvetkov, V. N. (1976). *Vysokomol. Soedin., Ser. A* **18**, 1621.

Tsvetkov, V. N., and Shtennikova, I. N. (1978). *Macromolecules* **11**, 306.

Valenti, B., Alfonso, G., Ciferri, A., Giordani, P., and Marrucci, G. (1981). *J. Appl. Polym. Sci.* **26**, 3643.

Van Luyen, D., and Strzelecki, L. (1980). *Eur. Polym. J.* **16**, 303.

Van Luyen, D., Liebert, L., and Strzelecki, *Eur. Polym. J.* **16**, 307.

Vilasagar, A., and Blumstein, A. (1980). *Mol. Cryst. Liq. Cryst. (Lett.)* **56**, 263.

Volino, F., Martins, A. F., Blumstein, R. B., and Blumstein, A. (1981). *C. R. Hebd. Seances Acad. Sci., Ser. 2*, **292**, 829.

Volksen, W., Dawson, B. L., Economy, J., and Lyerba, J. R. (1979). *Polym. Prepr., Am. Chem. Soc., Div. Polym. Chem.* **20**, 86.

Vorlander, D. (1923). *Hoppe-Seyler's Z. Physiol. Chem.* **105**, 211.

Walton, A. G., and Blackwell, J. (1973). "Biopolymers." Academic Press, New York.

Warner, M., and Flory, P. J. (1980). *J. Chem. Phys.* **73**, 6327.

Wee, E. L., and Miller, W. G. (1978). *In* "Liquid Crystals and Ordered Fluids" (R. Porter and J. Johnson, eds.), Vol. 3., p. 371. Plenum, New York.

Werbowyi, R. S., and Gray, D. C. (1980). *Macromolecules* **13**, 69.

White, J., Bheda, J., and Fellers, J. (1980). Polym. Sci. Eng. Rep. No. 150. Univ. of Tennessee, Knoxville.

Wirik, M. G., and Waldam, M. H. (1970). *J. Appl. Polym. Sci.* **14**, 579.

Wolfe, J. F., Loo, B. H., and Arnold, F. E. (1981). *Macromolecules* **14**, 915.

Wooten, W. C., Jr., McFarlane, F. E., Gray, T. F., Jr., and Jackson, W. J., Jr. (1979). *In* "Ultra-High Modulus Polymers" (A. Ciferri and I. M. Ward, eds.), p. 227. Appl. Sci. London.

Zippler, P., Krigbaum, W. R., and Kratky, O. (1969). *Kolloid Z. Z. Polym.* **235**, 1281.

Note added in proof: A few reports appeared after this chapter was completed.   W. J. Welsh, D. Bhaumik, and J. E. Mark (1981), *J. Macromol. Sci., Phys.* **B20**, 59, presented an interesting study of the conformation of molecular swivels able to reduce chain rigidity while maintaining extended conformations.   More data on thermotropic polyesters have been reported: R. W. Lenz and J. I. Jin (1981), *Macromolecules* **14**, 1405; C. Ober, J. I. Jin, and R. W. Lenz (1982), *Br. Polym. J.* **14**, 1; B. W. Jo, J. I. Jin, and R. W. Lenz (1982), *Eur. Polym. J.* **18**, 233; B. W. Jo and R. W. Lenz (1982), *Makromol. Chem., Rapid Commun.* **3**, 23; A. Blumstein, K. N. Sivaramakrishnan, R. B. Blumstein, and S. B. Clough (1982), *Polymer* **23**, 47; A. Blumstein, S. Vilasagar, S. Ponrathnam, S. B. Clough, R. B. Blumstein, and G. Maret (1982), *J. Polym. Sci. Polym. Phys. Ed.* **20**, 877.   A. Roviello and A. Sirigu (1982), *Makromol. Chem.* **183**, 895, reported even–odd effects on the isotropization entropy of homopolyesters with flexible spacers, which, they suggested, imply an anisotropic arrangement of the latter within the nematic phase.   A. Blumstein, G. Maret, and S. Vilasagar (1981), *Macromolecules* **14**, 1544, derived an opposite conclusion from magnetic birifringence data.   Spinning data on thermotropic polyesters, D. Acierno, F. P. LaMantia, G. Polizzotti, A. Ciferri, and B. Valenti (in press), *Macromolecules*, show that obtaining high modulus on these systems is more complex than anticipated.   Several new lyotropic systems have been reported: T. Kaneda, S. Ishikawa, H. Daimon, T. Katsura, and M. Ueda (1982), *Makromol. Chem.* **183**, 417, 434; M. Balasubramanian, M. J. Nanjan, and M. Santappa (1982), *J. Appl. Polym. Sci.* **27**, 1423; M. B. Polk, K. B. Bota, E. C. Akubuiro, and M. Phingbodhipakkiya (1981), *Macromolecules* **14**, 1626.   Schizophyllan was described by K. Van, T. Norisuye, and A. Teramoto (1981), *Mol. Cryst. Liq. Cryst.* **78**, 123; and DNA was described by A. A. Brian, H. L. Frish, and L. S. Lerman (1981), *Biopolymers* **20**, 1305.   A reanalysis of hydroxypropylcellulose solutions, G. Conio, E. Bianchi, A. Ciferri, A. Tealdi, and M. A. Aden (submitted), deals with the relative contribution to thermotropicity of anisotropic dispersion forces and the temperature dependence of the axial ratio, and with the correspondence between the latter and the Kuhn statistical segment.   Quantitative determination of the ternary phase diagram involving PBA and X-500 showed excellent agreement with theory, E. Bianchi, A. Ciferri, and A. Tealdi, (in press), *Macromolecules*.

4

Molecular Theories
of Liquid Crystals

P. J. Flory

Stanford University
Stanford, California

I. Introduction

Molecular asymmetry is a feature common to all liquid crystalline substances. Most nematic and cholesteric liquid crystals consist of rodlike molecules with axial ratios exceeding three. Platelike molecules or particles may also exhibit the mesomorphic characteristics associated with liquid crystallinity. Polymers occurring in the liquid crystalline state typically consist of comparatively rigid structures causing the chain backbone to be highly extended. Alternatively, rigid side chains appended to a flexible backbone may engage in the function of domains of mesomorphic order.

Asymmetry of molecular shape may be augmented by anisotropic intermolecular forces in promoting liquid crystallinity. This contributing factor is significant in aromatic nematogens (low molecular and polymeric) in which p-phenylene groups lie along the molecular axis. The polarizability anisotropy of the phenylene group causes the intermolecular London attractions to be anisotropic, with the result that parallel intermolecular alignment is favored. The principal molecular feature responsible for liquid crystallinity invariably is asymmetry of molecular shape, however.

Structures of liquids and solutions are dominated by influences of intermolecular repulsions. Intermolecular attractions have a comparatively minor

POLYMER LIQUID CRYSTALS

effect on the radial distribution function and, in the case of asymmetric molecules, on the orientational correlations as well. At the high density and close packing prevailing in the liquid state it suffices, therefore, to represent the molecules, in first approximation at least, as "hard" bodies of appropriate size and shape whose only interactions are the insurmountable repulsions that would be incurred if one of them should overlap another. Lattice treatments are well suited to the calculation of the intermolecular or "steric" part of the configuration partition function for such a system. The effects of intermolecular attractions may be appended to the thermodynamic functions thus derived after the manner of Bragg and Williams, or of regular solution theory.

II. Lattice Theory for Noninteracting Rods

Lattice methods can be adapted to treatment of a system of rodlike particles or molecules by the device illustrated in Fig. 1. The particle shown in Fig. 1a is oriented at an angle ψ from the domain axis, which we identify with one of the principal axes of the cubic lattice. It comprises x isodiametric segments, each equal in size to a cell of the lattice; x is also the axial ratio. In order to accommodate the particle thus inclined on the lattice, it is subdivided into y sequences of segments in Fig. 1b, with each sequence oriented along the preferred axis (Flory, 1956). The parameter y serves as a measure of disorientation of the rodlike molecule relative to the domain axis.

The analysis depends on evaluation of the number v_j of situations available to molecule j with disorientation y_j when $j - 1$ molecules have been added to the lattice previously. The average disorientation is $\bar{y}$. The number v_j is the product of expectations that sites required for occupation by suc-

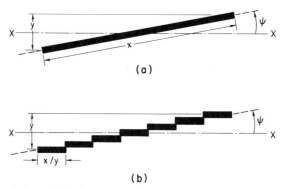

Fig. 1.   (a) Rodlike particle oriented at an angle ψ to the preferred axis, and (b) its subdivision into y sequences amenable to assignment to the cubic lattice. (From Flory and Ronca, 1979a.)

cessive segments are vacant. For a random distribution of orientations of the rods, this expectation can be approximated by the volume fraction of lattice vacancies. If the rods are preferentially oriented, this expectation is conditional on vacancy of the site required for the preceding segment. For perfect alignment it is the "number fraction" expressing the ratio of vacancies to the sum of vacancies and the number of rodlike molecules, or of rodlike sequences into which the molecules in the system are subdivided artificially as in Fig. 1. Thus each initial segment of a sequence requires a volume fraction factor and each succeeding segment a number fraction. The number fraction exceeds the volume fraction, hence v_j increases with orientation of molecule j, since fewer sequences entailing volume fraction factors for their initial segments are then required for lattice representation of this molecule. It increases also with the average orientation of other molecules since the number of sequences for their representation is decreased, with the result that the required number fraction of vacancies is increased.

The result obtained by combining factors, one for each segment, assigned in the prescribed manner is (Flory, 1956; Flory and Ronca, 1979a)

$$v_j = n_0 \left[\frac{n_0 - x(j - 1)}{n_0 - x(j - 1) + \sum_1^{j-1} y_i} \right]^x \left[\frac{n_0 - x(j - 1) + \sum_1^{j-1} y_i}{n_0} \right]^{y_j} \tag{1}$$

where n_0 is the total number of lattice sites. This result is central to the statistical mechanical treatment. That replacement of the rodlike particles by successions of sequences as in Fig. 1b can have little effect on the estimation of v_j is assured by intuition and analogies. Remaining steps in the analysis are straightforward.

The combinatory partition function Z_{comb} representing the number of accessible configurations for the system is obtained as the product of the v_j for each rodlike solute molecule added successively to the lattice; i.e.,

$$Z_{comb} = \frac{1}{n_x!} \sum_1^{n_x} v_j \tag{2}$$

where n_x is the total number of rodlike molecules. As expected from the preceding discussion, Z_{comb} increases with orientation; i.e., it decreases with increase in $\bar{y}$.

The complete partition function Z_M is the product of Z_{comb} and the function Z_{orient} that takes account of the orientational distribution, i.e.,

$$Z_M = Z_{comb} Z_{orient} \tag{3}$$

The latter function may be expressed by (Flory and Ronca, 1979a)

$$Z_{orient} = \prod_y \left(\frac{\omega_y n_x}{n_{xy}} \right)^{n_{xy}} \tag{4}$$

where n_{xy} is the number of rods with disorientation from the preferred axis denoted by y and ω_y is the fraction of solid angle associated with y. According to the "1979" treatment (Flory and Ronca, 1979a),

$$\omega_y = (\pi/4x)\tan\psi \tag{5}$$

with

$$\sin\psi = (\pi/4)(y/x) \tag{6}$$

In the earlier "1956" approximation (Flory, 1956), based on an arbitrary "square-well" distribution of uniform density over $\sin\psi$ with $|\psi| < \psi^*$, Z_{orient} reduces to the simple form

$$Z_{orient} = \bar{y}^{2n_x} \tag{7}$$

Whereas Z_{comb} decreases with disorientation ($\bar{y}$), Z_{orient} increases. For sufficiently small axial ratios x and concentrations, Z_{orient} dominates and Z_M increases monotonically as $\bar{y}$ increases from 1 to x. The completely disordered state with $\bar{y} = x$ is then stable. As the axial ratio is increased at a given concentration, the decrease in Z_{comb} with $\bar{y}$ becomes more marked. Eventually, Z_M exhibits a maximum and then a minimum with increase in $\bar{y}$, i.e., with decrease in order. Beyond a slightly larger x (or concentration) than is required for appearance of the extrema in Z_M, the value of $(Z_M)_{max}$ exceeds that for $\bar{y} = x$. Thereupon, $(Z_M)_{max}$ becomes the stable state. A discontinuous transition of first order is thus predicted with increase in concentration or in x. One of the coexisting phases at biphasic equilibrium is isotropic, the other is highly ordered (nematic), but not perfectly so. The latter may be only marginally more concentrated than the former.

Figure 2 shows compositions of the coexisting phases in a binary system consisting of hard rods and a solvent, calculated (Flory and Ronca, 1979a) according to theory as functions of the axial ratio x. The volume fractions v_x and v'_x in the isotropic and nematic phases, respectively, at equilibrium are given on the abscissa for the value of the axial ratio x plotted on the ordinate. In the limit $x = \infty$, $v'_x/v_x = 1.465$. As the concentration is increased, the two curves merge at $v_x = v'_x = 1.00$ and $x = 6.4$, the minimum axial ratio required for stable nematic order in a neat liquid consisting of hard rods.

The distribution of orientations in a nematic phase follows directly from Eq. (1) with j replaced by the total number n_x of rods. It is expressed by

$$n_{xy}/n_x = f_1^{-1}\omega_y\exp(-ay) \tag{8}$$

where

$$a = -\ln[1 - v'_x(1 - \bar{y}/x)] \tag{9}$$

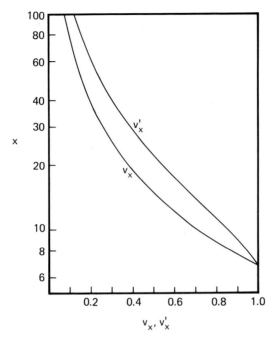

Fig. 2. Compositions of coexisting isotropic and nematic phases, expressed as volume fractions v_x and v_x', calculated as functions of the axial ratio according to theory for hard rods devoid of intermolecular attractions. (From Flory and Ronca, 1979a.)

and f_x^{-1} is a normalization factor; v_x' is the volume fraction of rodlike species. Substitution of Eq. (8) in (4) completes the evaluation of Z_{orient}.

The results yielded by the refined treatment (Flory and Ronca, 1979a) and described briefly above differ marginally from those obtained according to the earlier version Flory (1956). The limiting ratio $(v_x'/v_x)_{x=\infty}$ obtained in the 1956 approximation is 1.592, and $x_{\text{lim}} = 6.70$ for a stable nematic fluid of hard rods. The simplicity of the approximation embodied in Eq. (7), which was used instead of Eq. (4) for Z_{orient} in the 1956 treatment, greatly facilitates applications of the theory to a wide variety of systems. Included are treatments of polydisperse systems consisting of any arbitrary distribution of homologous rods differing in length (Flory and Abe, 1978; Abe and Flory, 1978; Flory and Frost, 1978; Frost and Flory, 1978), semiflexible particles consisting of rigid, rodlike members joined end-to-end by flexible joints (Flory, 1978a), mixtures of rods and random coils (Flory, 1978b), polymers comprising both rodlike and random-coiling sequences of segments (Matheson and Flory, 1981), and nonathermal solutions of rods subject to an (isotropic) exchange energy of interaction for contacting pairs of the solvent and solute (rod) components (Flory, 1956; Miller *et al.*, 1974a,b). Experiments on lyotropic nematic phases of polymers provide substantial confirmation for the various predictions of the lattice theory for systems of rods (Miller *et al.*, 1974a,b; Papkov *et al.*, 1974; Papkov, 1977; Morgan, 1977; Aharoni, 1979a,b, 1980; Aharoni and Walsh, 1979).

III. Orientation-Dependent Interactions

If the London dispersion forces of attraction between segments of neighboring molecules are anisotropic, they may confer added stability on the nematic state. Just as the dispersion energy between isotropic molecules depends on the polarizabilities α_1 and α_2 of the neighboring pair, the difference in energy between mutually aligned neighbors and uncorrelated ones depends on the anisotropies of their polarizabilities. If the polarizabilities of a pair of interacting segments are cylindrically symmetric, and if the interacting segments are identical, then the lowering of the energy with orientation is proportional to $(\Delta\alpha)^2$ (Flory and Ronca, 1979a), where $\Delta\alpha$ is the difference between the polarizabilities parallel to and perpendicular to the molecular axis.

Orientational energy from this source is invoked as the basis for the Maier and Saupe (1959, 1960) theory. Anisotropy of molecular shape and its influence on molecular packing in the fluid are not taken into account. Consequently, in order to reconcile this theory with experiments, it has been necessary to postulate unreasonably large orientation-dependent energies (Wulf, 1976) and, by implication, excessively large values of $\Delta\alpha$.

It is important to observe that the relevant interaction energy is that between contacting pairs of segments of neighboring molecules and not the energy for a pair of molecules, each considered as a whole (Flory and Ronca, 1979b). Obviously, the interaction between all of the segments of a chosen pair of rodlike molecules must increase with improvement of their mutual alignment, owing to the greater number of segments in contact. This increase in interaction occurs regardless of the anisotropy of the intermolecular forces; the anisotropy expressed by $\Delta\alpha \neq 0$ is not required. The enhancement of contacts between a given pair of molecules as they become mutually aligned is compensated by the concomitant elimination of contacts between other molecules (Flory and Ronca, 1979b). The compensation is exact at constant volume, so that the net change of energy on this score should be null if the incident dispersion forces are isotropic, i.e., if $\Delta\alpha = 0$. Since the volume change at the nematic–isotropic transition is very small, a contribution to the energy from this source may confidently be dismissed (Flory and Ronca, 1979b).

If, however, the anisotropy of the polarizability is substantial, then the segment–segment interaction energy becomes orientation dependent. This is the case for the numerous nematogens containing p-phenylene and other groups such as

$$-N=N-, \quad -N=N-, \quad -C\equiv C-, \quad \text{and} \quad -C\equiv N$$

with larger polarizabilities along their bond axes than perpendicular thereto (or along the p-phenylene axis compared to the lower polarizability perpendicular to the ring). In absence of a diluent, a segment inclined at an angle ψ with respect to the domain axis experiences an energy (Flory and Ronca, 1979b)

$$\varepsilon(\psi) = -kT^*s(3\cos^2\psi - 1)/2 \tag{10}$$

in the mean field of its neighbors. Here, k is Boltzmann's constant, T^* is the characteristic temperature that measures the strength of the orientation-dependent interaction, and s is the order parameter defined by

$$s = (3\langle\cos^2\psi\rangle - 1)/2 \tag{11}$$

The characteristic temperature is related to $\Delta\alpha$ according to

$$kT^* = \text{Const} \, r_*^{-6}(\Delta\alpha)^2 \tag{12}$$

where r_* is the characteristic distance between neighboring segments in contact.

The orientation-dependent energy for the system as a whole follows as

$$E_{\text{orient}} = -\tfrac{1}{2} xn_x kT^*s^2 \tag{13}$$

Multiplication of the partition function Z_M (Eq. (3)) for a system of rods without attractive forces by the Boltzmann factor for this energy yields the partition function for rods subject to orientation-dependent intermolecular interactions. By straightforward procedures, one may then deduce compositions of coexisting isotropic and nematic phases as functions of the reduced temperature $\tilde{T} = T/xT^*$ and the axial ratio x (Flory and Ronca, 1979b; Warner and Flory, 1980). The surfaces relating compositions of the respective phases to $\tilde{T}$ and x merge (Warner and Flory, 1980) to the line relating $\tilde{T}$ to x for the thermotropic fluid consisting of rods in absence of diluent (Flory and Ronca, 1979b). The calculated curve expressing this relationship is shown in Fig. 3. Starting at $x = 6.4$ for hard rods not subject to intermolecular attractions, the curve describes the increasing intermolecular attractions required to compensate a smaller axial ratio and thus maintain biphasic equilibrium. Observed transition temperatures for low molecular nematogens have been compared with theory on the basis of this functional relation-

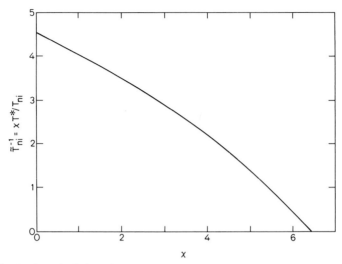

Fig. 3. Reciprocal of the reduced transition temperature plotted against the axial ratio. (From Flory and Ronca, 1979b.)

ship (Flory and Ronca, 1979b). The axial ratios are calculated from the molecular structure. The curve in Fig. 3 then gives the corresponding theoretical value of the reduced temperature. The observed transition temperature $T_{NI} = x\tilde{T}_{NI} T^*$ yields the characteristic temperature T^* for the given nematogen. Values of T^* thus obtained are mutually consistent with qualitative inferences from the molecular constitution.

For a more definitive test of the theory, Irvine and Flory (1982) determined the optical anisotropies of nematogens and analogous compounds from depolarized Rayleigh scattering measurements on their dilute solutions. From the value of $\Delta\alpha$ determined in this way, T^* can be estimated. Results for several compounds are compared in Table I with values of T^* deduced from T_{NI}, the axial ratios x, and the function shown in Fig. 3. The values (last column) estimated from the optical anisotropy $\Delta\alpha$ are subject to fairly large errors arising from uncertainty in the value of the constant in Eq. (12). They are consistently smaller than the T^* (penultimate column) calculated from the observed transition temperature T_{NI} and the curve in Fig. 3. The differences are within the range of the uncertainties mentioned, and the two sets of T^* values follow the same order.

Recently Wu and Flory (1982) investigated nematic–isotropic phase equilibria in low polymers

$$C_6H_5CO(OC_6H_4CO)_{x-2}OC_6H_5$$

obtained by random polycondensation of phenyl p-hydroxybenzoate and phenyl benzoate. The value of x is variable; its distribution is reliably

TABLE 1

Evaluation of T^{*a}

Nematogen	x	$T_{NI}(K)$	$T^*(K)$ From T_{NI}	$T^*(K)$ Estimated from $\Delta\alpha$
$H(C_6H_4)_5H$	5.0	718	200	130 ± 20
B-4[b]	5.2	529	125	75 ± 10
MBBA	4.2–4.5	319	120–160	85 ± 5
PAA	3.6	409	280	160 ± 20

[a] From Irvine and Flory (1982).
[b] $C_6H_5CO(OC_6H_4CO)_2OC_6H_5$

calculable from elementary statistical considerations. For $\bar{x} < 4$, melting points of the polymers do not exceed $\sim 200°C$ and experiments in the liquid range are consequently feasible. The molten polymers are quite fluid, hence the isotropic and nematic phases can be separated quantitatively. The observed fractional volumes occupied by the nematic phase are presented in Table II as functions of the average degree of polymerization $\bar{x}$ (Wu and Flory, 1982).

Values given in the last column were calculated according to theory for $T^* = 175$ K. The calculations require summations over all species present in significant amount (i.e., up to $x \approx 15$), orientation-dependent interactions being taken into account. The same value of T^* characterizing these interactions was used for all samples. The agreement is remarkably good. The

TABLE II

Effect of Average Degree of Polymerization on Volume Fraction of Mesophase[a]

	Volume Fraction of Nematic Phase at 191°C	
$\bar{x}$	Observed	Calculated ($T^* = 175$ K)
2.5	0	0
3.0	0.19	0.22
3.09	0.35	0.32
3.25	0.58	0.46
3.50	0.54[b]	0.66

[a] From Wu and Flory (1982).
[b] 22% of this sample separated as a crystalline phase.

value of T^* is somewhat larger than T^* deduced for the pure "tetramer" with $x = 4$, as recorded in Table I. Other experiments on the effect of temperature from 157 to 210°C are consistent with theory also.

IV. Summary and Conclusions

The principal determinant of nematic order is the axial ratio of the molecule in the case of low molecular nematogens and in polymers that yield lyotropic nematic phases it is the chain extension or "stiffness." Orientation-dependent intermolecular attractions may contribute also to the stability of the nematic state. They assume significance in molecules containing anisotropic groups such as phenylene.

Systems of rigid rods may be treated conveniently by adaptation to a lattice model. The partition function thus derived for hard rods devoid of orientation-dependent interactions depends on the composition, on the axial ratio x of the rods, and on a free thermodynamic variable y specifying the disorientation of a rod relative to the domain axis. Conditions for stability of nematic order are specified by theory, and equilibria thus predicted are in agreement with experiments.

The theory has been extended recently to comprehend systems in which orientation-dependent interactions are operative between rods. These interactions must be reckoned in terms of segments, instead of whole molecules, in contact. The strengths of these interactions deduced from the nematic–isotropic transition temperatures correlate well with calculations based on anisotropies of polarizabilities determined by depolarized Rayleigh scattering.

Acknowledgment

This work was supported by the Director of Chemical Sciences Air Force, Air Force Office of Scientific Research, Grant No. 77-3293-D, to Stanford University.

References

Abe, A., and Flory, P. J. (1978). *Macromolecules* **11**, 1122.
Aharoni, S. M. (1979a). *Macromolecules* **12**, 94, 537.
Aharoni, S. M. (1979b). *J. Polym. Sci., Polym. Phys. Ed.* **17**, 683.
Aharoni, S. M. (1980). *J. Polym. Sci., Polym. Phys. Ed.* **18**, 1439.
Aharoni, S. M., and Walsh, K. E. (1979). *Macromolecules* **12**, 271.
Flory, P. J. (1956). *Proc. R. Soc. London, Ser. A* **234**, 73.
Flory, P. J. (1978a). *Macromolecules* **11**, 1141.
Flory, P. J. (1978b). *Macromolecules* **11**, 1138.
Flory, P. J., and Abe, A. (1978). *Macromolecules* **11**, 1119.

Flory, P. J., and Frost, R. S. (1978). *Macromolecules* **11**, 1126.
Flory, P. J., and Ronca, G. (1979a). *Mol. Cryst. Liq. Cryst.* **54**, 289.
Flory, P. J., and Ronca, G. (1979b). *Mol. Cryst. Liq. Cryst.* **54**, 311.
Frost, R. S., and Flory, P. J. (1978). *Macromolecules* **11**, 1134.
Irvine, P. A., and Flory, P. J. (1982). In preparation.
Maier, W., and Saupe, A. (1959). *Z. Naturforsch., A* **14A**, 882.
Maier, W., and Saupe, A. (1960). *Z. Naturforsch., A* **15A**, 287.
Matheson, R. S., Jr., and Flory, P. J. (1981). *Macromolecules* **14**, 954.
Miller, W. G., Rai, J. H., and Wee, E. L. (1974a). *In* "Liquid Crystals and Ordered Fluids",
    (J. F. Johnson and R. F. Porter, eds.), Vol. 2, pp. 243–255. Plenum, New York.
Miller, W. G., Wu, C. C., Wee, E. L., Santee, G. L., Ray, T. H., and Goebel, K. (1974b). *Pure*
    Appl. Chem. **38**, 37.
Morgan, P. W. (1977). *Macromolecules* **10**, 1381.
Papkov, S. P. (1977). *Vysokomol. Soedin., Ser. A* **19**, 3.
Papkov, S. P., Kulichikhin, V. G., Kalmykova, V. D., and Malkin, A. Y. (1974). *J. Polym.*
    Sci., Polym. Phys. Ed. **12**, 1753.
Warner, M., and Flory, P. J. (1980). *J. Chem. Phys.* **73**, 6327.
Wu, D., and Flory, P. J. (1982). In preparation.
Wulf, A. (1976). *J. Chem. Phys.* **64**, 104.

5

Mechanical Properties
of Nematic Polymers

P. G. de Gennes

Collège de France
Paris, France

I. General Aims

In fluid polymer systems, mechanical measurements have been extremely helpful in delineating certain basic microscopic processes, such as backflow effects around dilute coils, or entanglement effects in concentrated systems (Ferry, 1970; Graessley, 1974). For conventional nematics, the three elastic constants (Fig. 1) or the five friction constants (Fig. 2) should also provide us with a wealth of information. However, because the molecules concerned are not very very large, the mechanical properties are probably sensitive to the details of the chemical structure and are not very revealing.

115

POLYMER LIQUID CRYSTALS

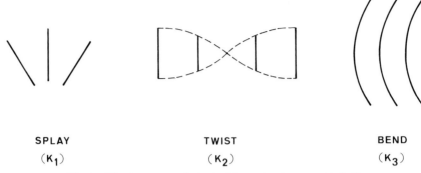

SPLAY   TWIST   BEND

(K_1) (K_2) (K_3)

Fig. 1.   The three types of elastic deformation in a nematic fluid.

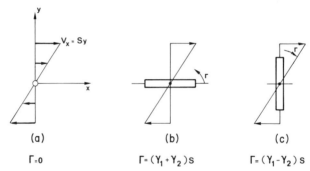

(a)   (b)   (c)

$\Gamma = 0$ $\Gamma = (Y_1 + Y_2)s$ $\Gamma = (Y_1 - Y_2)s$

Fig. 2.   The three geometries for measurement of the Miesowicz viscosities $(\eta_a, \eta_b, \eta_c,)$ in a nematic fluid. Also indicated are the torques Γ due to the flow acting on the molecules. On the whole we are dealing with five friction coefficients.

For instance, a careful analysis of the elastic constants in certain homologous series of Schiff bases shows that no existing theoretical scheme is completely adequate (Leenhouts and Dekker, 1981).

With nematic polymers, however, the situation may be somewhat simpler: systematic variations with molecular weight can be looked for. Of course, if we go to very high molecular weights, we deal with highly viscous materials, which are difficult to probe. But there are certain ways of circumventing this difficulty:

(a)   Systematic studies on a few oligomers may suffice to establish certain laws.

(b)   Dilution in an isotropic solvent can be helpful. This is now classical for rodlike molecules: we shall discuss this case only briefly in Section II.

(c)   Dilution in a conventional nematic is another interesting procedure. Certain mechanical measurements on dilute solutions may give information on the conformation of the solute chain.

The problem of these conformations is in fact rather complex.

(1)   Completely flexible chains, which do not contain any nematogenic groups, are poorly soluble in a conventional nematic (Kronberg *et al.*, 1978; Dubault *et al.*, 1980; Brochard, 1979b, 1980).

(2)   Completely rigid, passive rods in isotropic solvents represent an ideal limiting case, which we present in Section II. But the practical situations are often complicated by anisotropic interactions or by flexibility effects.

(3)   Semirigid chains may fall into a variety of conformational types, depending on their intrinsic flexibility and on the properties of the surrounding medium; we insist on these differences in Section IV. In Section V we try to define the mechanical properties of the various conformational types, and to see how they will vary with the molecular weight of the chains.

Our approach is mainly qualitative. The aim is to discuss concepts rather than numbers, and also to maintain a style that should remain accessible to both physicists and chemists.

II. Rigid Rods in Isotropic Solvents

Long passive rods tend to align parallel to each other when their number density v is above a certain limit $v_{NI} \sim L^{-2}d^{-1}$ (L = length, d = diameter) (Onsager, 1949; Flory, 1956, 1978a; see also Flory, Chapter 4, this volume). A detailed discussion of the elastic energies for these nematic systems has been given within mean field theory (Priest, 1973; Straley, 1973). We can summarize some of the results in the following way. An elastic constant K is of the form

$$K = vJl^2$$

where J is a coupling energy and l measures the range of the couplings. For the rod systems with purely steric interactions, $J \sim kT$ and $l \sim L$. This implies that at the nematic threshold ($v = v_{NI}$)

$$K \sim kT/d$$

i.e., that the elastic constants should be essentially independent of molecular weight. The existing experiments do not yet enable us to confirm or disprove this prediction (DuPré, Chapter 7, this volume.)

A few words now on the friction properties: Here interesting effects occur, not only in the nematic phase, but also at lower concentrations. Doi and Edwards (1978) noticed that the rod–rod interactions become negligible at $v \ll v_{NI}$ for all static properties, but remain essential for the dynamics: the rotational relaxation of one rod is hindered by its neighbors, to concentrations as low as $v = v^* \sim L^{-3}$. The concentration v^* is much lower than v_{NI}.

At $v = v^*$, the viscosity η of the isotropic solution is comparable to the viscosity η_0 of the pure solvent. But at $v > v^*$, $\eta(v)$ increases rapidly because of the rotational hindrance.

Thus, just below the isotropic nematic transition, η/η_0 is large; this effect can be most inconvenient for certain extrusion processes. It may be worthwhile to notice that this difficulty is removed if, instead of hard rods, we use a suspension of *hard disks* (diameter L, thickness d). With such disks the two concentrations v^* and v_{NI} become comparable: $v^* \sim v_{NI} \sim L^{-3}$. Thus if we want to extrude a solution of platelets, we expect to find a viscosity at the onset of nematic order that is not much larger than the solvent viscosity.

Returning to the rod system in the high concentration nematic phase ($v > v_{NI}$), we find detailed predictions regarding the friction parameters (Doi, 1980; Marrucci, 1981). The main practical feature is that the viscosity in simple shear decreases with increasing v: one interesting contribution to this effect is due to an increase of rotational rates when the nematic order increases (each rod is now restricted to a rather small solid angle, but can explore it freely).

III. Dilute Rods in a Nematic Solvent

A. Antagonists and Agonists

It is quite difficult to prepare a stable suspension of hard objects inside a conventional nematic; apart from the usual trend toward precipitation due to long range van der Waals forces, there are also specific complications related to the existence of long range order (characterized by an order parameter S) in the matrix. The suspended material may be "antagonist" or "agonist" to S.

An agonist solute will increase S in its neighborhood, while an antagonist will depress it (Fig. 3). An important consequence of this effect is that an agonist molecule M will attract another agonist M': since M' prefers to be in regions of high S it will tend to stick to M. Similarly, an antagonist P attracts another antagonist P' because P' prefers to be in regions of low S and finds such a region near P.

This effect is omnipresent: for instance, Flory (1978b) shows that random coils tend to segregate out of a suspension of passive rods. Among other effects, each coil acts as an antagonist for the nematic order of the rod system.[†]

[†] Another source of attraction between suspended bodies in a two-component liquid is a modification of the composition of the liquid near the suspended body. This also plays a role in the coil–rod–solvent system. Similarly, hard spheres in an isotropic solution of rods (no S effect) can have a significant attraction (Auvray, 1981).

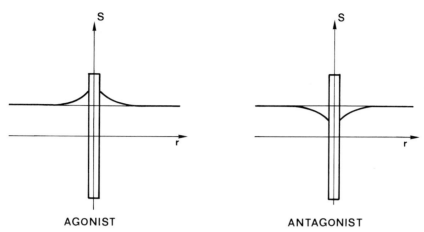

Fig. 3. Deformation of the nematic order parameters in the vicinity of a solute object. The drawing ignores the tensorial character of S and is highly idealized.

Thus to stabilize a colloid of rods in a nematic matrix, the rods should be neither antagonist nor agonist to S, but neutral. One plausible pathway for rod stabilization is shown in Fig. 4. Here each rod is coated with an adsorbed layer of nematic polymer; the nematogenic unit of this polymer is chosen to be quite similar to the matrix. The spacer is assumed to adsorb

Fig. 4. One possible method for stabilization of a long particle in a nematic matrix.

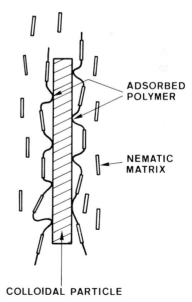

preferentially on the rod surface. The adsorbed clouds from two adjacent rods should create a repulsive (stabilizing) force when they overlap.

All this discussion on suspensions of rods inside a nematic liquid may seem exceedingly remote from known facts. However, we should remember that suspensions of elongated magnetic particles, if they can be made, should have admirable properties (Brochard and de Gennes, 1970). For instance, with a cholesteric matrix a color change could be induced by a magnetic field of order 1 G.

B. Effects on Elastic Constants

When we put one (neutral) rod in a nematic solvent it may decide to align parallel to the director, or normal to it, or even at some fixed nontrivial angle ψ. However, the case of parallel alignment is probably rather frequent (Brochard and de Gennes, 1970) and we shall concentrate on this situation only.

The effects of one rod on elastic deformations are shown in Fig. 5. We expect very weak effects for splay and twist, and strong effects for bend $K_3 \rightarrow K_3 + \delta K_3$. If the director is well anchored to the rod, we expect

$$\delta K_3 = \frac{(\text{const}) \cdot K v L^3}{\ln(L/d)} \tag{1}$$

where K is an average of splay and bend constants. The essential feature (expressed by the factor $v L^3$) is that a large volume ($\sim L^3$) around the rod is resisting the bend constraint.

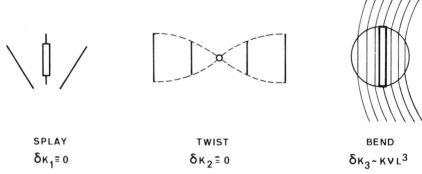

SPLAY	TWIST	BEND
$\delta K_1 \cong 0$	$\delta K_2 \cong 0$	$\delta K_3 \sim K v L^3$

Fig. 5. Effects of dilute rods on the elastic constants of a nematic. The rods lie parallel to the local director. Strong anchoring is assumed.

C. *Effects on Friction Coefficients*

The three basic geometries are shown in Fig. 2. For a thin rod that aligns parallel to the director, it is clear that

$$\delta\eta_a = \delta\eta_b = \delta(\gamma_1 + \gamma_2) \cong 0 \tag{2}$$

but two coefficients are expected to be seriously modified:

$$\delta\eta_c \cong \delta(\gamma_1 - \gamma_2) \cong \nu L^3 \eta \tag{3}$$

where η is some average of the matrix viscosities and where, for simplicity, we ignore all logarithmic factors.

Thus both for elasticity and friction we expect to find corrections on certain coefficients that scale as νL^3 (where ν is the number density) or equivalently as ϕL^2 (where ϕ is the volume fraction of solute).

IV. Semirigid Molecules: the Conformation Problem

What we have in mind here is the chain shown in Fig. 6a, with a nematogenic backbone separated by spacers. Depending on the size of the spacers, we may go from very rigid to very flexible chains. But it is not only the chain that matters. The surroundings are also important: thus we shall discuss first the case of one chain embedded in a conventional nematic; and then turn to the more complex, but important, case where the medium itself is made of other identical chains.

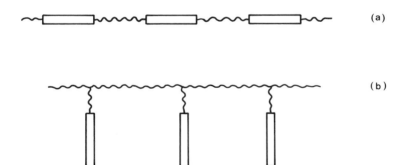

Fig. 6.   Two types of nematic polymers: (a) nematogenic backbone, (b) lateral nematogen. Most of the discussion in the text is concerned with (a).

A. *One Chain in a Low Molecular Weight Nematic*

Following our terminology for hard rods, we assume that the solute chain is "neutral" (neither antagonist nor agonist) to the surrounding matrix. This

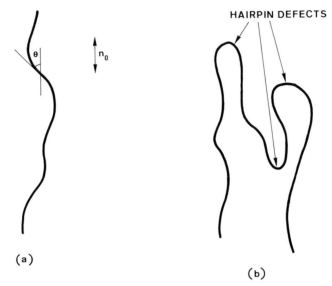

HAIRPIN DEFECTS

(a)

(b)

Fig. 7. Chains (a) without hairpins and (b) with hairpins. In case (a) the chain size along the nematic axis increases linearly with molecular weight M. In case (b) the size increases as $M^{1/2}$.

will be necessary to avoid precipitation. Depending on the chain flexibility, we find two important limiting cases:

1. Without Hairpins (Figure 7a)

Locally, the tilt angle θ (between the chain and the overall nematic axis) is everywhere small. Our starting point in discussing such a case is a free energy of the form

$$F = F_0 + \int ds \, \frac{1}{2} \left[Q\theta^2 + B\left(\frac{d\theta}{ds}\right)^2 \right] \tag{4}$$

where s is the curvilinear abscissa along the chain. The coefficient Q has the dimensions of a force and measures the coupling between chain alignment and matrix alignment.

The very existence of a coefficient Q is not obvious for the following reason: the nematic director $\mathbf{n}$ around the chain is perturbed by the local tilts θ, and this perturbation is long range. This leads to a renormalization of Q, which is discussed in the appendix. The result is that, strictly speaking, Q is not a constant but depends on the wavelength $(2\pi/q)$ of the distortion considered for $\theta(s)$. Fortunately, however, this dependence is weak (logarithmic) and for all practical purposes Eq. (4) with a constant Q is a reasonable starting point.

The direct (unrenormalized) Q is probably proportional to the matrix order parameter S:

$$Q^{\mathrm{dir}} = Q_0 S \tag{5}$$

On the other hand, the correction Q^{el} due to elastic effects in the matrix is proportional to the Frank constants K of the matrix, and is thus expected to scale as S^2:

$$Q^{\mathrm{el}} = Q_1 S^2 \tag{6}$$

When both effects are present, the rule is that the compliances Q^{-1} are additive, as shown in the appendix:

$$1/Q = (1/Q^{\mathrm{dir}}) + (1/Q^{\mathrm{el}}) \tag{7}$$

Thus the dependence of Q on S may be complex but in all cases we expect Q to increase with S.

Returning to Eq. (4), let us comment on the second phenomenological coefficient B. It has the dimensions energy · length, and measures the local flexibility of the chain. In a first approximation, we might assume that B is independent of the surrounding medium, i.e., independent of S.

The essential conclusion drawn from Eq. (4) is then the existence of a correlation length

$$l = (B/Q)^{1/2} \tag{8}$$

Quantitatively, the chain is then expected to behave as a succession of rigid "pieces," each of length l. Note thate if the spacers are small and rigid, l may be much larger than the monomer size. Successive pieces have different orientations. The average tilt of any piece has a mean square value

$$\langle \theta^2 \rangle = \tfrac{1}{2} kT(BQ)^{-1/2} \tag{9}$$

and this is assumed to be much smaller than unity. The form of Eq. (9) may be understood by saying that the tilt energy $\tfrac{1}{2} Q l \theta^2$ for one rigid piece is comparable to the thermal energy kT.

Let us now assume that the contour length L of the chains is much longer than l. Then the overall end-to-end dimensions of one chain are simply

$$\langle R_{\parallel}^2 \rangle = L^2(1 - \langle \theta^2 \rangle) \tag{10}$$

$$\langle R_{\perp}^2 \rangle = LkT/Q \qquad (L \gg l) \tag{11}$$

Projected along the directions x (or y) normal to the director, the chain is Gaussian and is a sequence of units, each of projected size $b \cong (l^2 \langle \theta^2 \rangle)^{1/2}$.

The number of these units is L/l and

$$R^2 \cong (L/l)b^2 \cong Ll\langle\theta^2\rangle$$

in qualitative agreement with (11).

In practice, one major problem will be to assess the values of l and $\langle\theta^2\rangle$, or equivalently of B and Q, and in particular to find out how B and Q depend on the spacer length. We shall see later that mechanical measurements can help to determine B and (possibly) Q.

2. With Hairpins (Figure 7b)

If the molecular weight of the chain is very high, the chances of finding a hairpin defect become significant. One essential parameter is then the average (contour) distance between consecutive hairpins, which we call h. Another related quantity is the energy ε_h required to build up one hairpin. We expect a relation between them of the form

$$1/h = (\text{const})(1/l)\exp(-\varepsilon_h/kT) \qquad (12)$$

The value of ε_h depends on the detailed molecular conformation of the hairpin. It may also depend on the surrounding medium, for various reasons:

(a) If the medium is made of identical chains, their packing is perturbed in the entire vicinity of the hairpin. Similar effects (for a free chain end) are discussed by Meyer (Chapter 6, this volume).

(b) Even if the medium is a low molecular weight nematic, it tends to enforce a certain alignment in all parts of the chain, including the spacers. Thus ε_h may be a function of S. This statement can be made precise in one particular limit, which is probably not very physical, but which is conceptually instructive: namely, when each hairpin involves many spacers (Fig. 8). Then the continuous model of Eq. (4) can be used, provided that we make the replacement

$$\tfrac{1}{2}Q\theta^2 \to \tfrac{1}{2}Q\sin^2\theta \qquad (13)$$

which is a natural extension for large θ.

The shape of a hairpin is easily found by minimization of the modified energy F. The result is

$$\theta = 2\tan^{-1}[\exp(s/l)] \qquad (14)$$

(with $\theta = 0$ for $s \to -\infty$, and $\theta = \pi$ for $s \to +\infty$). The contour length of the hairpin is $\sim 2l$ and the hairpin energy is

$$\varepsilon_h = 2Ql = 2(BQ)^{1/2} \qquad (15)$$

For instance, if B is independent of S, and Q is linear in S, ε_h would then be proportional to $S^{1/2}$. The calculation is not very realistic because the con-

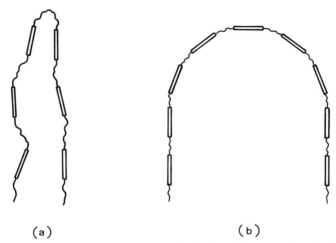

(a)   (b)

Fig. 8.   (a) Sharp hairpin versus (b) slow hairpin. Case (a) is expected to be found with long spacers. Case (b) would be expected for short spacers, but is of high energy and thus rather improbable.

tinuous limit of Eq. (4) requires a rather large l, leading to high values of ε_h: hairpins may then become unobservable in practice. But the main merit of Eq. (15) is to show us that ε_h may be strongly dependent on temperature (through S).

Returning to the conformations of a long chain, we see that the crucial parameter is:

$$L/h = (L/l)\exp(-\varepsilon_h/kT) \tag{16}$$

If $L/h > 1$, we expect a random coil. The end-to-end size along the nematic axis is no longer given by Eq. (8), but is rather the size of a coil with persistence length h:

$$R_{\parallel}^2 = (\text{const})Lh \tag{17}$$

Equation (17) neglects all excluded volume effects; they are probably weak for our locally rigid chains.

B. Melts of Semirigid Chains

These melts have a remarkable "fibrous" structure usually represented as in Fig. 9a for the limit of very few hairpins. The mechanical consequences of the fibrous structure have been discussed by de Gennes (1977) and Meyer (Chapter 6, this volume); the most spectacular effect is that splay deformations become very difficult.

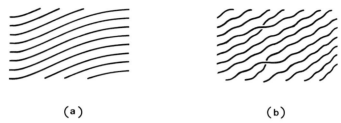

Fig. 9. "Fibrous" structures for melts of nematic polymers, assuming very few hairpins: (a) the conventional representation; (b) an improved representation, showing how each chain may deviate from the local director.

From the point of view of the conformations of one (labeled) chain, the representation of Fig. 9a may be somewhat misleading. By assumption, our melt is a nematic, not a crystal. The order parameter S is large, but remains definitely smaller than unity. Thus, although the chains do (on the average) follow the local director, they can have some deviations from it, as shown in Fig. 9b.

This remark implies that a description of the conformations for one labeled chain in the melt may still be based on deviations $\theta(s)$ with an associated energy of the form shown in Eq. (4). There is one minor difference: the value of the coupling constant Q to be used here is probably $Q = Q_{direct}$. There should be no renormalization of Q due to deformations of the director field around a wavy chain: all the required deformations would have a splay component, and splay is quenched in our system. Thus the chain fluctuates in a frozen director field.

Provided that we put $Q = Q_{direct}$, most of the analysis of Section IV.A.1 is expected to remain valid—with or without hairpins.

V. Semirigid Molecules: Mechanical Properties

A. Entanglements in Melts

In most (present) practical cases with relatively low molecular weights, we expect that chains do not make knots. For conventional polymer melts this is achieved whenever the degree of polymerization N is below a certain entanglement threshold $N_e \sim 300$ (Graessley, 1974). We do not know anything about N_e for semiflexible nematic chains:

(a)   In the absence of hairpins, entanglements may still exist: two chains may wind slowly around each other without ever making large angles with the nematic axis. Knots of this type would generate a rubber elasticity for stresses normal to the optical axis, but they appear inefficient for stresses along the optical axis.

(b)   With hairpins, all entanglements are allowed and we can generate a plateau modulus $G_\parallel$ for stresses applied along the optical axis. It is not clear, however, whether $G_\parallel$ should be proportional to the density of hairpins (i.e., to h^{-1}) or to its square (h^{-2}).

In practice it is, of course, possible to enhance considerably the entanglement density by chain branching. In nematic melts (as opposed to conventional melts), relatively short lateral groups (with one or two nematogenic units) may be sufficient to establish firm entanglement points—because each unit is rather strongly locked by the aligning potential.

It is also possible to prepare gels. For side chain nematogens this has been described by Finkelmann (Chapter 2, this volume). The case of backbone nematogens should give more spectacular mecano-optic couplings. This is discussed qualitatively by de Gennes (1975) and more quantitatively by Jarry and Monnerie (1979). The basic equation describes the free energy as a function of the nematic order parameter S, of the mechanical deformation $\lambda - 1 = e$, and of the applied stress σ

$$F = \tfrac{1}{2}A(T)S^2 - USe + \tfrac{1}{2}\mu e^2 - \sigma e \qquad (18)$$

for small S and e. The coefficient $A(T)$ describes the entropy and the interactions of the nematogens,

$$A(T) \cong k(T - T^*)/V \qquad (19)$$

where V is the volume per monomer and T^* is a virtual nematic–isotropic transition point. Both the elastic modulus μ and the coupling coefficient U are proportional to the density of cross-links (or entanglements)[†]

$$\mu \sim U \sim kT/N_e V. \qquad (20)$$

Various physical effects follow from Eq. (18). (a) Starting from the isotropic phase, one finds a strong stress–optical coefficient. (b) The nematic–isotropic transition is shifted by the stress,

$$\Delta T/T_c \sim \sigma V/\Delta H \sim e/N_e \qquad (21)$$

where ΔH is the enthalpy of the N–I transition. (c) The existence of a stress-induced transition should lead to very nonlinear plots of e versus σ. These effects may have weak counterparts in conventional rubbers where a small nematic coupling can exist between elastomers (Jarry and Monnerie, 1980).

[†] The estimate of U by de Gennes (1975) is incorrect.

For our really nematic systems they could also be observed without cross-linking (in transient regimes) when entanglements are present. One complication however is that N_e might depend strongly on S.

B. Nonentangled Regimes ($N < N_e$)

For a melt of chains without hairpin defects, the Frank elastic constant K_3 for bend is expected to be normal:

$$K_3 = B\Sigma^{-1} \tag{22}$$

(where Σ is the cross-sectional area per chain normal to the optical axis). On the other hand, as already mentioned, the splay constant K_1 is large. I have suggested (de Gennes, 1977)

$$K_1 = EL^2 \tag{23}$$

where E has the dimensions of an elastic modulus (but is mostly related to the fluctuations of S). For the twist constant K_2 there are no definite predictions, but it is plausible to assume that K_2 is independent of L.

Let us now turn to the friction constants. Here we can deal simultaneously with the cases with and without hairpins, provided that the chains are not entangled. A very simple but useful model has been constructed by F. Brochard for the friction effects of dilute neutral chains in a conventional nematic matrix (Brochard, 1979a). Recently we realized that the same phenomenological analysis can be applied to nonentangled melts. The basic idea is to describe an anisotropic dumbbell with two size parameters ($R_{\parallel}$ and $R_{\perp}$) and two relaxation rates ($\tau_{\parallel}$ and $\tau_{\perp}$). All friction coefficients can then be written in terms of these four parameters. The results are:

(1) Without hairpins, two friction coefficients are particularly large (just as in the case of rods): namely, η_c and $\gamma_1 - \gamma_2$. Both are expected to scale as the square of the molecular weight (M^2).

(2) With hairpins, the anisotropies are less spectacular, and the most crucial effect is that now all friction coefficients should scale as M.

(3) With or without hairpins, there is a remarkably simple prediction for the extinction angle in simple shear flow (the "rest angle" ψ, as it is sometimes called). This is the angle between the director and the flow lines and the prediction for it is

$$\cos 2\psi = (R_{\parallel}^2 - R_{\perp}^2)/(R_{\parallel}^2 + R_{\perp}^2) \tag{24}$$

independent of $\tau_{\parallel}$ and $\tau_{\perp}$.

Equation (24) is interesting, because it allows us (in principle) to relate a macroscopic measurement to the microscopic structure of the chains.

We end by emphasizing once more the difference with the entangled regime: when entanglements are dominant, conventional reptation arguments (de Gennes, 1979) would suggest that the friction coefficients scale as M^3.

Appendix.   Shape Fluctuations of a Semirigid Chain in a Nematic Matrix of Small Molecules

The situation is represented in Fig. 10.

Our nematic matrix has an unperturbed director n_0 parallel to $0z$. But the presence of the fluctuating chain induces distortions in the director field $n = n_0 + \delta n$. The local tilt of the chain is described by two components $\theta_x(s)$, $\theta_y(s)$ and the free energy is

$$F = \int ds \tfrac{1}{2} Q_0 (\delta n(s) - \theta)^2 + \int dr \tfrac{1}{2} K (\nabla \, \delta n(r))^2 \tag{25}$$

where $\delta n(s)$ is the value of the distortion very near the chain (at a radius a), and where we use a single elastic constant K. We select one Fourier component for θ

$$\theta(s) = \theta \cos(qs) \tag{26}$$

Outside of the chain region, the distortion δn satisfies an equation of the form

$$\nabla^2 \, \delta n = 0 \tag{27}$$

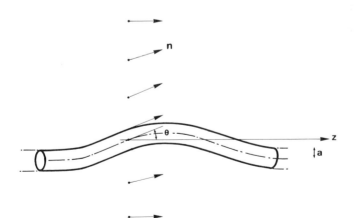

Fig. 10.   A chain in a nematic medium. The chain (of radius a) induces certain distortions in the director field around it.

giving

$$\delta \mathbf{n}(s, \rho) = \alpha \boldsymbol{\theta}(s) \frac{K_0(q\rho)}{K_0(qa)} \tag{28}$$

where $K_0(x)$ is a Bessel function, decaying exponentially at large arguments, but diverging logarithmically at small arguments.

The parameter α is then the ratio of $\delta \mathbf{n}(a, s)$ to $\boldsymbol{\theta}(s)$ and will be determined later.

After integration over the radial distance ρ, we find (per centimeter of length s):

$$F/\theta^2(s) = \tfrac{1}{2}Q_1\alpha^2 + \tfrac{1}{2}Q_0(\alpha - 1)^2 \tag{29}$$

$$Q_1 = 2\pi K/\ln(1/qa) \qquad (qa \ll 1) \tag{30}$$

Optimizing (29) with respect to α we obtain:

$$\alpha = Q_0/(Q_0 + Q_1) \tag{31}$$

and

$$F = \tfrac{1}{2}Q\theta^2(s) \tag{32}$$

where

$$1/Q = (1/Q_0) + (1/Q_1) \tag{33}$$

We may say that two elastic compliances work in series: one described by $1/Q_0$ coupling the chain to its immediate environment and one described by $1/Q_1$ coupling this environment to the distant unperturbed nematic.

Acknowledgments

This chapter was written during the conference on polymer liquid crystals at Santa Margherita Ligure in 1981 and has benefited from discussions with many participants, in particular F. Brochard, R. Meyer, and P. Pincus.

References

Auvray, L. (1981). *J.Phys. (Orsay, Fr.)* **42**, 79.
Brochard, F. (1979a). *J. Polym. Sci., Polym. Phys. Ed.* **17**, 1367.
Brochard, F. (1979b). *C. R. Hebd. Seances Acad. Sci., Ser. B* **289**, 299.
Brochard, F. (1980). *C. R. Hebd. Seances Acad. Sci., Ser. B* **290**, 485.
Brochard, F., and de Gennes, P. G. (1970). *J. Phys. (Orsay, Fr.)* **31**, 691.
de Gennes, P. G. (1975). *C. R. Hebd. Seances Acad. Sci., Ser. B* **281**, 101.
de Gennes, P. G. (1977). *Mol. Cryst. Liq. Cryst. (Lett.)* **34**, 177.
de Gennes, P. G. (1979). "Scaling Concepts in Polymer Physics." Cornell Univ. Press, Ithaca, New York.

Doi, M. (1980). *Ferroelectrics* **30**, 247.

Doi, M., and Edwards, S. F. (1978). *J.C.S. Faraday II* **74**, 568,918.

Dubault, A., Casagrande, C., Veyssie, M., and Deloche, B. (1980). *Phys. Rev. Lett.* **45**, 1645.

Ferry, J. (1970). "Viscoelastic Properties of Polymers." Wiley, New York.

Flory, P. (1956). *Proc. R. Soc. London, Ser. A* **234**, 73.

Flory, P. (1978a). *Macromolecules* **11**, 1119.

Flory, P. (1978b). *Macromolecules* **11**, 1138.

Graessley, W. (1974). *Adv. Polym. Sci.* **16**, 1.

Jarry, J. P., and Monnerie, L. (1979). *Macromolecules* **12**, 316.

Jarry, J. P., and Monnerie, L. (1980). *J. Polym. Sci.* **18**, 1879.

Kronberg, B., Bassingnana, D., and Patterson, D. (1978). *J. Phys. Chem.* **22**, 171.

Leenhouts, F., and Dekker, A. (1981). *J. Chem. Phys.* **74**,1956.

Marrucci, G. (1981). *Mol. Cryst. Liq. Cryst* **72L**, 153.

Onsager, L. (1949). *Ann. N.Y. Acad. Sci.* **51**, 627.

Priest, R. G. (1973). *Phys. Rev. A* **7**, 720.

Straley, J. P. (1973). *Phys. Rev. A* **8**, 2181.

6

Macroscopic Phenomena in Nematic Polymers

Robert B. Meyer

Martin Fisher School of Physics
Brandeis University
Waltham, Massachusetts

I. Introduction

A. Macroscopic Phenomena

Much of the excitement of liquid crystal physics has grown out of the striking and unique macroscopic phenomena that result from the combination of fluid and crystalline properties in these materials. Complex textures, field-induced structural changes, and instabilities produced by flow are examples of phenomena so rich in variety that they are still being explored actively after years of study. There is every reason to believe that polymer nematics and cholesterics will prove to be just as interesting as the low molecular weight materials. For example, the successful spinning of

POLYMER LIQUID CRYSTALS

high modulus fibers from nematic melts or solutions depends on a number of macroscopic processes that are only vaguely understood at this time.

The microscopic view of these materials is fascinating in its own right, and even for one mainly interested in macroscopic phenomena it is clear that the understanding and possible control of the fundamental macroscopic material parameter lies at the molecular level. In this chapter some of the consequences of constructing a nematic phase out of extremely long semi-rigid molecules are explored. Essentially the only molecular information used is the great length of the molecules. This alone has very important effects at the macroscopic level, and a number of specific examples will be discussed. For the most part, systems with a high degree of orientational order and fully extended rather than kinked chains will be considered, but a few comments will be made on the possible role of molecular kinks and the dependence of material parameters on the degree of orientational ordering.

B. Low Energy Deformations

The curvature elastic deformations of an ordinary nematic involve extremely small energies because they can be achieved without changing the density of the system. This is the first rule to be applied to understanding similar deformations in the polymer systems. The second, stated most drastically, is that molecules must not be broken. A similar restriction is that a system with fully extended unkinked molecules at equilibrium should not have kinks introduced by the deformations; producing kinks has results similar to breaking the molecules.

In the following two sections these simple ideas will be applied to the study of static deformations and elementary flow processes.

II. Statics

A. Elastic Coefficients

The basic idea of this section was pointed out by de Gennes (1977), who noted that splay deformation becomes difficult in a nematic composed of long molecules. The situation in two dimensions is sketched in Fig. 1. To achieve splay at constant density requires that the gaps opened up between molecules be filled by the ends of neighboring molecules. For ordinary nematics with a length to diameter ratio L/d of 3 or 4, this is hardly a notice-able effect, and the splay elastic constant k_{11} may be determined as much by specific molecular attractive forces as by repulsive effects. However, as the molecules become long there are fewer ends available to fill the gaps, making

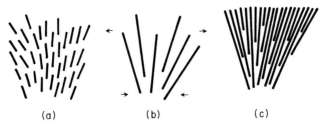

Fig. 1. Splay deformation (a) in a low molecular weight nematic, (b) in a nematic of long rods, indicating the coupling of splay to relative compression and dilation, and (c) showing efficient packing of chain ends to achieve high density.

splay difficult. For infinite chains it cannot occur at all if the density is to remain constant. In this limit the coupling between splay deformation and density gradients is precise:

$$\text{splay} = \mathbf{V} \cdot \mathbf{n} = -\mathbf{n} \cdot \mathbf{V}\rho/\rho \qquad (1)$$

ρ is the number density of chains in a plane normal to the nematic director $\mathbf{n}$.

Bend and twist deformations are not constrained in this way. For molecules that are sufficiently flexible, bend deformation is in no way inhibited as molecular length increases; each molecule bends to follow the director pattern. As shown in Fig. 2, this is not true for rigid rods; as their length grows the ends of each molecule extend further into regions in which they are at odds with the average alignment. For any real "rigid" molecule, however, there must be a length beyond which it bends continuously, or perhaps sharply, to follow the local curvature.

Twist is a more subtle case since, for perfectly ordered rods or semiflexible chains, it would appear that there is no inhibition to twist deformation arising from purely repulsive forces; each rod or chain remains straight and closely packed between its neighbors, no matter how long it is. Successive layers of molecules are simply stacked on one another in a helicoidal fashion. However, if we introduce any degree of orientational disorder, then individual molecules will not be in perfect layers, but each will have some component

Fig. 2.   Bend deformation (a) in a flexible chain system, and (b) in a rigid rod system.

of its orientation parallel to the twist axis. As with bend deformations, this causes rigid rods to enter regions in which they are misaligned, an effect that becomes more serious as rod length increases. Once again, flexible molecules will bend to follow the local orientation and thus the twist elastic constant will also remain finite as chain length grows.

To summarize, as molecular length increases from a small value all the elastic constants will probably show some initial increase, but in the very long chain limit in which all real molecules must be thought of as flexible, only the splay elastic constant will increase without limit.

One important point about this statement remains to be clarified. If there are chain ends available, one can always draw a picture (like Fig. 1c) in which any degree of splay is accommodated by arranging the ends of even very long molecules to fit precisely between one another while keeping the density constant. This may take time to occur in a real system, but as a truly static deformation it looks as if splay should not be limited by molecular length. However, note that the pattern in Fig. 1c is achieved by collecting the "bottom" ends of molecules together; the "top" ends are excluded from this region. This lowers the entropy of the system and thus increases the free energy at any finite temperature.

To make this more precise, consider a well ordered nematic composed of identical molecules of length L, each occupying cross-sectional area d^2 in a plane normal to the director. If we label the top and bottom ends of each molecule relative to the director in some region, and consider the spatial distribution of top and bottom ends described by number density functions $\rho_t(r)$ and, $\rho_b(r)$, respectively, then by sketching the geometry as in Fig. 3 we can use Gauss's theorem to relate splay to ρ_b and ρ_t at any point:

$$\nabla \cdot \mathbf{n}(\mathbf{r}) = d^2[\rho_b(\mathbf{r}) - \rho_t(\mathbf{r})] \tag{2}$$

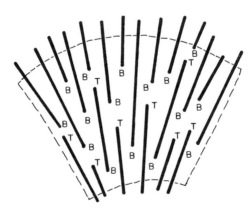

Fig. 3. Splay is proportional to the difference between the numbers of "bottom" (B) and "top" (T) ends of molecules in a volume element.

We can also label the center of each molecule and describe the number density of molecular centers by $\rho_c(\mathbf{r})$. These densities are all related by the fact that each molecule has a top, a center, and a bottom, so that again taking splay into account we can write

$$\rho_c(\mathbf{r} + \tfrac{1}{2}L\ \mathbf{n}) = \rho_b(\mathbf{r})(1 - \tfrac{1}{2}L\ \mathbf{V} \cdot \mathbf{n}) \qquad (3a)$$

$$\rho_c(\mathbf{r} - \tfrac{1}{2}L\ \mathbf{n}) = \rho_t(\mathbf{r})(1 + \tfrac{1}{2}L\ \mathbf{V} \cdot \mathbf{n}) \qquad (3b)$$

In the absence of splay $\rho_c = \rho_b = \rho_t = (d^2L)^{-1}$. In a region of dimensions much larger than L in which the splay is constant ρ_c, ρ_b, and ρ_t will also be constant and related by the following equations:

$$\Delta\rho_b = -\Delta\rho_t = \rho_b - \rho_c = \frac{L}{2} \rho_c\,\mathbf{V} \cdot \mathbf{n} \qquad (4)$$

$$\rho_c = \frac{1}{2}(\rho_b + \rho_t) - \frac{L}{4d^2}(\mathbf{V} \cdot \mathbf{n})^2 \qquad (5)$$

We see that ρ_b and ρ_t change linearly with splay, while ρ_c has a quadratic correction term. Thinking of the chain ends as ideally dissolved in the liquid crystal and noninteracting, the density changes produced by splay create a change in the free energy density, ΔF:

$$\Delta F = \frac{1}{2}\rho_c kT \left(\frac{\Delta\rho_b}{\rho_b}\right)^2 = \frac{kT}{2d^2L}\left(\frac{L}{2}\mathbf{V} \cdot \mathbf{n}\right)^2$$

$$= \frac{kT}{8d}\left(\frac{L}{d}\right)(\mathbf{V} \cdot \mathbf{n})^2 \qquad (6)$$

Looking at this in parallel with the Frank free energy (Frank, 1958), the splay elastic constant is

$$k_{11} = \frac{kT}{4d}\left(\frac{L}{d}\right) \cong 10^{-7}\left(\frac{L}{d}\right) \quad \text{dynes} \qquad (7)$$

for d of the order of a few angstroms at room temperature. Nonlinear corrections to k_{11} occur due to changes in ρ_c,

$$\Delta\rho_c/\rho_c = \tfrac{1}{4}L^2(\mathbf{V} \cdot \mathbf{n})^2 \qquad (8)$$

These clearly become very large when $\mathbf{V} \cdot \mathbf{n}$ is of order $1/L$.

In his 1977 paper, de Gennes estimated the magnitude of k_{11} by an argument based on elastic energy rather than entropy. The idea is that the splay is produced in discrete steps by molecules of finite size; between the ends of molecules there must be regions of relative dilation or compression, the energy of which accounts for the splay elasticity.

The difficulty of such a model is deciding just how the chain ends and the associated elastic strains are distributed. De Gennes pictured each molecule embedded in a continuous elastic medium with no other chain ends next to it along its entire length. In the presence of splay the center of the molecule is in a region of equilibrium density, and the dilative strain given by Eq. (1) varies linearly with distance z from the center toward the chain ends. Integrating over the chain length, de Gennes concludes that

$$k_{11} = E\overline{z^2} = (1/12)EL^2 \tag{9}$$

in which E^{-1} is the compressibility or osmotic compressibility of the medium. The problem with this model is the use of L as the characteristic length over which one finds no chain ends. In a nematic composed of chains all of length L there must be other chain ends along the length of any one molecule. De Gennes's calculation is precisely correct only for a kind of "supersmectic," in which all the chains are aligned into layers of thickness L. A nematic will always achieve lower splay energy than this by interpenetrating molecules among one another. Is there a correct characteristic $\overline{z^2}$ to put into Eq. (9)? From Eq. (2) we see that there is a characteristic volume V for each excess chain end introduced by splay,

$$V = d^2/|\mathbf{V} \cdot \mathbf{n}| \tag{10}$$

But the correct shape for this volume and therefore the linear dimension parallel to the director that enters Eq. (9) is determined by both the spatial arrangement of the chain ends and the bending elasticity of the individual molecules, dictating how they conform to the discontinuity created by the end of a chain. This latter problem is discussed further below. The arrangement of the chain ends is a difficult problem depending on chain end interactions and entropic effects. Except for the special case of the supersmectic, the length L does not play any obvious role in a model of this kind for estimating the purely energetic contribution to k_{11}.

De Gennes's conclusion that measurements of k_{11}/k_{33} should distinguish between rigid rod and flexible chain systems remains correct, but in the latter case k_{11}/k_{33} should increase as L, not L^2.

B. The Infinite Chain Limit[†]

An instructive exercise for understanding the possible differences between low molecular weight and polymer nematics is to consider the infinitely long chain limit. With no chain ends and no hairpin defects present splay deformation is strictly impossible at constant density.

[†] The work in Sections II.B, II.C, and II.D.1 was done in collaboration with Victor Taratuta.

One way to deal with this case is to take seriously the limit of no density changes and no splay and try to think of all possible textures and defect structures consistent with those rules. The analogous exercise for smectics (no bend or twist allowed), incorporating only line disclinations, leads to the theory of focal conic domains (Friedel, 1922; Bouligand, 1972). A similar exercise for discotic phases (no twist or splay allowed) has helped to understand the observed textures in those materials (Kleman, 1980). The corresponding study for the limiting case of interest here has not been done. The possibilities should be somewhat richer than either of the other cases, since only one kind of curvature rather than two has been eliminated. We shall return to this question below.

Another way of thinking about this limit is to examine small deformations of a single crystal. This is geometrically much simpler than the study of textures, and it allows one to consider the effects of both density changes and orientational deformations at the same time. The analogous study of mixed elasticity in smectics has been very fruitful (Martin *et al.*, 1972). It allows one to understand properties of the fluctuation phenomena and wave propagation that are unique to those systems.

A single crystal of an infinite chain polymer nematic consists of fully extended chains, all parallel to one another, with constant density. A cross section taken normal to the director would look like a two-dimensional liquid. The fundamental deformations of this system consist of displacements of small volume elements from their original positions. The elastic free energy density is a function of the various spatial derivatives of the displacement pattern. Deriving its correct form is mainly dependent on the symmetry of the system. One major simplification is made by noting that all displacements parallel to the director result in no rotations of the director or changes in the density of packing of the chains. Gradients in these displacements parallel to the director correspond to stretching or compressing the chains, an important elastic process, but one that can be described equivalently by adding a bulk compressibility term to the ones we consider below. Because of this, a complete description of the elasticity can be made using only displacements normal to the director. The analogous description of smectics was first made by de Gennes (1974, Section 7.2).

Using the Cartesian coordinate system sketched in Fig. 4, with the z axis parallel to the director, the two components of the transverse displacement are $u(\mathbf{r})$ in the x direction and $v(\mathbf{r})$ in the y direction. The director components are given for small displacements ($n_x, n_y \ll 1$) by

$$n_x = \partial u / \partial z, \qquad n_y = \partial v / \partial z \qquad (11)$$

The fact that the director can be described this way is a fundamental difference between this system and an ordinary nematic. In a small molecule

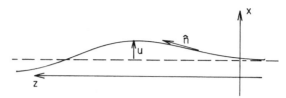

Fig. 4. The coordinate system with the z axis parallel to the average director. The director n at any point is described by gradients of displacements u and v.

nematic director rotations can be accomplished by rotating each molecule about its center, with no continuous displacements of volume elements. However, when the chains are infinitely long all rotations of the director must be described by gradients of displacements (de Gennes, 1974, Section 5.1.2).

For the rest of the development here, we shall be concerned only with static orientational deformations. The elastic free energy density can be written as follows (a partial derivative is indicated by a subscript, $\partial u/\partial x = u_x$):

$$f = \tfrac{1}{2}E(u_x + v_y)^2 + \tfrac{1}{2}k_{11}(u_{xz} + v_{yz})^2 + \tfrac{1}{2}k_{22}(u_{yz} - v_{xz})^2 + \tfrac{1}{2}k_{33}(u_{zz}^2 + v_{zz}^2) \tag{12}$$

The first term describes first order elastic compressibility due to changes in ρ, the number of chains per unit area in a plane normal to the director. If the chains are already fully extended, this is a high energy process very much like a bulk compression. However, for less than fully extended chains or equivalently for a nematic order parameter less than one, changes in the lateral packing density of the chains are coupled to changes in order parameter or chain extension (de Gennes, 1977). In terms of continuum elasticity this means that a decrease in the cross-sectional area of a volume element is coupled to an increase in its length along the director, resulting in almost no bulk compression, and therefore being a moderately low energy process. Therefore E^{-1} can be much larger than a typical compressibility for an organic liquid. If one is dealing with a polymer solution, the low energy processes being pictured involve displacements of the chains relative to the solvent, keeping the overall density essentially constant, and only changing the concentration; E^{-1} will typically be larger for a solution than for a melt. For numerical estimates below we shall guess arbitrarily that E is 10^7 ergs/cm^3.

The remaining terms in f are the usual Frank elastic terms (Frank, 1958). Note that k_{11} here is due to local free energy changes and is similar to k_{11} for a low molecular weight nematic; it is not the same as the k_{11} for a system with a finite density of chain ends discussed earlier. The values k_{22} and k_{33} also arise from local interactions with k_{33} strongly reflecting molecular rigidity. We guess that the Frank constants are of order 10^{-7} dyne.

Note that the correct description of the director as derivatives of the displacements leads naturally to Eq. (1):

$$-\mathbf{n} \cdot \frac{\nabla \rho}{\rho} = \frac{\partial}{\partial z}(u_x + v_y) = \frac{\partial}{\partial x}(u_z) + \frac{\partial}{\partial y}(v_z) = \nabla \cdot \mathbf{n} \tag{13}$$

This means that for any long wavelength processes the splay energy is negligible compared to the E term in f. The splay energy becomes comparable to the E term for wavelengths in the z direction shorter than

$$\lambda_s = 2\pi(k_{11}/E)^{1/2} \tag{14}$$

which is of the order of molecular dimensions in many cases, but may be larger if E is particularly small. There are other terms that could be added to f, such as

$$\left\{ \left[\frac{\partial}{\partial x}(u_x + v_y) \right]^2 + \left[\frac{\partial}{\partial y}(u_x + v_y) \right]^2 \right\} \tag{15}$$

that are of the same order of elasticity as the splay term, and equally unimportant compared to the E term. The only terms that play an important role are the k_{22} and k_{33} terms, since they can have finite magnitudes while the E term is zero. Finally, for certain surface interaction problems one must construct the terms in f equivalent to the k_{24} term in nematics. A case will be mentioned below in which this is important.

Once the free energy is formulated one of its uses is the solution of boundary value problems. If one knows how the liquid crystal interacts with the surfaces enclosing a sample, one can solve for the internal structure of the sample. Using calculus of variations, one can derive from f the equilibrium equations for u and v:

$$-E(u_{xx} + v_{yx}) + k_{11}(u_{xxzz} + v_{xyzz}) + k_{22}(u_{yyzz} - v_{xyzz}) + k_{33}u_{zzzz} = 0 \tag{16}$$
$$-E(u_{xy} + v_{yy}) + k_{11}(u_{xyzz} + v_{yyzz}) - k_{22}(u_{xyzz} - v_{xxzz}) + k_{33}v_{zzzz} = 0$$

C. Interaction with a Grooved Surface

As an illustration of the use of the infinite chain model, consider the interaction of a polymer nematic with a grooved surface as sketched in Fig. 5. This kind of model has been proposed as the explanation of the orienting effect that rubbed surfaces or obliquely evaporated thin films have on ordinary nematics (de Gennes, 1974, Section 3.1.4.2; Berreman, 1972).

We must assume that the amplitude of the grooves is small compared to their wavelength. If the director lies in the xz plane, exactly normal to the grooves, then $v = 0$, $\partial/\partial y = 0$, and there can be no twist. Then Eq. (12)

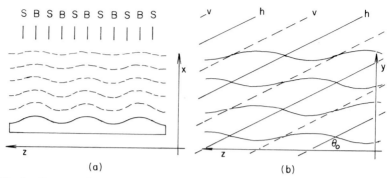

Fig. 5. Response of a nematic to a grooved surface. (a) Cross section normal to the grooves, showing the alternating regions of bend (*B*), and splay (*S*) that is needed to relax the bend. (b) Planar view of the surface, with the mean director at angle θ_0 to the grooves. In plane undulations of the director are indicated, relative to the hills (h) and valleys (v).

reduces to

$$f = Eu_x^2 + k_{11}u_{xz}^2 + k_{33}u_{zz}^2 \tag{17}$$

and Eq. (16) to

$$-Eu_{xx} + k_{11}u_{xxzz} + k_{33}u_{zzzz} = 0 \tag{18}$$

Assuming the director is parallel to the surface and becomes constant far from the surface, the boundary conditions are

$$u(0, z) = u_0 \sin qz \quad \text{and} \quad u(\infty, z) = 0 \tag{19}$$

The solution of Eq. (18) is then

$$u(x, z) = u_0 \sin qz \exp\left[-x\left(\frac{k_{33}q^4}{E + k_{11}q^2}\right)^{1/2}\right] \tag{20}$$

We see here explicitly that as long as q is small ($\ll 2\pi/\lambda_S$), we can ignore $k_{11}q^2$ compared to E. In that case the ripples in the director at the surface die away in the bulk with a characteristic length

$$x_0 = (q^2\lambda_B)^{-1}, \qquad \lambda_B = (k_{33}/E)^{1/2} \tag{21}$$

This is highly reminiscent of the behavior of smectic layers on an undulating surface (with the roles of splay and bend reversed). The relaxation length x_0 depends quadratically on the width of the grooves. If λ_B is of order 60 Å, for instance, and $q = 2\pi \times 10^3$ cm^{-1} (10-μm grooves), then $x_0 \cong 1$ cm. The elastic energy per unit area of surface due to the undulations is

$$F = \tfrac{1}{4}u_0^2q^2[k_{33}(E + k_{11}q^2)]^{1/2} \tag{22}$$

By comparison, if we do the same problem for an ordinary nematic (simply set $E = 0$), the solution for the director is

$$n_x = \partial u / \partial z = u_0 q \cos qz \exp[-qx(k_{33}/k_{11})^{1/2}] \tag{23}$$

Now the relaxation length x_0 varies linearly with the width of the grooves, so the ripple caused by 10-μm grooves will die out in about 10 μm instead of 1 cm. The energy F is

$$F(E = 0) = \tfrac{1}{4} u_0^2 q^3 (k_{33} k_{11})^{1/2} \tag{24}$$

For long wavelengths the ratio of this surface energy to the surface energy for infinite chains is

$$F(E = 0)/F(E \neq 0) = \lambda_s q \ll 1 \tag{25}$$

This can be understood qualitatively by seeing that the bend distortion imposed by the surface can only be relaxed by splay deformation (see Fig. 5). When splay is strongly inhibited by the first order elastic effects in the infinite chain limit, the bend distortion propagates a long way into the bulk, with a consequently higher energy.

There is a further subtlety to this problem that weakens the conclusion of the preceding paragraph somewhat and also points out that the degrees of freedom available to the infinite chain nematic can allow it to behave more like an ordinary nematic than might at first be expected. To complete the above model for the orienting effect of grooves, one should consider the behavior of the director as its mean direction changes all the way from parallel to perpendicular to the grooves. When the director is parallel to the grooves, clearly $F = 0$. For an oblique orientation of the average director to the grooves one might think that one could assume a planar configuration again, with the director parallel to the xz plane, which is now oriented at some angle θ_0 to the grooves, rather than perpendicular to them. Carrying through this calculation for the ordinary nematic with the added assumption that $k_{11} = k_{22} = k_{33} = \bar{k}$, one finds simply

$$F(E = 0) = \tfrac{1}{4} u_0^2 q^3 \bar{k} \sin^2 \theta_0 \tag{26}$$

clearly indicating a minimum for $\theta_0 = 0$.

For the infinite chain case, one still finds the smectic-like response to the undulations with energy

$$F(E \neq 0) = \tfrac{1}{4} u_0^2 q^2 |\sin \theta_0| (E\bar{k})^{1/2} \tag{27}$$

exhibiting an interesting singularity at $\theta_0 = 0$. Unfortunately, the simplifying assumption of a planar solution eliminates an essential part of the actual response to the undulating surface. For oblique alignments undulations of the director *in* the plane of the surface appear, and relieve some of the strains of the undulations normal to the surface (see Fig. 5). For the ordinary nematic this effect occurs and changes some features of the solution, but may not be very important energetically. For the infinite chain case this effect is crucial. The in-plane undulations can be thought of physically as the director tending to run along the "valleys" and then turning to go over the "hills" rather abruptly. This adds extra material to the valleys, effectively filling them up. Another way of putting it is that there is now an in-plane component of splay equal and opposite to the out-of-plane component of splay needed to relax the bending caused by the undulating surface. The result is that for the oblique case the infinite chain nematic behaves again almost like an ordinary nematic rather than like a smectic. The mathematics of the solutions is very complex and proper accounting for free rotation of the director at the surface involves the k_{24} term in the elastic free energy (Wolff *et al.*, 1973). When the average director is nearly perpendicular to the grooves, the in-plane undulations disappear, and the smectic-like solution given above is correct. Therefore, although for small misalignments relative to the grooves the infinite chain case is much like an ordinary nematic, the maximum misalignment energy is much larger.

The important points here are the restricted freedom of the infinite chain nematic leading to smectic-like behavior, and the two-dimensional character of splay deformations, such that apparent splay in one plane can be canceled by splay in a perpendicular direction. Such cancelation effects do not exist in smectics for the forbidden bend and twist deformations.

D. Chain Ends

To the extent that one might think of the infinite chain nematic as the "perfect" nematic polymer, the chain end should be thought of as the fundamental defect in such a system. As is true for fundamental defects in many other systems, chain ends are essential for a number of important properties of nematic polymers. We have seen above their role in determining the splay elastic constant. In this section we shall consider other aspects of chain ends.

1. Elastic Distortions around a Chain End

The abrupt termination of a molecule causes distortions in the surrounding material. Assuming that the chain end is an isolated defect in a medium of otherwise infinite chains, Eqs. (12) and (16) give the elastic free energy density and the differential equation for the equilibrium structure. The geometry of the problem is sketched in Fig. 6. Since there is cylindrical symmetry around a chain end, if we assume that the displacements R are all in the

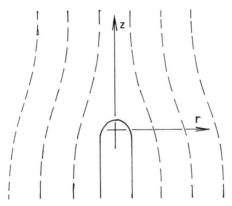

Fig. 6.   Distortions of the director field around a single chain end.

radial direction, we can write these equations in cylindrical coordinates as

$$f = \tfrac{1}{2}E(R_r + R/r)^2 + \tfrac{1}{2}k_{33}(R_{zz})^2 \tag{28}$$

and

$$-E(R_{rr} + R_r/r - R/r^2) + k_{33}R_{zzzz} = 0 \tag{29}$$

There is no twist distortion and the splay term is unimportant. Solution of Eq. (29) is an exercise in Fourier analysis and Bessel functions. Qualitatively, one can say that the combination of first order elasticity in the radial direction and second order elasticity in the z direction means that the ripple caused by the chain end extends a long distance perpendicular to the director and dies out quickly along the director. This means that chain ends interact with one another most strongly if they lie in the same plane normal to the director. An isolated chain end will have associated with it an elastic distortion energy. If a top and bottom end approach one another, their strain fields overlap and partially cancel one another, reducing this energy. If they join together, effectively making a continous chain, the elastic strain energy disappears. The consequences of such end-to-end binding are now discussed.

2.   Chain End Interactions

The elastic distortions around the end of a chain provide one mechanism for chain ends to interact with one another. There may be other mechanisms of interaction as well. For instance, the polypeptide α-helix molecules are effectively giant electric dipoles whose ends interact with one another electrostatically. The elastic interaction is such that the top end of one molecule attracts the bottom end of another, and repels other top ends. For the electrostatic interaction, depending on the orientation of the dipole moments of a given pair of molecules, the top end of one may attract or repel the bottom end of another. In the nematic, however, there should be as many molecules with their dipoles "up" as "down," so there are always pairings possible in which the elastic and electrostatic forces act together to bind molecules end to end.

In any particular system there may be other kinds of forces between molecules. For instance, tobacco mosaic virus particles in water form a nematic in which each virus particle has a net electrostatic charge, so there are long-range repulsive forces tending to prevent aggregation. However, if during preparation of the virus the conditions are right, they tend to bind together tightly end to end due to some particular short-range interactions (Oster, 1950; Kreibig and Wetter, 1980).

Dealing only with the case discussed above in which there is an attractive force tending to bind the bottom of one molecule to the top of another, there are two extreme possibilities. If the binding energy is much greater than the thermal energy kT, then the ends are mostly bound in pairs and the molecules are effectively much longer than their fundamental length. The opposite extreme, with the binding energy much less than kT, produces almost all unbound ends.

In the mostly bound case there is an exponential distribution of effective molecular lengths, with mean effective length $\tilde{L} = Le^{E/kT}$, where E is the energy needed to break an end-to-end bond. Putting this into our calculation of k_{11}, we find in this case

$$k_{11} = \frac{kT}{4d}\left(\frac{\tilde{L}}{d}\right)e^{E/kT} \tag{30}$$

The above description is straightforward if the molecules are at least moderately long to begin with. When the molecules are very short we do not even think about them being bound together end to end. If we did think of calculating the elastic strain energy around the end of one molecule, we would not use the infinite chain model for the description of the surrounding material but rather the ordinary curvature elasticity of a nematic. This would give us a much lower energy associated with a chain end. For moderate length molecules we could use the infinite chain model to describe the elastic strains fairly near the chain end, but farther away, at a characteristic distance for encountering other chain ends, this would be incorrect. This means that the elastic contribution to the energy of a chain end depends on the density of free (unbound) chain ends; $E = E(\rho_f)$. Similar effects occur for electrostatic interactions of chain ends. We can now write for the density of free ends (top or bottom)

$$\rho_f = \rho_c e^{-E(\rho_f)/kT} \tag{31}$$

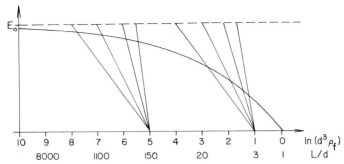

Fig. 7. Chain binding energy $E(\rho_f)$, and $-kT\ln(\rho_f/\rho_c)$, plotted as functions of $\ln(d^3\rho_f)$, for two values of L/d (3 and 150), and four temperatures, $kT = E_0/3$, $E_0/2$, E_0, and $2E_0$, from left to right, for each value of L/d.

Recall that $\rho_c = (d^2L)^{-1}$ is the density of molecules. Writing Eq. (31) as

$$-kT\ln(\rho_f/\rho_c) = E(\rho_f) \qquad (32)$$

we can sketch a graphical representation of this equation, plotting both sides as a function of $\ln(d^3\rho_f)$. For example, as a simple model calculation, if the chain ends interact with a $1/r$ potential cut off at a characteristic radius $r_f = \rho_f^{-1/3}$, then $E = E_0(1 - r_0\rho_f^{1/3})$, where $r_0 \cong d$, corresponding to very short molecules. As seen in Fig. 7, if $L \gg d$, the solution for ρ_f corresponds to an energy E near E_0 no matter what the temperature. However, for smaller L, as temperature is raised the equilibrium ρ_f changes as it evolves from large to small values of E. The same thing can be observed by changing L at fixed temperature. As the equilibrium values of ρ_f and E change, the effective length $\tilde{L}$ and k_{11} change too. This nonlinear dependence of k_{11} on L would be seen most clearly in systematic studies of series of oligomers if chain end binding is significant at large L.

3. Surface Alignment Effects

How a nematic aligns at a surface is an important subject. In ordinary nematics alignment effects are poorly understood at the fundamental level. The director can be either parallel or perpendicular, and even occasionally oblique, at surfaces.

For infinite chain polymer nematics, the director must be parallel to the surface. Perpendicular alignment implies that the ends of molecules are at the surface; infinite molecules do not have ends. For finite but long molecules condensing many ends at the surface costs a great deal of entropy and is therefore unlikely unless there is a strong binding energy for ends at the surface.

To clarify this, imagine the conceptual process of creating an interfacial surface in a liquid crystal. For an interface parallel to the director construct a mathematical plane of the right orientation cutting through the ordinary bulk configuration of the molecules. Those molecules that are intersected by this plane, and other nearby molecules, must be rearranged to create the final interface. This involves changes in both the local energy and entropy. For a parallel interface these local changes do not depend in any obvious way on molecular length, at least for molecule longer than some lower limit. The range for significant rearrangements probably extends no more than a few molecular diameters into the bulk.

Performing the same conceptual process for an interface perpendicular to the director involves more drastic rearrangement. Each molecule intersected by the plane must be moved to position its end at the surface. Assuming this positioning of the end has a precision of about a molecular diameter, this reduces the configurational entropy of the molecule by

$$\Delta S = k \ln(L/d) \tag{33}$$

If there is a binding energy E_s of the chain at the surface, the net free energy change for the molecule is

$$\Delta f = -E_s + kT \ln(L/d) \tag{34}$$

Each chain end bound to the surface also experiences other adjustments in its energy and entropy, Δf_0, which may be similar in magnitude to the adjustments that occur for each monomer in the case of a chain lying parallel to the surface, but do not depend on L. For chains that are long enough, the configurational entropy term given above must dominate, although its logarithmic dependence on length means that it may not force parallel alignment to occur except for very long molecules. The contribution of that term to the anisotropy of the surface tension is $\Delta \sigma = (kT/d^2)\ln(L/d)$, which is of the order of $[10\ln(L/d)]$ ergs/cm^2 for $d^2 = 4 \times 10^{-15}$ cm^2. This is not a negligible effect.

When we consider cooperative effects among the molecules there is a further complication to the surface alignment problem. For the energetics one may argue that one contribution to E_s is the elastic deformation energy around a free chain end. By condensing a uniform layer of chain ends at the surface, this energy is eliminated, favoring perpendicular alignment. However, for this term to be effective by itself it must be much larger than kT. If it is this large, then chain ends in the bulk will already be mostly bound in pairs, so there will be no energy to be gained by binding ends to the surface.

For cooperative contributions to the entropy, in one case there is an important consideration. If the molecules are all the same length, then forming a close packed layer of them perpendicular to the surface simply forms a new surface a distance L into the bulk, at which the next layer of molecules must be properly arranged, further reducing the entropy. If the molecules are already strongly bound end to end, this effect is not important, since that entropy is already gone. Also, if the molecules are not all the same length but broadly distributed in length, the ordering effect of the external interface rapidly dies out. But for identical length molecules not bound end to end, this effect is important and further disfavors perpendicular alignment at the boundary. However, there is a way around this effect.

If the molecules adopt an oblique orientation at the surface and then bend (and splay) toward perpendicular alignment as in Fig. 8, gaps between them open up, into which other molecules can penetrate, restoring a random distribution of chain ends in the bulk and avoiding the cooperative entropy reduction problem. Because of their oblique alignment, fewer molecules have their ends bound to the surface, raising the surface energy. The compromise between binding energy and entropy minimizes the surface free energy. Remember, however, that the prerequisite for this oblique structure is that, for each molecule bound to the surface, the Δf of Eq. (34) must be negative. Just how far the molecules bend toward perpendicular alignment is controlled by curvature elastic energy, which varies as $\sim \Delta\theta^2/L$, in which $\Delta\theta$ is the net rotation in the boundary layer. For large L this becomes small, resulting in nearly perpendicular orientation on the interior side of the boundary layer. It will be hard to see such a boundary layer directly, but observation of nearly perpendicular alignment would be good evidence for its existence. Formally, one can say that the surface curvature results from a surface induced spontaneous splay, an effect thought of already for ordinary nematics; for polymer nematics we now have a clear physical mechanism for it.

Experimentally, for poly(γ-benzyl-L-glutamate) in dioxane (Uematsu and Uematsu, 1979) or mixed dioxane and methylene chloride (Lonberg

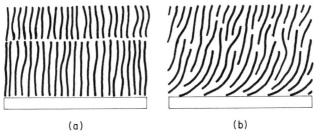

(a)   (b)

Fig. 8.   Chain end binding at a solid surface. (a) Close packing of identical length chains, indicating surface induced layering. (b) Oblique curved boundary layer that lowers interfacial entropy.

and Meyer, 1982) as solvents, for a degree of polymerization up to several hundred at least, perpendicular alignment is observed at glass and indium oxide surfaces. This is probably due to some specific surface binding of the polymer ends. At an air interface and the interface with the isotropic phase the director is parallel. This makes sense, since there is nothing for the chain ends to interact with at such surfaces. The alignment also appears to be parallel at a Teflon surface (Lonberg and Meyer, 1982), so glass or indium oxide might be coated with Teflon to produce parallel alignment at those surfaces, which are important in many experiments.

For poly(p-benzamide) published photographs suggest that the director is parallel to glass and fluid surfaces (Morgan, 1977; Panar and Beste, 1977).

E. Defects and Textures

1. General Considerations

A systematic study, either theoretical or experimental, of the possible defects and textures in nematics in the limit of k_{11} becoming either large or infinite has not been made. Some general ideas and a few specific cases will be discussed here, without any illusions that the most important points have been included.

The complete freedom of curvature of a nematic allows it a remarkable ability to adapt to any boundary conditions in a container of any shape, so that the director field is continuous except for a few point and line singularities. Experience with smectics has shown that they are also very adaptable, but their "splay only" rule of curvature for the director leads to high densities of disclinations to satisfy the boundary conditions in all but a few sample geometries. The strict "no splay" limit for polymer nematics may lead to similar results. For example, consider the two-dimensional problem. If the director lies in a plane, with only bend curvature allowed, and we look for structures in elastic equilibrium that involve only point disclinations, the fundamental solution is a disk of director lines forming concentric circles (see Fig. 9). As soon as we try to build a texture with more than one disk unit we encounter a problem: to fill the spaces between disks we need an infinite series of even smaller disks. Such problems are well known in smectics (Bidaux *et al.*, 1973), where it is evident that this "infinite" series is not present, but the splay only rule is violated below some characteristic length scale at which dislocation distributions must appear to allow some splay.

In three dimensions infinite chain polymer nematics may be more adaptable than smectics, since they can incorporate both bend and twist. If we think of the possibilities for line disclinations, the simplest one to picture

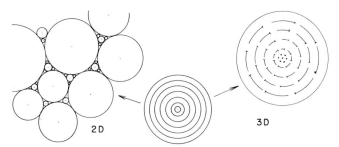

Fig. 9. Pure bend disclination, showing the two-dimensional packing of a texture of disks and the three-dimensional twist relaxation that eliminates the singular core.

is again based on a cross section with the director lines making concentric circles around the core. As for ordinary nematics, this kind of singular disclination core can be removed by allowing the director to rotate toward parallelism with the line formed by the core (see Fig. 9) (Meyer, 1973; Cladis and Kleman, 1972). This requires twist and bend, but no splay. Since this seems to be the only line disclination involving no splay, one might conclude that the textures formed by infinite chain nematics would contain no singular disclination lines. If the truly infinite chain limit can be attained in real materials, it will be interesting to see what its low energy textures look like.

The more realistic case to consider is that of large but finite k_{11}. Topologically this is equivalent to any other nematic, but the elasticity is so anisotropic that new textures may appear, tending to minimize splay energy at the expense of more complex bend–twist configurations. The resulting textures will often contain large areas of splay with bend and twist confined to boundary layers and discontinuities. This may sound backwards but it results from two causes. First, many sample configurations involving a large rotation of the director can only be accomplished with a combination of splay and other curvatures. Second, because the curvature energy density is quadratic in the curvatures, energy is minimized by spreading the curvature over the largest dimension possible; if splay is the highest energy curvature, then splay will occupy the largest part of the sample relative to bend and twist.

As an example of this consider the simple problem of a layer of nematic with the director perpendicular at one surface and parallel at the opposite one, requiring a 90° rotation in the bulk. For a planar solution (see Fig. 10) described by a single rotation angle θ, such that $\theta = 0$ at the perpendicular boundary, the free energy density is

$$f = \tfrac{1}{2}(k_{11}\sin^2\theta + k_{33}\cos^2\theta)(d\theta/dz)^2 \tag{35}$$

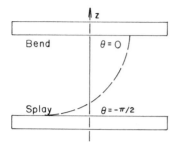

Fig. 10. A splay–bend layer produced by hybrid boundary conditions.

and the Euler–Langrange equation is

$$(k_{11}\sin^2\theta + k_{33}\cos^2\theta)(d^2\theta/dz^2) + (k_{11} - k_{33})\sin\theta\cos\theta(d\theta/dz)^2 = 0 \quad (36)$$

For $k_{11} = k_{33}$ the solution is

$$\theta = (\pi/2)(z/d - 1) \quad (37)$$

with bend $(= \cos\theta\, d\theta/dz)$ at one surface evolving continuously into splay $(= \sin\theta\, \partial\theta/\partial z)$ at the other surface. For unequal elastic constants, in the limit $k_{11}/k_{33} \to \infty$, the solution is

$$\theta = \cos^{-1}(z/d) \quad (38)$$

For this solution the splay is uniform and the bend is concentrated as much as possible, becoming infinite at the perpendicular surface:

$$|\text{splay}| = \frac{1}{d}, \quad \text{and} \quad |\text{bend}| = \frac{1}{d}\cot\theta = \frac{z}{d(d^2 - z^2)^{1/2}} \quad (39)$$

Notice that even though we have taken the limit of $k_{11}/k_{33} \to \infty$ we implicitly assume that splay is possible and described by second order elasticity, so this is not the infinite chain limit. For finite k_{11} the solution will be a little different with only finite bend at the perpendicular surface.

2. Disclinations

For the large k_{11} limit the basic question is whether the change in the energetics relative to ordinary nematics will result in the appearance of different kinds of disclinations for long chain nematics. Since all the well-known line disclinations other than the one discussed above involve splay in combination with other curvatures, there is no obvious reason to favor one structure over others. Since the pure bend disclination of Fig. 9 can be converted to a continuous structure by introducing twist, it is not clear that it should play any special role either. It may well be that the singular lines seen in polymer nematics (Arpin *et al.*, 1977; Millaud *et al.*, 1978) are the familiar twist disclinations seen in ordinary nematics, for which the director **n** rotates to −**n** in one circuit along any closed path around the disclination (Friedel and de Gennes, 1969).

The best way to test this idea is to look at a region of the sample in which the polarized light optics can be analyzed clearly enough to determine how the director rotates around a disclination. For instance, in a sample with parallel boundary conditions, a twist disclination line running parallel to the plane of the sample with its twist axis perpendicular to the plane will not exhibit any distortion of the birefringence of the sample, except within a distance from its core on the order of a few wavelengths of light. In highly birefringent materials, if there is a disturbance of the birefringence around a disclination, it shows up as strong focusing effects for the extraordinary wave. In less birefringent samples examination with monochromatic light will provide a sensitive test of this condition. For other than pure twist disclinations in this geometry the region of distorted birefringence around a disclination will extend for a distance of the order of the thickness of the sample.

The other common geometry exhibiting simple disclinations is the schlieren texture or texture "with nuclei." The singular points representing a 180° rotation of the director in the plane of the sample (two brushes, seen with crossed polarizers) must represent singular lines running normal to the plane of the sample. If they are pure splay–bend lines with the axis of rotation for the director normal to the plane of the sample, the director will be everywhere parallel to the plane of the sample (Frank, 1958) and again there will be no change of birefringence near the disclination. If they tend to be twist lines in the bulk, there will be some asymmetric birefringence pattern around the nuclei. For the 360° rotation nuclei (four brushes seen with crossed polarizers), there are three main possibilities. If as in ordinary nematics these represent point disclinations at the surfaces with no singular core connecting them (Meyer, 1972), then the director is normal to the plane of the sample right at the nucleus, becoming planar far from it, and thus producing a very symmetric birefringence pattern seen as rings with monochromatic light if the sample is thick enough (de Gennes, 1974, Fig. 4.16b). If these are true singular splay–bend disclinations, there will be no birefringence pattern. If they are singular twist disclinations, there will be an asymmetric birefringence pattern. This simple guide to disclination identification should help in qualitative examination of new materials.

In one case in poly(γ-benzyl-L-glutamate) simple 180° rotation twist disclinations are observed (Lonberg and Meyer, 1982). The sample is a frustrated cholesteric; when a sample with perpendicular boundary conditions is made thin enough (less than one helix pitch) the helix disappears. In the otherwise uniform vertical alignment there are occasional metastable 180°

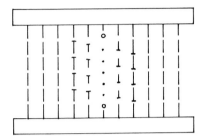

Fig. 11. Cross section of a 180° twist wall in a frustrated cholesteric, indicating twist disclinations (○) near the surfaces.

twist walls that terminate near the sample surfaces with twist disclinations. A cross section of such a wall is shown schematically in Fig. 11. This sample geometry is simple enough to make optical analysis relatively easy. The disclination cores are easily seen because they are decorated with dust particles and tiny droplets of isotropic material.

3. Textures

By a texture in liquid crystals one generally means a pattern of distortions or an array of defects that is usually not repeatable in detail but has qualitative features that are characteristic of some sample preparation or treatment. Understanding the structure of a texture and the mechanism of its formation often contributes to a fundamental understanding of the material under examination.

An example of such a texture is found in thin layers of poly(γ-benzyl-L-glutamate) solutions between glass and air interfaces (Lonberg and Meyer, 1982). At the glass the director is perpendicular, while at the air it is parallel to the interface. This produces a splay–bend deformation as shown in Fig. 12a. Because the parallel boundary is a fluid, the director is free to rotate

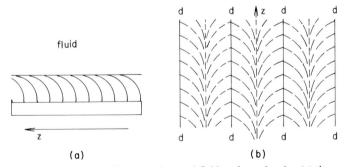

Fig. 12. A splay–bend layer between glass and fluid surfaces showing (a) the splay bend structure in a vertical cross section, and (b) a planar view of the texture consisting of splay domains separated by disclinations (d———d).

in the plane of the surface. This allows the director to adopt a splay deformation in the plane of the surface opposite in sign to the splay in the direction normal to the surface, so that the total splay approaches zero. This can be accomplished at the cost of introducing more bend and twist energy. Moreover, the geometry of splay deformation is such that this compensated splay structure cannot be uniform or continuous. It breaks up into quasi-periodic domains separated by disclinations, as shown in Fig. 12b. For the optimum splay compensation the width of the domains must be of the order of twice the sample thickness. The precise configuration depends on the structure and energy of the disclinations. Above a critical thickness the splay energy is too small to drive this domain formation, and it does not appear. For very thin samples the domains are visible down to the resolution limit of the microscope. This analysis of the observed domain pattern is for now only a proposal; if we can confirm that k_{11} is very large in this material, it will be more likely to be a correct explanation.

III. Dynamics

A. Anisotropic Viscosity

If we follow the rule proposed in the introduction, that low energy processes cannot break molecules, and consider the infinite chain limit, then clearly certain deformations are forbidden. Any flow that stretches (or shortens) the molecules is impossible. Likewise, a velocity perpendicular to the director with a gradient along the director produces a shear that tends to cut the molecules and is therefore forbidden. However, shears that simply slide the chains parallel to one another or "roll" them over one another are not inhibited by the infinite molecular length—as long as the chains are almost perfectly parallel to one another and not entangled. In terms of the Meisowicz viscosities, defined in Fig. 2 of de Gennes's chapter (this volume, p. 116), η_a and η_b are finite while η_c (corresponding to cutting across the molecules) is infinite. In the ordinary nematic the coupling between shear flow and rotation of the director is described by two viscosity coefficients since rotation and shear flow are fundamentally independent variables. In the infinite chain limit, as stressed earlier, these are no longer independent vairables and the rate of rotation of the director is described exactly by velocity gradients. This means that processes accomplished in an ordinary nematic by rotating each molecule around its center of mass with no accompanying macroscopic flows become impossible for infinite chains. Therefore, although a simple cholesteric helical structure is possible in the infinite chain limit, and changing the pitch of the helix (unwinding it or winding it up) can conceptually be accomplished without breaking any molecules, the required fluid flow pattern would involve shear rates that diverge with the size of the sample. Twist distortions with finite wavelength along the director with their accompanying fluid flows are still possible.

In terms of the Leslie viscosity coefficients, the infinite chain limit requires that $\gamma_1 \to \infty$, $\gamma_2 \to -\infty$, $\alpha_1 \to \infty$, $\alpha_2 \to -\infty$, $\alpha_5 \to \infty$, while α_3, α_4, and α_6 remain finite. In terms of the Harvard group's description, $v_1 \to \infty$, $\gamma_1 \to \infty$, $\lambda \to 1$ and v_2 and v_3 remain finite. For a unified discussion of all the coefficients see de Gennes (1974, Chapter 5).

As an example of the properties of the infinite chain limit, consider the twist–bend fluctuation of the director in a nematic single crystal (de Gennes, 1974, Sections 3.4 and 5.2.5). Referring to Fig. 13, if the wave vector has components q_x and q_z, then this fluctuation of the director is described by

$$n_y = n_y^0 \sin(q_x x + q_z z) \qquad (40)$$

Such a fluctuation relaxes with a time dependence controlled by an effective viscosity η_2 described in the infinite chain limit by

$$\eta_2 = \eta_b + (q_x^2/q_z^2)\eta_a \qquad (41)$$

When $q_x = 0$ this fluctuation involves only bend distortion and $\eta_2 = \eta_b$, the Meisowicz viscosity for sliding the chains along one another. For a finite value of q_x twist deformations appear accompanied by shear flows transverse to the director with corresponding viscosity η_a. Then as q_z goes to zero the fluctuation becomes pure twist, and the effective viscosity becomes infinite. This is just the limit described above in which the shear rates required for rotation of the director diverge with the sample size (or the wavelength along the director). Observations of this fluctuation mode by light scattering will be especially interesting, since for $q_x = 0$ one will see

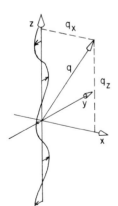

Fig. 13. A twist–bend fluctuation, showing the orientation of the wavevector **q** relative to the fluid displacements and director undulations in the yz plane.

behavior similar to that of an ordinary nematic, while as pure twist is approached one will see the relaxation time diverge.

For finite chain lengths the viscosities that are infinite in the infinite chain limit must become large as L increases. To see how this occurs, consider, for example, η_c. As shown in Fig. 14, the center of each molecule moves with the average fluid velocity at that point. Every point in the molecule moves with that same velocity. Therefore a monomer along the chain a distance δ from the center is moving with velocity $\delta \partial v/\partial z$ relative to its surroundings. The power p dissipated at each monomer along the chain is proportional to the local velocity difference squared times a friction coefficient v:

$$p = v\delta^2(\partial v/\partial z)^2 \tag{42}$$

The mean power dissipated at any monomer in the sample is then

$$\bar{p} = v\overline{\delta^2}(\partial v/\partial z)^2 = v(L^2/12)(\partial v/\partial z)^2 \tag{43}$$

Since the power dissipated in shear flow is also proportional to $(\partial v/\partial z)^2$, we see that $\eta_c \sim L^2$ for fully extended chains. The same model applies to the other divergent viscosities, γ_1 (rotation of molecules about their centers), or v_1 and α_1 (elongational flow parallel to the director).

Once the infinite chain limit is not strictly achieved, the polymer nematic is no different from any other nematic on the basis of symmetry for processes that are slow enough not to depart far from equilibrium and of sufficiently long wavelength not to encounter nonlinearities. They will have large anisotropies that can be measured. For instance, the pure twist–bend fluctuation discussed above will exhibit viscosity $\eta_2 \cong \eta_b$ for the pure bend geometry and, for the pure twist mode $\eta_2 = \gamma_1$, which will now be finite, since finite length molecules can rotate around their centers with no coupled shear flows.

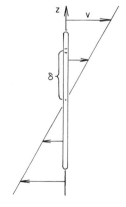

Fig. 14.   A single molecule in a transverse shear flow, indicating the mismatch between the velocity of a monomer and its surroundings, growing linearly with distance δ from the center of the molecule.

B. Nonlinear Effects

One should be aware that the true hydrodynamic behavior for long chain nematics discussed in the last paragraph may require the use of very small forces and a great deal of patience to observe. In many cases, even with forces that are small in some absolute sense, nonlinearities will occur in which the system will respond almost as if it were an infinite chain system. These are nonlinearities involving structural changes, i.e., reorientations of the director. The general rule is that for sufficient driving force the response that most quickly relaxes the applied stress will be the one observed. It must be emphasized that these nonlinearities are macroscopic phenomena rather than the microscopic effects familiar in the study of isotropic polymer solutions. In the processes considered here the response in any small volume element is quite linear. Some examples will clarify this situation.

1.   The Frederiks Transition

This example is a little unusual in that the Frederiks transition is already a nonlinear response of the kind discussed above. However, thinking of the well-known Frederiks transition as the analog of a linear response, we shall see that if we drive it with a sufficiently strong force, a new kind of Frederiks transition appears with a much faster response time.

The case being studied is sketched in Fig. 15. The initial configuration has the director perpendicular to the sample surfaces, which are transparent electrodes. The electric field applied parallel to the director drives the transition, because of the negative dielectric anisotropy of the material.

The geometry of the usual (mode-one) Frederiks transition is shown in Fig. 15a. Above a threshold voltage V_1 a bend deformation appears and grows in amplitude as voltage is increased. As a static deformation the equilibrium structure at any voltage is well understood. However, if starting from the

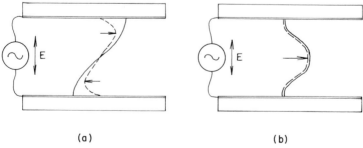

(a)   (b)

Fig. 15.   Electric field induced Frederiks transition: (a) the usual mode-one and (b) mode-two, indicating the director field (——) and the fluid displacement field (---) for each.

undeformed sample one suddenly applies a voltage above V_1, the response is a dynamic one. A small perturbation in the sample is amplified and grows exponentially with a rate that depends on the square of the scaled voltage $(V/V_1)^2$ and the appropriate viscosities of the material. Because of the coupling between director rotation and fluid flows, the dynamic deformation can be quite different in appearance from the static equilibrium structure at the same voltage. Pieranski et al. (1973) have done a careful theoretical and experimental study of this subject in an ordinary nematic.

When we consider the long-chain limit the mode-one Frederiks transition of Fig. 15a encounters a serious problem. Thinking of the director pattern as composed of continuous molecules, we see that the ends of the molecules at one surface must be displaced with respect to those at the other surface. However, the boundary conditions for fluid flow at solid surfaces demand no molecular displacements at the surfaces. In the infinite-chain limit, therefore, the mode-one deformation cannot occur. For long chains it can only occur very slowly.

In an effort to see something quickly, one increases the voltage, and above a second threshold V_2 ($=2V_1$) there is a rapid response. The mode-two deformation corresponding to this response is shown in Fig. 15b. It is just like two of the mode-one deformations on top of one another. It is not normally seen because it has twice the threshold voltage and because in ordinary nematics the Frederiks transition is usually studied as a quasi-static phenomenon. The important feature of this second mode of deformation is that the displacements required at the surfaces are zero. The fluid flow accompanying the deformation is similar to Poiseuille flow except that there is no velocity gradient at the surfaces. In fact, the exact dynamic equations show that the director rotation and fluid flows are compatible, so that the shape of the deformation is independent of the driving voltage and the response rate is linear in $[(V/V_2)^2 - 1]$, which is proportional to the driving force. In the infinite chain limit this is a pure bend mode again with effective viscosity η_b, as expected. The dominance of this second mode of deformation at large driving force follows exactly the general prescription given above for nonlinear effects.

This phenomenon has been observed (Lonberg and Meyer, 1982) in preliminary studies of poly(γ-benzyl-L-glutamate) in a dioxane/methylene chloride solvent mixture (to eliminate the cholesteric helix). A high frequency (300 kHz) electric field was used to have a negative dielectric anisotropy and to avoid electrohydrodynamic instabilities. For a degree of polymerization of about 300, a concentration of about 20% by weight, and a sample about 100 μm thick, the relaxation times for the two modes discussed above were of the order of seconds for mode two and minutes for mode one. Optically the modes can be distinguished from one another in conoscopic observation since mode one produces an initial rotation of the average optical axis while mode two produces only a biaxial distortion. Precise measurements will provide an easy means of determining some of the highly anisotropic viscosities of these materials. Using the rough data for the fast mode and guessing $k_{33} = 10^{-7}$ dynes, η_b is not much more than the solvent viscosity, while γ_1 is much higher.

An interesting footnote to this work is that, being aware of the possibility of different modes of Frederiks transition with different response times, if one looks at low molecular weight nematics again, one finds that they are already anisotropic enough to exhibit the same effect. There are even some published data (Pieranski *et al.*, 1973) that suggest that this has already been observed without being recognized. This is an example of what will probably be a common occurrence—ideas and knowledge coming from the study of nematic polymers will prove very useful in understanding ordinary nematics as well.

2. An Elongational Flow Instability

It is occasionally noted in articles on syntheses of nematogenic polymer materials that at some point in the reaction opalescence is seen in the material while it is being stirred (Morgan, 1977; Panar and Beste, 1977). This is probably an example of another flow induced structural nonlinearity. A possible mechanism for this phenomenon is proposed here, using a geometry simpler than that of a stirrer in a resin pot!

Picture a long chain polymer nematic flowing in a cylindrical tube of cross-sectional area A with the director parallel to the tube axis. This simple Poiseuille flow will be characterized by viscosity η_b, one of the Meisowicz viscosities that remains small in the infinite-chain limit. What happens if, along the direction of flow, the cross-sectional area of the tube increases gradually? To maintain a simple Poiseuille flow, there must be an elongational component to the velocity, $\partial v_z / \partial_z < 0$ (see Fig. 16). In the infinite-chain

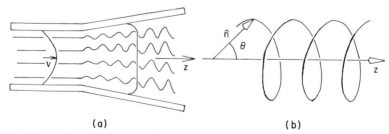

Fig. 16. An instability in elongational flow showing (a) the growth of the instability with increasing tube diameter and (b) the helicoidal nature of the director pattern.

limit this is not possible since the elongational flow viscosity diverges with chain length. For finite chains elongational flow is possible and a simple Poiseuille flow can be stable but only at very low velocities. Above a critical velocity or pressure gradient an instability can occur that allows the cross-sectional area of a flowing volume element to increase without elongational flow.

The basic mechanism is the tilting of the director by an angle θ relative to the axis of the tube, which increases the effective cross-sectional area per molecule in a plane normal to the tube axis by a factor sec θ. For no elongation, $A(z)\cos\theta(z)$ must be constant. Also, to maintain the macroscopic average director and flow direction parallel to the tube axis there must be some oscillatory behavior of this deformation. The only oscillation that can occur without modifying θ is a helicoidal distortion around the z axis as shown in Fig. 16. What determines the pitch of the helix? The director pattern requires (in the infinite-chain limit) shear deformations that produce a corrugated surface on the liquid crystal, which is incompatible with the smooth surface of the tube. The amplitude of the corrugations is proportional to the helix pitch, so a short pitch reduces this incompatibility at the expense of extra bend energy in the helix. The corrugations can also be removed by elastic compressive and dilative strains described by the E term in Eq. (12). This situation is similar to that for a smectic undulation instability (Clark and Meyer, 1973; Delaye et al., 1973) in which the balancing of compressional and curvature energy leads to a distortion wavelength, or pitch p in our case,

$$p = \alpha(\lambda_B d)^{1/2} \tag{44}$$

in which $\lambda_B = (k_{33}/E)^{1/2}$ (Eq. (21)), d is the diameter of the tube, and α is a numerical coefficient of order one. Since λ_B may easily be as small as 10^{-7} cm, then for a 1-cm-diameter tube p is of the order of 10 μm, giving a nice opalescence. This pitch is correct for the infinite-chain limit and is a lower limit for the pitch if the molecular length is finite. For finite chains suddenly applying a high pressure gradient would produce a pitch approaching this value, while lower pressures would allow some elongational flow, resulting in a longer pitch. Incidently, the helix will be right- or left-handed purely by accident.

The critical velocity or pressure gradient for this instability to appear depends on how rapidly the tube cross section changes. In the infinite chain limit this instability becomes a static deformation, the threshold for which is described by a change in the tube diameter equal to a small factor times λ_B. Thus a change in the tube diameter of 100 Å could initiate the instability. Opalescence would become evident at larger dilations. In the finite chain case this translates into a very small threshold velocity in most practical cases. This instability is also nonlinear in another sense: a decrease in tube diameter does not cause an instability. In fiber processing this instability would occur as the nematic exits from the die, but during the succeeding drawing of the fiber under tension the undulation should disappear again reversibly.

In the presence of this helical instability the continuing flow in the tube will no longer be Poiseuille flow. As the pressure gradient forces molecules through the helical structure they follow helical paths, undergoing rapid rotational motion. This dissipates energy in a way similar to Helfrich's "permeation" flow for cholesterics (Helfrich, 1969; de Gennes, 1974, Section 6.3.2), except that γ_1 in the power dissipation expression for the cholesteric is replaced by something like $\eta_b + \eta_a \tan^2 \theta$, which is finite (for $\theta < 90°$) even in the infinite-chain limit (Eq. 41). The velocity profile in the tube will not be parabolic but uniform except for thin boundary layers. The pressure gradient needed to maintain the flow will be very large.

To relate this model to stir opalescence we have only to establish the effective dimension d that enters into the expression for the pitch that will be correct in the rapid stirring limit. It will be of the order of the average dimension of the stirring tool in a direction normal to its velocity.

This proposed mechanism is speculative and experiments are being prepared to test it. Moreover, this instability may be only the first of a cascade of processes leading to more rapid flow mechanisms at higher pressures.

3. Nonlinearities at the Molecular Level

Two kinds of nonlinearities at the molecular level are easily imagined; they are closely related. First, in response to the appropriate stresses, chain kinks may be created. A double kink in a chain effectively introduces another molecule into the system. Second, in systems with significant end-to-end chain binding, these bonds can be broken. The former process can relax stresses tending to cause dilations normal to the director but not chain stretching stresses. The latter process is effective for relaxing both signs of elongational stress.

IV. Conclusion

This chapter contains a few facts and many speculations. In concept, polymer liquid crystals are very stimulating to the imagination. In the laboratory they should prove equally interesting. As with ordinary liquid crystals, the empirical approach will probably yield a richer variety of phenomena than any amount of theoretical speculation will ever conjure up!

Acknowledgments

This work has benefited greatly from the collaboration of two graduate students at Brandeis University, Franklin Lonberg and Victor Taratuta. Their specific contributions are cited in the text. Detailed publications of their work are in preparation.

This research was supported in part by the National Science Foundation, grant number DMR 79-15149 A02, by the Army Research Office, contract number DAAG 29-89-K-0050, and by the Martin Fisher School of Physics at Brandeis University.

References

Arpin, M., Strazielle, C., and Skoulios, A. (1977). *J. Phys. (Orsay, Fr.)* **38**, 306.

Berreman, D. (1972). *Phys. Rev. Lett.* **28**, 1683.

Bidaux, R., Boccara, N., Sarma, G., de Seze, L., de Gennes, P. G., and Parodi, O. (1973). *J. Phys. (Orsay, Fr.)* **34**, 661.

Bouligand, Y. (1972). *J. Phys. (Orsay, Fr.)* **33**, 525.

Cladis, P. E., and Kleman, M. (1972). *J. Phys. (Orsay, Fr.)* **33**, 591.

Clark, N., and Meyer, R. B. (1973). *Appl. Phys. Lett.* **22**, 493.

de Gennes, P. G. (1974). "The Physics of Liquid Crystals." Oxford Univ. Press, London and New York.

de Gennes, P. G. (1977). *Mol. Cryst. Liq. Cryst. (Lett.)* **34**, 177.

Delaye, M., Ribotta, G., and Durand, G. (1973). *Phys. Lett. A* **44**, 139.

Frank, F. C. (1958). *Discuss. Faraday Soc.* **25**, 1.

Friedel, G. (1922). *Ann. Phys. (Leipzig)* **18**, 273.

Friedel, J., and de Gennes, P. G. (1969). *C. R. Hebd. Seances Acad. Sci., Ser. B* **268**, 257.

Helfrich, W. (1969). *Phys. Rev. Lett.* **23**, 372.

Kleman, M. (1980). *J. Phys. (Orsay, Fr.)* **41**, 737.

Kreibig, U., and Wetter, C. (1980). *Z. Naturforsch., C: Biosci.* **35C**, 750.

Lonberg, F., and Meyer, R. B. (1982). To be published.

Martin, P., Parodi, O., and Pershan, P. (1972). *Phys. Rev. A* **6**, 2401.

Meyer, R. B. (1972). *Mol. Cryst. Liq. Cryst.* **16**, 355.

Meyer, R. B. (1973). *Philos. Mag.* **27**, 405.

Millaud, B., Thierry, A., and Skoulios, A. (1978). *J. Phys. (Orsay, Fr.)* **39**, 1109.

Morgan, P. W. (1977). *Macromolecules* **10**, 1381.

Oster, G. (1950). *J. Gen. Physiol.* **33**, 445.

Panar, M., and Beste, L. F. (1977). *Macromolecules* **10**, 1401.

Pieranski, P., Brochard, F., and Guyon, E. (1973). *J. Phys. (Orsay, Fr.)* **34**, 35.

Uematsu, Y., and Uematsu, I. (1979). *Polym. Prepr., Am. Chem. Soc., Div. Polym. Chem.* **20**, 66.

Wolff, U., Greubel, W., and Kruger, H. (1973). *Mol. Cryst. Liq. Cryst.* **23**, 187.

7

Techniques for the Evaluation of Material Constants in Lyotropic Systems and the Study of Pretransitional Phenomena in Polymeric Liquid Crystals

Donald B. DuPré

Department of Chemistry
University of Louisville
Louisville, Kentucky

I. Introduction

Very little is known about the three major elastic constants k_{ii} ($i = 1, 2, 3$) of an important class of liquid crystals formed by macromolecules. These constants, representative of the forces between molecules, are fundamental measurable properties of the liquid crystal and determine the way the material is held together and its response to a disturbance.

Measurements of the k_{ii} are possible by optical and laser light scattering methods that monitor specific magnetic (or electric) field distortions imposed on the liquid crystal and spontaneous hydrodynamic fluctuations. The Frederiks transition is familiar in liquid crystal physics and yields values of the elastic constants by the optical detection of the onset of a field-induced deflection of the director in opposition to special boundary surface orientations. In the case of spontaneously twisted (cholesteric) liquid crystals, it is possible to unwind the twisted organization by application of a sufficiently strong field. Following this cholesteric-to-nematic transition to a critical field can yield the twist modulus k_{22}. Splay, twist, and bend modes also occur in natural thermal excitation and may be sensed by light scattering. Quantitative values of the k_{ii} and corresponding viscosity coefficients η_i, may be obtained from the intensity, frequency spectrum, and time correlation functions of scattered light. These procedures are known to work well with small molecule, thermotropic liquid crystals, but little work has been published on their application to polymeric systems. Our present understanding of similarities, differences, and difficulties in dealing with macromolecules by these methods will be discussed in the following sections.

Theoretical developments that relate macroscopic viscoelastic properties of liquid crystals to microscopic molecular factors will also be reviewed to the extent that they relate to large molecule assemblies.

Our results to date on the experimental determination and theoretical correlation of the k_{ii} of polypeptide liquid crystals will then be examined. Pretransitional ordering of these helical macromolecules has also been detected at concentrations of polymer just below that required for liquid crystal formation.

II. Lyotropic Polymer Liquid Crystals

It is now more widely appreciated that a number of high molecular weight polymeric substances can form a liquid crystal or mesomorphic state (Samulski and DuPré, 1979; DuPré and Samulski, 1978; Black and Preston, 1979; Ciferri and Ward, 1979; Blumstein, 1978). This new interest is due in part to recently recognized practical applications of polymer liquid crystals. Noteworthy are new technological advances in the fabrication of temperature-resistant and ultrahigh-strength fiber materials that depend upon a liquid crystal precursor organization of macromolecules. This application alone has provided considerable impetus to further consideration and characterization of the properties of large molecule liquid crystals.

Polymers in fact would seem predisposed to liquid crystallinity since they may adopt an extremely elongated shape necessary for the long-range orientational order of these special fluids. It is possible in some cases by design of the monomer to optimize polymer rigidity while retaining useful thermal or solubility properties. These optimal factors occur most noticeably in the rodlike structure of helical polypeptides, and the linear extended chain conformation of some polyarylamides. These polymers form lyotropic liquid crystals at sufficiently high concentration in appropriate solvents. Thermotropic polymer liquid crystals have also been synthesized (Blumstein, 1978; Barrall and Johnson, 1979) but will not be discussed here.

The object material of this chapter will be liquid crystal solutions of poly(benzyl glutamate) (PBG), a synthetic polypeptide of the following basic repeating unit:

$$\left(\!NH\!-\!CH\!-\!CO\!\right)_n$$
$$|$$
$$CH_2$$
$$|$$
$$CH_2$$
$$O\!=\!C \diagdown_O \diagup^{CH_2} \!-\!\!\bigcirc\!\!-\!H$$

In solvents that support the rodlike α-helical conformation, this polymer has been shown to exhibit the mesomorphic behavior of nematic and cholesteric (spontaneously twisted) structures. Because the α carbon on the chain backbone is a chiral center, polyglutamates usually form cholesteric liquid crystals. The sense of the cumulative twist is solvent dependent. The tertiary structure of the side chains of this polymer is sensitive to solvent interactions and may be right or left handed without modification of the backbone helical sense. Nematic liquid crystals may be formed by compensatory effects on the pitch in binary solvent mixtures of opposite chiral influence. Equimolar mixtures of the D and L isomer of this polymer are also compensated and form nematic liquid crystals. Coupling between polymer chains at high concentration in nonheliogenic solvents may also induce chain rigidity leading to a liquid crystal in unexpected solvents. This macromolecule is in general an excellent model system, since many

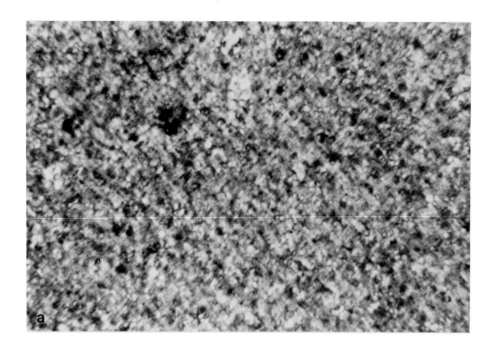

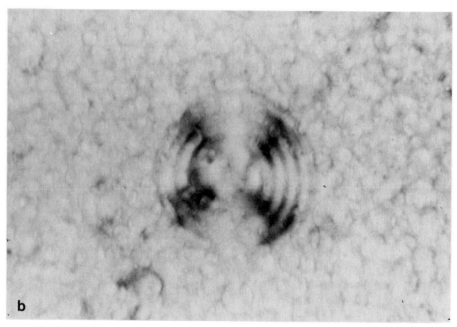

Fig. 1. (a) Photomicrograph of the pretransition of PBLG (molecular weight = 100,000)/ CH$_2$Cl$_2$ liquid crystals. (b) Conditions have been adjusted to the **A** point and a cholesteric spherulite has begun to form. View is between crossed polars at a magnification of 50x.

variables (molecular weight, supporting solvent character, concentration, temperature) may be easily controlled for study over wide ranges.

A. Formation of the Liquid Crystal and Pretransitional Effects

The formation of lyotropic liquid crystals from concentrated solutions of rod-shaped macromolecules proceeds in several identifiable stages. Theoretical investigations (Onsager, 1949; Flory, 1956) of the possibility of self-ordering of rodlike solutes in concentrated solution predict that, as the concentration of elongated particles is increased, a point will be reached beyond which randomness of orientation is no longer possible. Above this limiting concentration, referred to as the **A** point, theory predicts that the solution will separate into two phases: a dilute phase, isotropic in the arrangement of solute particles, and a more concentrated phase, which is anisotropic. This phase separation is entropically driven and is a consequence of the rodlike shape of the solute particles; no specific attractive forces or solvent interactions are required. At a higher concentration of total solute, the entire solution becomes anisotropic at what is known as the **B** point.

Features of this liquid crystal formation have been observed in solutions of poly(benzyl glutamates) (Robinson, 1956, 1961, 1966; Robinson and Ward, 1957; Robinson et al., 1958), where the predicted inverse relation between axial ratio of the macromolecule and polymer volume fraction for the onset of orientational ordering has been verified (Straley, 1973a). When the polypeptide concentration exceeds the critical volume fraction at the **A** point the solution separates into two phases. With care, conditions of temperature and concentration can be adjusted so that solutions of polyglutamates remain indefinitely just below the critical point (**A** point) for the appearance of anisotropic spherulitic structures. In this pretransition state some light is seen to pass through the sample when viewed between crossed polars and a uniform grainy texture is observed, without any evidence, however, of phase separation. A photomicrograph of this region is presented in Fig. 1a. A small change (0.5 to 1%) in polymer concentration will cause the solution to turn completely isotropic (dilution of polymer) or will precipitate the formation of cholesteric spherulites (concentration of polymer to the **A** point). A rise in temperature will also destory embryonic cholesteric ordering and result in an isotropic solution. In Fig. 1b conditions have been adjusted to the **A** point and a cholesteric spherulite has begun to form. As the **A** point is passed (Fig. 2), the spherulites grow in size with increasing concentration of polymer and phase separation can occur if the density of the solvent is sufficiently different from that of the polymer. Liquid crystal spherulites can be seen to move in the gravitational field, either up or down depending on the solvent density. The morphology and magnetic field

Fig. 2. Fully formed spherulites suspended in isotropic liquid phase. PBLG (molecular weight = 100,000) in tetrahydrofuran, 18.5% wt/vol.

orientation of these spherulitic structures has been discussed elsewhere (Patel and DuPré, 1980a). With continued concentration of polymer, the spherulites coalesce and the anisotropic phase grows at the expense of the isotropic component until the **B** point is reached. The fully formed liquid crystal is characterized by regions of regular striations arranged in a swirl-like or fingerprint pattern (Robinson *et al.*, 1958; Robinson, 1961) and unusually high optical rotatory power (as high as 10^5 °/mm). The distance between these striations, which is of the order of $5-200$ μm for these polyglutamate solutions, is a measure of the pitch of the cholesteric structure.

In Fig. 3 we plot the birefringence versus volume fraction of polymer for a poly(benzyl-L-glutamate) (PBLG) solution in *p*-dioxane. The birefringence (Δn) is, of course, zero below the **A** point but jumps in the **A–B** biphasic region. In the single phase, anisotropic region Δn is larger and increases slowly with added polymer. We have also found in the study of a series of solvents and molecular weights that the polymer adds a small anisotropic increment to the refractivity of the supporting solvent and that the optical birefringence does not depend significantly on polymer molecular weight (DuPré and Lin, 1981).

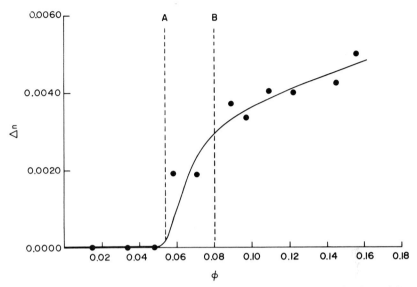

Fig. 3. Concentration dependence of the birefringence of PBLG (molecular weight = 240,000) solutions in *p*-dioxane. The vertical line labeled A indicates the volume fraction at which phase separation just begins; the line labeled B, the volume fraction after which the medium is totally anisotropic. Between A and B the solution is biphasic and the liquid crystal has not fully formed (Fig. 2).

We have made extensive measurements (Patel and DuPré, 1980a,b; DuPré and Patel, 1980) of the optical rotatory dispersion (ORD) of dilute solutions of PBLG and the enantiomer PBDG in various solvents with the concentration of polymer and temperature of solution held just below the **A** point as described above. Of the various combinations of solvent and polyglutamate isomers studied we find pretransition ordering effects to be most dramatic in PBDG/CHCl$_3$ solutions. The chirality of the liquid crystal phase of this polymer/solvent system is opposite in sense to that of the dilute solution. Dilute solutions ($\sim 5\%$ wt/vol) of PBDG in CHCl$_3$ display small positive rotations in the ORD curve. Figure 4 is a plot of ORD curves of this solution concentrated to 12.5% wt/vol and held, in the lowest curve, just below the **A** point at 19.6°C. In this condition PBDG in CHCl$_3$ displays large negative rotations, which indicates the emergence of the higher form optical rotatory power of the liquid crystal. As the temperature is increased in the family of ORD curves labeled 1 to 7, the specific rotation decreases in absolute value and eventually changes to the sign of the dilute solution (isotropic phase). The course of this thermal inversion of $[\alpha]_{324}$ is shown in

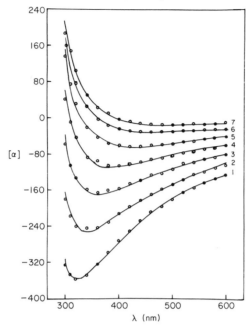

Fig. 4. ORD curves of PBDG (150,000)/CHCl₃ solution (12.5% wt/vol) in the pretransition region. Curve $1 = 19.6°C$; $2 = 21°C$; $3 = 22°C$; $4 = 23°C$; $5 = 24°C$; $6 = 27°C$; $7 = 32°C$.

Fig. 5. It is seen that differences in the optical rotatory power in the pre-transition region over and above that found in dilute solution (▲ in Fig. 5) persist for over a decade on the temperature scale. The optical rotatory power of a solution of uncorrelated polyglutamate macromolecules is only weakly temperature dependent for moderate temperature variations far from the intramolecular helix–coil transition (Fasman, 1967). This enhanced optical rotatory power with strong temperature dependence is a manifestation of the emergence of short-range chiral ordering in the isotropic phase of this lyotropic liquid crystal. An analogous effect has been reported in the isotropic phase of thermotropic cholesteric liquid crystals (Cheng and Meyer, 1972, 1974) and has been similarly interpreted.

We are examining the possibility that the speckled appearance of PBLG solutions below the **A** point (Fig. 1a) is due to an ordered lattice arrangement such as occurs in the blue phase of thermotropic cholesterics. If such an arrangement does exist, we would expect the unit cell dimensions to be very much larger than with small molecule thermotropics that exhibit this curious structure.

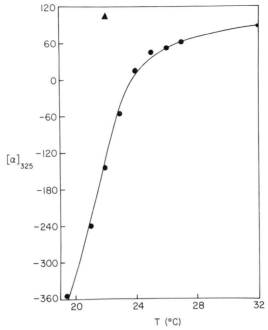

Fig. 5.   Specific rotation at 325 nm, $[\alpha]_{325}$, as a function of temperature in the pretransition of PBDG (150,000)/CHCl$_3$ solutions (12.5% wt/vol); ▲ represents the value of $[\alpha]_{325}$ for a completely isotropic solution (5% wt/vol) of the polymer in CHCl$_3$ at 22°C. Chiral orientational ordering is thus not completely destroyed even after more than a decade of temperature elevation at the higher concentration of the polymer.

III. Elastic Moduli of Liquid Crystals: k_{11}, k_{22}, k_{33}

Liquid crystals are unusual in that, among other things, as fluids they have the ability to sustain a shear and transmit a distortion. There exists no counterpart of this elastic phenomenon in normal liquids. In true solids, the elastic moduli are high and deformations are strongly resisted. The nature of molecular alignment in liquid crystal media is frequently discussed in terms of the disposition of a unit vector **n**, the director, which refers to the preferred direction of the long axes of more or less parallel molecules in a small local volume of the material. In an ideal "single liquid crystal" **n** would be uniform over the entire sample. In most situations, however, **n** will vary in space. The macroscopic condition of the liquid crystal is thus characterized by the vector field of this director, i.e., **n(r)** at each spatial position. If changes in molecular orientation and hence $\mathbf{n(r)}$ vary slowly in space relative to a molecular dimension, the material may be discussed and treated as an elastic continuum. It has been shown that physical distortions of a liquid crystal of the nematic category lead to an excess free energy that in general is composed of only three independent terms (de Gennes, 1974):

$$F_d = \tfrac{1}{2}k_{11}(\text{div } \mathbf{n})^2 + \tfrac{1}{2}k_{22}(\mathbf{n} \cdot \text{curl } \mathbf{n})^2 + \tfrac{1}{2}k_{33}(\mathbf{n} \times \text{curl } \mathbf{n})^2 \qquad (1)$$

where F_d is the distortion free energy per unit volume and k_{11}, k_{22}, and k_{33} are elastic constants of modes of splay, twist, and bend, respectively. A representation of these three fundamental distortions is given in Fig. 6. The interesting feature of liquid crystals is that the moduli k_{ii} are extremely weak, typically some 22 orders of magnitude less than those associated with corresponding deformations in true solids. (In crystals the moduli considered here are masked by the more significant constants of the stress–strain tensor and are therefore never discussed.) The low values of k_{ii} are responsible for (or a reflection of) the unusual sensitivity of liquid crystals to external stimuli and boundary conditions. The fact that only three moduli are of consequence considerably simplifies the description of material anisotropy and response. The k_{ii} occur in almost all liquid crystal theories (de Gennes, 1974; Priestly *et al.*, 1975; Chandrasekhar, 1977; de Jeu, 1980). A knowledge of their quantitative values is therefore of fundamental importance from both theoretical and experimental points of view. Measurements of the k_{ii} ($i = 1, 2, 3$) have been made on a number of small molecule liquid crystals. As yet only the twist modulus k_{22}, has been examined extensively in polymer systems (Duke and DuPré, 1974a,b; DuPré and Duke, 1975; Duke *et al.*, 1976; Guha-Sridhar *et al.*, 1974; DuPré *et al.*, 1977a,b; Patel and DuPré, 1979a,b, 1980a,b).

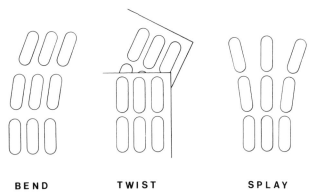

BEND   **TWIST**   **SPLAY**

Fig. 6. A molecular representation of modes of bend, twist, and splay in the liquid crystal.

IV. Dependence of the k_{ii} upon Molecular Factors: Theoretical Considerations, Large Molecules

Values of the elastic moduli found in small molecule liquid crystals are generally of the same order of magnitude ($\sim 10^{-7}$ dyne) and vary little (except at points of certain phase transitions). The understanding of molecular factors that influence the elastic constants for these materials is rather limited, although the effect of overall molecular dimensions and the inclusion of flexible groups within the molecule has been analyzed in a few recent systematic studies (de Jeu and Claassen, 1977; Leenhouts et al., 1979; Schadt and Muller, 1979; Schad et al., 1979; Schad and Osman, 1981).

Theoretical considerations pertinent to macromolecular liquid crystals indicate a clear and dramatic influence of molecular parameters that is amenable to test by experimental procedures.

Straley (1973a,b) has made calculations for the k_{ii} of a suspension of long hard rods at concentrations sufficient to model liquid crystal formations of rigid polymers (theory of Onsager). For a specified form for the distribution of rod orientations, the k_{ii} have the form

$$k_{ii} = c \cdot \phi^2 L^4 DkT \qquad (2)$$

where ϕ is the volume fraction of rods of length to breadth ratio L/D and $c \simeq 0.02$–0.44 is a constant dependent on the orientational distribution function of the long axes of the rods. A distinct molecular length and concentration variation is thus predicted. The ratio of the splay to twist elastic modulus is, however, predicted to remain constant: $k_{11}/k_{22} = 3$, independent of any assumption about the form of the rod orientation distribution. The ratio $k_{33}/k_{11} \simeq 5$ but varies with a parameter of the distribution function.

Priest (1973) has also developed an expansion for the k_{ii} applicable to long rod-shaped molecules. His theory incorporates the averages of the even numbered Legendre polynomials. Calculations truncated at the first two terms $\bar{P}_2$ and $\bar{P}_4$ yield

$$k_{11}/\bar{k} = 1 + \Delta - 3\Delta'(\bar{P}_4/\bar{P}_2)$$
$$k_{22}/\bar{k} = 1 - 2\Delta - \Delta'(\bar{P}_4/\bar{P}_2) \qquad (3)$$
$$k_{33}/\bar{k} = 1 + \Delta + 4\Delta'(\bar{P}_4/\bar{P}_2)$$

where $\bar{k} = \frac{1}{3}(k_{11} + k_{22} + k_{33})$. The quantities Δ and Δ' are constants dependent on molecular parameters. For cylindrically shaped particles with spherical caps that interact via hard core repulsions:

$$\Delta = (2R^2 - 2)/(7R^2 + 20); \qquad \Delta' = \tfrac{9}{16}(3R^2 - 8)/(7R^2 + 20) \qquad (4)$$

where $R = (L - D)/D$. The variation of the $k_{ii}/\bar{k}$ ratios is illustrated in Fig. 7 as a function of axial ratio R for a fixed $\bar{P}_4/\bar{P}_2$ value.

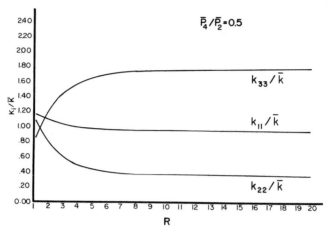

Fig. 7. Calculations for $k_{ii}/\bar{k}$, where $\bar{k} = \frac{1}{3}(k_{11} + k_{22} + k_{33})$ from the theory of Priest (1973) as a function of axial ratio R. $\bar{P}_4/\bar{P}_2$ is taken to be 0.5 here.

For positive values of $\bar{P}_4/\bar{P}_2$ the values of k_{33}/k_{11} increase with the length/width ratio in a regular manner. A decrease is possible, however, if $\bar{P}_4/\bar{P}_2 < 0$. The liquid crystal order parameter $\bar{P}_2$ has been measured by a number of techniques for many thermotropics (de Gennes, 1974) and some polymers (Murthy *et al.*, 1976; DuPré *et al.*, 1977b). Data on $\bar{P}_4$ in liquid crystals is more limited, but under some conditions negative values have been obtained (Jen *et al.*, 1977; Chapoy and DuPré, 1979). Priest's theoretical results have not been adequately tested at this point.

Most recently de Gennes (1977) has considered theoretical features connected with the elasticity of polymer liquid crystals composed of partially flexible extended chains. He finds that the bend elastic constant k_{33} is dominated by the chain rigidity, whereas k_{22} is primarily a function of the interaction between chains. The splay constant k_{11} is predicted to be very large, increasing as the square of the molecular weight. The measurement of the ratio of k_{11}/k_{33} is suggested as a discriminant of polymer chain flexibility in these associated systems. It was pointed out that the case of partially flexible polymer chains is quite novel: the coupling between chains may react on the effective rigidity of each. Indeed, polypeptides that are flexible in dilute solutions in nonhelicogenic solvents (such as dichloroacetic acid) become stiff and liquid crystalline at higher concentration (Robinson *et al.*, 1958; de Gennes and Pincus, 1977).

In short, there is ample though largely untested theory indicating that the elasticity of polymer liquid crystals is expected to be unusual, quite variable, and controllable by variation of molecular parameters.

Since these bulk moduli are fundamentally a reflection of an interaction potential among a collection of elongated molecules, one may consider the effects of variations of microscopic parameters and in what way, if any, these parameters alter the macroscopic response of liquid crystals. Polymer liquid crystals—particularly the lyotropic phases—provide excellent systems for controlling, in many cases in a quantitative manner, such molecular factors as molecular weight, size and shape, concentration, and polymer–polymer and polymer–solvent interactions. The latter factor includes the wide variability of dielectric constant, dipolar strength, and chemical binding specificity of solvent and polymer segment. The variations mentioned are not easily obtained in small molecule thermotropics. (That is, the synthetic route to obtain a certain degree of interaction among mesogens is in general more complicated in low molecular weight thermotropics.)

Our results to date (Duke and DuPré, 1974a,b; DuPré and Duke, 1975; Duke *et al.*, 1976; Guha-Sridhar *et al.*, 1974; DuPré *et al.*, 1977a,b; Patel and DuPré, 1979a,b, 1980a,b) on PBLG liquid crystals have shown that molecular parameters are important in the variation of at least one liquid crystal elastic modulus, k_{22}. In some cases the trends are understandable on the basis of theory; in other cases only empirical variations (or lack thereof) can be stated. Only limited information on k_{11} and k_{33} is available at this time. These results will be discussed in Section VI.

V. Experimental Methods

Numerous experimental techniques that yield the elastic constants of liquid crystals appear in the literature. We shall focus here on optical and light scattering methods that are capable of monitoring specific magnetic (or electric) field distortions imposed upon the material and spontaneous hydrodynamic fluctuations governed by the viscoelastic constants. Measurements of the k_{ii} are possible by optical detection of a critical electric or magnetic field necessary to just induce the distortions of splay, twist, or bend. This is referred to as the *Frederiks transition* in liquid crystal physics. In the case of spontaneously twisted (cholesteric) liquid crystals it is possible to "unwind" the twisted organization by application of a sufficiently strong field. Following this cholesteric-to-nematic transition by determining the critical field can yield the twist modulus k_{22}. The splay, twist, and bend modes are also in natural thermal excitation and may be sensed by light scattering from resultant fluctuations in the refractive index (or dielectric constant) of the material. The theory of light scattering of liquid crystals is sufficiently well developed that quantitative values of the k_{ii} and corresponding viscosity coefficients η_i, may be obtained from measurements of the intensity, frequency spectrum, or intensity fluctuation time correlations of scattered light.

We shall indicate in the subsections below the physical basis behind the plan of approach of each selected optical procedure. These methods are known to work well with small molecule thermotropic liquid crystals, but little work has been published on their application to polymeric systems. Our present understanding of similarities, differences, and difficulties in dealing with macromolecules with these techniques will be discussed where appropriate.

One of the objects of our research, other than obtaining quantitative information on the viscoelastic parameters, has been to determine to what extent these optical procedures carry over to large molecule liquid crystals, frequently concentrated polymer solutions. The accrued data are interpreted where possible in terms of theoretical developments such as those described in Section IV.

A. The Frederiks Transition: k_{11}, k_{22}, k_{33}

It is possible to generate deformations of pure splay, twist, or bend by the imposition of an electric or magnetic field of sufficient strength on strongly anchored liquid crystals. The detection of the onset of such director-field distortions may be related to the elastic moduli.

One of the curiosities of liquid crystals is that these fluids may be anchored to a surface in certain special alignments that may carry through the entire bulk of the sample, thus creating "single liquid crystals." Two important types of such nematic single crystals are a uniform perpendicular alignment with respect to parallel glass walls (homeotropic texture) and a planar uniaxial parallel alignment (homogeneous texture). Liquid crystal molecules are thought of as falling into an "easy axis" at the boundaries where the surface contact free energy is minimized and made directional as a result of rubbing, scratching, or molecular depositions. If such strong anchoring conditions prevail, the boundary conditions for the director field may be regarded as fixed.

In the presence of an external field the anisotropic medium will become distorted in ways that may be contradictory to the alignments mentioned above. An external magnetic field **H** will add a term of the form $F_m = \frac{1}{2}\chi_a(\mathbf{n} \cdot \mathbf{H})^2$ to the elastic free energy, Eq. (1), where $\chi_a = \chi_{||} - \chi_{\perp}$ is the anisotropy of the diamagnetic susceptibility of the medium. ($_{||}$ and $\perp$ are defined with respect to the director axis **n**). For nematic liquid crystals χ_a is usually positive and the total free energy will thus be minimized by a rotation of **n** to become parallel with the direction of the imposed field. For certain geometries, competition of field and wall effects will result in curved conformations within the liquid crystal.

The special configurations of Fig. 8 will induce transitions of splay, twist, and bend, respectively, in appropriately oriented nematic liquid crystals. It

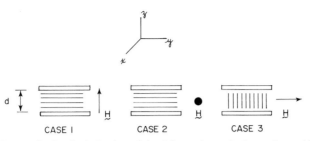

Fig. 8.   Geometries for the induction of the deformations of splay, twist, and bend by the Frederiks transition. The magnetic field strength **H** in the illustrations is below the threshold $H_{c,i}$. Lines represent the pattern of the undistorted director field. Wavy underscore denotes vector quantities.

is found that a critical magnetic field strength H_c must be reached before the respective distortions occur, and that (Gruler *et al.*, 1972)

$$H_{c,i} = (\pi/d)(k_{ii}/\chi_a)^{1/2} \tag{5}$$

where $i = 1, 2,$ or 3 for the situations depicted in Fig. 8, cases 1, 2, and 3, respectively. Hence an accurate measurement of the critical field strength $H_{c,i}$ for the onset of the distortion case i may be used to determine the elastic constant k_{ii} if d (the sample thickness) and χ_a are separately known.

In principle, any physical quantity that has different values along and perpendicular to the long axis of the molecule can be used to detect these sudden variations of the director field. A convenient method relies on the anisotropy of the refractive index to detect the transition optically.

If we consider the geometry of case 1 of Fig. 8 with incident light propagating normal to the glass plates, a phase difference δ of emergent light results, which is given by (Saupe, 1960; Gruler *et al.*, 1972; de Jeu *et al.*, 1976)

$$\delta = \frac{1}{\lambda} \int_0^d (n_o - n_{eff}) \, dz \tag{6}$$

where $n_{eff} = n_e n_o (n_e^2 \sin^2 \theta + n_o^2 \cos^2 \theta)^{-1/2}$. Here n_o and n_e are the ordinary and extraordinary refractive indices, respectively. The optical path of the component of light polarized along the x axis depends on the ordinary refractive index n_o for all values of H. The effective refractive index n_{eff} for light polarized along the y axis depends on the deformation angle θ of the director. This angle in turn is a function of applied field strength. Under no distortion $(H < H_c)$, $\theta = 0$ and the phase difference remains constant. As H increases the intensity of light transmitted will undergo a series of maxima and minima beginning at the critical field H_c. These oscillations are illustrated

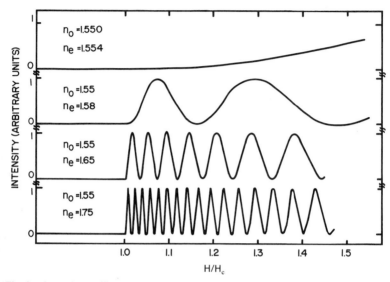

Fig. 9. Intensity oscillations as a function of reduced magnetic field strength H/H_c calculated on the basis of Eqs. (12) and (13) with $k_{11} = 5.19 \times 10^{-7}$ dyne, $k_{33} = 6.49 \times 10^{-7}$ dyne for various values of the birefringence Δn. Note that as Δn decreases, oscillations above H_c are reduced.

in Fig. 9 and correspond to $\delta(H)$ going through integer and half-integer values.

In principle, both k_{11} and k_{33} could be obtained from a single experiment such as this in the "splay mode geometry." The threshold value $H_{c,1}$ determines k_{11}. The ratio k_{33}/k_{11} may also be obtained from a fit of the experimental intensity oscillations to a complete theoretical curve of δ versus H. Accurate values of the refractive indices are required since the fit is very sensitive to the birefringence Δn. For this reason, and other complications involving weak anchoring of the liquid crystal to the surfaces of the cell, confirmation of k_{11} and k_{33} in both geometries 1 and 3 is preferred. (A similar discussion applies to the homeotropic case 3, with k_{11} and k_{33} interchanged.) There are some special problems with the optical detection of the threshold for the twist mode (case 2), which may be overcome with conoscopic (divergent light) observation. Details of this case are discussed in the source references for the Frederiks effect (Saupe, 1960; Cladis, 1972; Gruler *et al.*, 1972; de Jeu *et al.*, 1976).

It should be emphasized that all of the procedures described above depend on the strong anchoring of the liquid crystal, with the easy axis exactly normal or parallel to the slab. If other boundary conditions obtain, the moduli determined will be in error. It can be expected (Rapeni and

Papoular, 1969), for example, that weak boundary anchoring will result in lower values of the $H_{c,i}$ and hence the k_{ii}. At the beginning of this research little was known concerning surface anchoring of polymer liquid crystals. Progress to date with this aspect of handling large molecule liquid crystals is discussed in Section VI.A. We have found some unexpected results in testing surface preparation methods traditionally used with thermotropics.

Figure 9 also illustrates how the oscillatory patterns of $I(H)$ vary with changes in Δn of the material (everything else being held constant). We note that as Δn decreases there are fewer maxima and minima in the trace. This has caused problems with the polypeptide liquid crystals where Δn is quite small (two orders of magnitude less than that of small molecule thermotropics). A principal elastic modulus can be obtained from the first inflection of the curve, but as we shall see below (Section VI.B), other extrema do not occur. We have devised a procedure based on the rate of change of $I(H)$ to extract elastic constant ratios, which is also discussed below.

Experimentally, to locate the threshold fields and possible intensity oscillations, an appropriately surface-aligned polymer specimen is positioned between the poles of a continuously variable electromagnet to correspond to the geometries of Fig. 8. The transmission intensity of a small polarized laser beam incident normal to the plane of the preparation is observed after passing through a polarization analyzer (see Fig. 10). Transmission of the light beam undergoing a phase retardation $\delta(H)$ is recorded as a function of magnetic field strength. To enhance the signal to noise ratio the laser beam may be chopped and detected by a photosensitive diode in synchronous

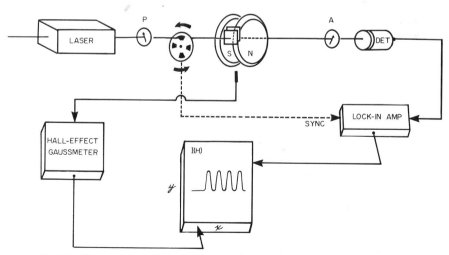

Fig. 10.   Block diagram of the apparatus for the detection of the Frederiks distortion.

amplification. A calibrated Hall-effect probe measures the field strength and sweeps the x axis of a recorder. Separation d of the plates may be determined interferometrically and χ_a must also be known.

B. Field-Induced Cholesteric-to-Nematic Transition: k_{22}

A very convenient method for measurement of the twist modulus k_{22} is available that does not require a uniformly anchored liquid crystal. If the liquid crystal can be prepared in the spontaneously twisted (cholesteric) structure, the pitch P of the supramolecular organization may be measured optically. In many cases this may be done by simple observation of the striations of average separation S in a polarization microscope equipped with a micrometer eyepiece. The striations (Robinson, 1956; Robinson *et al.*, 1958) are caused by periodic changes in the refractive index that follow the helical molecular organization and $S = P/2$. If the liquid crystal is not cholesteric, it may be induced to be so by adding a small amount of a chiral molecule that imparts a cumulative twist to the director field. Polypeptide liquid crystals are naturally cholesteric in most solvents with striations usually greater than or equal to the wavelength of light. Polyarylamides may be made cholesteric by the addition of suitable chiral dopants (Panar and Beste, 1977) or by modifications of the main chain with small quantities of optically active residues (Krigbaum *et al.*, 1979).

If the diamagnetic and dielectric anisotropies (χ_a, ε_a, respectively) of the macromolecule are positive, the application of an electric or magnetic field will act to "unwind" this helical superstructure. The process can be followed by the change in the pitch (or striation separation S). The process is found to be slowly varying at first, then to proceed rapidly as a critical field is approached. It has been shown that the critical field for the magnetic case is inversely proportional to the pitch P_0 of the undisturbed (zero field) structure, and is given by (de Gennes, 1968b; Meyer, 1968)

$$H_c = (\pi^2/2)(k_{22}/\chi_a)^{1/2}(1/P_0) \tag{7}$$

(An analogous relationship occurs for the critical electric field E_c where χ_a is replaced by ε_a.) A functional form for the dependence of the cholesteric pitch on field strength has also been derived (de Gennes, 1968b; Meyer, 1968), and is plotted in reduced form as the solid line of Fig. 11. The value of the twist modulus can be obtained from the critical field at which the pitch diverges to infinity, using Eq. (7) with independent measurements of P_0 and χ_a. The data points of Fig. 11 are one such determination from our studies on PBLG liquid crystals (DuPré and Duke, 1975). In dealing with polymer systems it is necessary to perform pitch–field measurements slowly

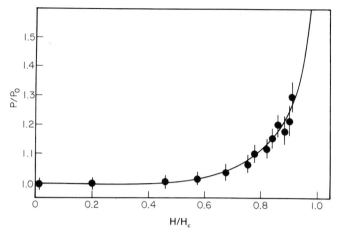

Fig. 11.   Dependence of the pitch P on magnetic field strength **H** for liquid crystal PBLG of molecular weight 310,000 in dioxane (20% wt/vol). Reduced plot where P_0 is the pitch in the absence of the field and H_c is the critical field (Duke and DuPré, 1975). $P_0 = 30 \pm 1 \, \mu$, $H_c = 5.06 \pm 0.02$ kG. The solid curve is after theory (de Gennes, 1968b; Meyer, 1968).

to assure equilibrium. The kinetics of the transition is considerably slower than in small molecule cholesterics.

The values of k_{22} obtained by this method in PBLG liquid crystals have been successfully correlated with theoretical predictions on the influence of molecular parameters as discussed in Section VI.C.

C. Light Scattering

Strong depolarized light scattering was an early historical observation in liquid crystal research. It is now recognized that the light scattering phenomenon is due to long wavelength spontaneous fluctuations of the director field in the continuous medium. Such fluctuations modulate the anisotropic refractive index at a local level in the material and hence divert the path of oncoming light. The fluctuations have been described theoretically in terms of elastic curvatures of the liquid crystal medium regarded as a continuum. The extent of fluctuation depends on the elastic constants k_{ii}. De Gennes (1974) has shown that the differential scattering cross section per unit solid angle, $d\sigma/d\Omega$, resulting from orientational fluctuations in a nematic liquid crystal has these important elements:

$$\frac{d\sigma}{d\Omega} = \left(\frac{\varepsilon_a \omega^2 V}{4\pi c^2}\right)^2 \sum_{\alpha = 1,2} \langle |\delta n_\alpha(\mathbf{q})|^2 \rangle (\mathbf{i}_\alpha \mathbf{f}_z + \mathbf{i}_z \mathbf{f}_\alpha)^2 \tag{8}$$

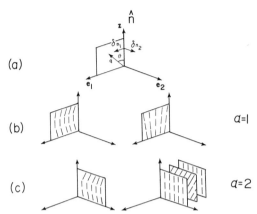

Fig. 12. Representation of modes of orientational fluctuations leading to light scattering. (a) The two uncoupled modes δn_1 and δn_2. (b) Components of the deformation in the δn_1 mode: bend and splay. (c) Components in the n_2 mode: bend and twist.

where

$$\langle |\delta n_\alpha(\mathbf{q})|^2 \rangle = (kT/V)(k_{\alpha\alpha}\mathbf{q}_\perp^2 + k_{33}\mathbf{q}_{//}^2)^{-1}, \qquad \alpha = 1, 2 \tag{9}$$

In the above, $\mathbf{q}$ is the scattering vector, $\mathbf{i}$ and $\mathbf{f}$ are unit vectors specifying the polarization of the incident and exit beams, and ε_a is the dielectric anisotropy.

The scattering amplitude depends upon Fourier components of fluctuations in the local director, $\langle |\delta n_\alpha(\mathbf{q})|^2 \rangle$, in two independent modes designated by $\alpha = 1$ and 2 in the sum above. The first mode ($\alpha = 1$) is a mixed deformation of splay and bend of the director; the second ($\alpha = 2$) involves twist and bend. Components of the deformations are illustrated in Fig. 12. By suitable choice of polarizations and scattering geometries all three elastic moduli, ratios of elastic moduli, and ratios of individual moduli to viscosity coefficients of the material may be extracted from a light scattering experiment. The selection rules on the experimental geometries are more complicated than those shown above for the Frederiks transition but are explained in references cited below. The procedures involve the measurement of the total intensity of scattered light or the statistical correlation of intensity fluctuations in this beam. Observations are performed on the Rayleigh line (central or inelastic scattering component).

1. Intensity and Angular Dependence of Light Scattering

A proper choice of the geometry of the scattering experiment and polarization states of an incident and exit laser beam with respect to the director of a well aligned liquid crystal can isolate one or the other of the two modes pictured in Fig. 12.

The strongest scattering occurs when the polarized laser beam and its analyzer are crossed at 90°. Examination of the intensity variations as a function of scattering angle has been shown (de Gennes, 1968a; VanEck and Perdeck, 1978) to yield values of ratios of k_{11}/k_{22}, k_{33}/k_{11} and k_{33}/k_{22} with the **q** vector properly oriented with respect to the director (easy axis) of homeotropic or homogeneous preparations.

We should remark here on an aspect of our experience with polypeptide liquid crystal solutions that have an extraordinary optical clarity (weak light scatterers). This can be a result of a combination of physical factors unusual in this system. The intensity of scattered light depends upon the squares of the birefringence and amplitude of orientational fluctuations. The birefringence of PBLG liquid crystals is noted to be about two orders of magnitude smaller than that found in small molecule thermotropics, considerably reducing the scattered light intensity from this factor alone. The amplitude of director-field fluctuations is controlled by the elastic moduli, which may differ greatly in these polymer solutions. Furthermore, enhanced viscous interactions among these large molecules may result in overdamped oscillations with a restricted maximum displacement in a mode of fluctuation. Experimentally this facet of polymer liquid crystals can have positive and negative implications. On the one hand, stray light problems frequently encountered with strong scatterers will be reduced. Reduced scattering, however, results in weaker signals to process. Modern photon-counting and other electronic signal averaging instrumentation are available, however, to overcome the latter difficulty.

2. Anisotropy of Turbidity

Integration of Eq. (8) over all scattering angles yields an expression for the total light scattered or turbidity of the sample. In three selected geometries considered by Langevin and Bouchiat (1975; Langevin, 1974) the extinction coefficients (σ_i) of the remaining forward-directed transmitted beam yield the separate moduli k_{11}, k_{22} and k_{33}. It would serve no purpose to reproduce these lengthy equations here, and we simply note that the expressions are coupled integral equations that involve the three moduli separately (not as ratios). The integrals we mention are not analytic functions and require computer methods to evaluate for the possibilities of the k_{ii}. Accurate measurements of the σ_i in all three of the proposed geometries are necessary as input into three equations with the three sought unknowns (k_{ii}: $i = 1, 2, 3$). The ordinary (n_o) and extraordinary (n_e) refractive indices of the sample must also be known in this procedure.

The only report that we are aware of that utilizes this turbidity method is that of the originating authors (Langevin, 1974; Langevin and Bouchiat, 1975). They were quite successful in reproducing previously known values of the k_{ii} for MBBA, a well studied thermotropic liquid crystal. It is noteworthy that this method does not require uniformly surface anchored liquid crystal preparations. A simple magnetic field serves to define the major director axis. Effects of static surface defects on the anisotropic turbidity have been noted, however, and a method devised to take these defects into account (Langevin, 1974).

The procedures described above are essentially static (time integrated) measurements, i.e., with the sample between crossed polars, a total intensity transmitted or scattered to a given angle is monitored. Modification of an incident laser beam is also a result of the dynamics of the director-field fluctuations in the liquid crystal. The dynamic properties of these modes involve viscous interactions as well as elastic restorative forces.

Relatively new spectroscopic methods that depend on the monochromaticity and coherence properties of the laser provide quantitative information on the frequency distribution and statistical time correlations of light scattered in conjunction with molecular relaxation mechanisms. These methods are referred to as quasi-elastic or laser light beating spectroscopy (Chu, 1974; Cummins and Pike, 1974; Berne and Pecora, 1976). Since each of the $\alpha = 1, 2$ modes is dissipative, their relaxation is reflected in a broadening in frequency (as well as a change in intensity) of the outgoing laser beam. For a nonpropogating mode the optical shifts are to higher and lower frequencies about the central laser frequency and fall off in proportion to the mobility or diffusivity of the disturbance sensed by the light beam. From the point of view of a photon train emerging in time from the scattering volume, relaxation processes destroy statistical correlations among the pulses of the coherent beam. The decay of such correlations is a measure of the relaxation time τ_c of the interacting molecular mode. One experimental procedure is to measure directly the time decay and relaxation time using photon-count time-correlation spectrometry. This type of spectral analysis makes rapid statistical correlations among individually counted (though discriminated) photon pulses that strike the PMT of the spectrometer. The digital nature of the spectrometer considerably simplifies the acquisition and analysis of data and allows superior signal averaging and noise reduction capabilities. Correlators are available that have a minimum resolution limit of 100 nsec.

3. Magnetic Field Quenching of Intensity Fluctuations

Imposition of a magnetic (or electric) field on an aligned liquid crystal has the effect of stabilizing the sample against orientational fluctuations, reducing the scattering intensity, and decreasing the damping time (τ_c) associated with the two modes $n_\alpha(\mathbf{q})$. The field strength appears in the theory as an additional factor in the term of Eq. (9) for the director fluctuations (de Gennes, 1968a; Martinand and Durand, 1972):

$$\langle|\delta n_\alpha(\mathbf{q})|^2\rangle \sim (k_{\alpha\alpha}\mathbf{q}_\perp^2 + k_{33}\mathbf{q}_{//}^2 + \chi_a H^2)^{-1} \tag{10}$$

The electric field case is similar. For selected geometries (wave vector presentations with respect to the director), a measure of I^{-1} or τ^{-1} versus H^2 can yield the elastic constants. The bend elastic constant of MBBA has been measured by this method with results in good agreement with other independent measurements (Martinand and Durand, 1972).

4. Viscoelastic Ratios from Inelastic Light Scattering

Dynamic light scattering measurements as a function of wave vector (variation of scattering angle in specific geometries and polarization states) can be interpreted in terms of purely viscoelastic relaxations of the Fourier components of the two liquid crystal eigenmodes: $n_1(\mathbf{q})$ and $n_2(\mathbf{q})$. The relaxation time of each mode is given by (Orsay Liquid Crystal Group, 1969)

$$\tau_\alpha^{-1}(\mathbf{q}) = k_{\alpha\alpha}(\mathbf{q})/\eta_\alpha^{\text{eff}}(\mathbf{q}), \qquad \alpha = 1, 2 \tag{11}$$

where η_α^{eff} is an effective viscosity of the mode, in general a complicated function of a number of viscous torque parameters of the material. There are scattering geometries of special interest that decrease the complexity of η_α^{eff} and yield k_{11}/η_1, k_{22}/η_2, and k_{33}/η_3, where η_1, η_2, η_3 are viscosity coefficients for pure splay, twist, and bend, respectively. Ratios of a fundamental elastic constant and its viscous drag constant may thus be obtained. These viscosities have yet to be measured for polymer liquid crystals. Values of the k_{ii} obtained from one of the above independent optical methods would yield the η_i in conjunction with these measurements.

VI. Progress to Date with Poly(benzyl glutamate) Liquid Crystals

A. Surface Anchorage

Strong anchorage of liquid crystals to glass surfaces in the homeotropic and/or homogeneous alignments is essential for the successful execution of the Frederiks distortion and inelastic light scattering measurements. We have prepared PBLG liquid crystals in thin cells in the homeotropic alignment (molecular axes perpendicular to the surface). Curiously, all of the standard procedures for enhancing sympathetic surface attachments in thermotropic liquid crystals (rubbing, mechanical scribing, cleaning, chemical depositions of organic and polymer films) have yielded the homeo-tropic texture only (Fernandes and DuPré, 1981).

The most satisfactory homeotropic preparations of PBLG liquid crystals have been obtained by the following procedure. Glass plates are soaked in chromic acid overnight, rinsed in water, followed by a short soaking in a weak solution of sodium bicarbonate to remove excess acid. This is followed by a thorough washing in running water for about 30 min and a rinsing in deionized distilled water. The plates are then dried, first by blotting up excess water using filter paper, followed by about three hours in an oven at about 100°C. Sample cells are prepared by transferring PBLG solutions onto one of the plates, placing the other plate on top and allowing the solution to flow to fill the entire area. Mylar spacers are used to define the sample thickness. About three to four hours are allowed for the sample to align itself in homeotropic attachments. Preparations degrade for sample thickness greater than $\sim 10 \ \mu$m.

The solutions we have dealt with here are cholesteric liquid crystals in bulk. Uniform nematic textures are found only when specimens are confined in thin cells. Such a boundary induced cholesteric-to-nematic phase change has been observed in a series of chirally doped thermotropics (Harvey, 1977) where the plates were treated to favor homeotropic attachment. The transition was shown to occur when the separation of the plates was less than or equal to the undisturbed pitch of these large pitch twisted nematics. A similar nematization of PBLG and poly(n-amyl glutamate) polypeptide liquid crystals has also been reported (Uematsu and Uematsu, 1979). In this case no special treatment of the surfaces was attempted.

The origin of the stand-up-straight homeotropic tendency of these polypeptides may be in electrostatic interaction of the charged terminal (N^+, COO^-) residues with ionic charges on the surface of the glass plates.

B. Frederiks Distortion

Having achieved suitable homeotropic preparations of PBLG liquid crystals, we are able to observe the magnetooptical distortion in the geometry of case 3 of Fig. 8. Since the birefringence of this lyotropic liquid crystal is quite small (DuPré and Lin, 1981), the maximum path difference $\Delta n \cdot d$ that can be introduced between the ordinary and extraordinary components of the laser beam is less than one wavelength of the incident light source (He–Ne laser, 6328 Å) for accessible values of the sample thickness d. The transmitted beam does not exhibit an oscillatory behavior as sketched in the lower traces of Fig. 9 but is more like the relatively featureless top trace. The onset of the transition is evident however at a critical field $H_{c,3}$ but no extrema at higher fields can be found. Because of the large viscous component of these polymer solutions, the magnetic field must be swept very slowly (~ 20 G/min). A room temperature measurement on a 6-μm-thick sample of 18% solution of PBLG (mol wt $= 70,000$) in dioxane yields a critical field of 2 kG (Fernandes and DuPré, 1981). Using previous measurements (DuPré and Duke, 1975; Duke et al., 1976) of χ_a, we obtain a k_{33} of 6.77×10^{-10} dyne for the preparation (Fernandes and DuPré, 1981).

Failing to observe intensity oscillations with this liquid crystal, we have devised an alternate method to extract k_{11} from the Frederiks distortion data in this geometry (Fernandes and DuPré, 1981). The method involves the change of transmitted intensity above the critical field and applies even in the absence of extrema. The maximum deformation angle θ_m occurs at the center of the cell and is related to the field strength H by the following equation (Gruler et al., 1972) central in the development of the theory of the Frederiks distortion:

$$\frac{Hd}{2}\left(\frac{\chi_a}{k_{33}}\right)^{1/2} = \int_0^{\pi/2} dx \left[\frac{1+\kappa \sin^2 \theta_m \sin^2 x}{1 - \sin^2 \theta_m \sin^2 x}\right]^{1/2} \tag{12}$$

where $\kappa = (k_{11} - k_{33})/k_{33}$, $\sin\theta = \sin\theta_m \sin x$, and θ is the angle of the director with respect to the z axis. k_{11} occurs in the integrand and may be interpreted as an additional restraint [independent of surface anchorage and therefore related to $\theta(z)$] resisting the deformation. The transmitted intensity $I = I_0 \sin^2(\delta/2)$ in this geometry depends on the phase angle through:

$$\delta = \frac{2\pi n_0 d}{\lambda}\left\{1 - \frac{2H_c}{\pi H}\int_0^{\pi/2} dx \left[\frac{1+\kappa \sin^2 \theta_m \sin^2 x}{(1 - \sin^2 \theta_m \sin^2 x)(1 + v \sin^2 \theta_m \sin^2 x)}\right]^{1/2}\right\} \tag{13}$$

with $v = (n_0^2 - n_e^2)/n_e^2$, and contains information on k_{11}, even if an oscillatory pattern is not observed. The derivative of the above equation with respect to H [which includes a derivative of $\theta_m(H)$ inside of the integrand] presents an equation that may be numerically fitted to the derivative of the experimental intensity variations to extract κ and hence k_{11}.

Figure 13 is an illustration of this derivative trace obtained from differentiation of Eqs. (12) and (13). Comparison of the featureless $I(H)$ curve for a small sample birefringence with its derivative shows that dI/dH more accurately locates the critical field. The sensitivity of both functions to variations in the k_{33}/k_{11} ratio is also shown. Experimentally the derivative trace may be obtained by on-line computer processing of the output of the photodetector or by a numerical differentiation of the complete $I(H)$ curve.

We also note that at large fields ($H \gg H_c$) complete deformation of the sample is obtained ($\theta_m = \pi/2$) and δ assumes its maximum value δ_{max}. The intensity I does not change with further increase in H. This "steady state"

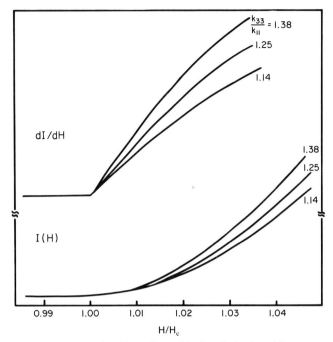

Fig. 13. Plot of the transmitted intensity and its first derivative with respect to magnetic field strength in the Frederiks distortion experiment. Here the birefringence Δn (0.004) of the liquid crystal is taken to be quite small, precluding detailed intensity oscillations above H_c. Identification of H_c from the derivative trace is facilitated. The sensitivity of both traces to variations of the k_{33}/k_{11} ratio is also illustrated. k_{33} is held constant at 6.49×10^{-7} dyne. Derivatives where performed on Eqs. (12) and (13) of the text.

intensity I_s is given by $I_s = I_0 \sin^2(\delta_{max}/2)$. Since δ_{max} may be calculated from Eq. (13) for an assumed value of k_{11}, I_0, which is a complex function of the optical setup, may be determined. Values for δ may now be obtained at intermediate values of the field from the experimental intensity $[\delta = 2 \sin^{-1}(I/I_0)^{1/2}]$ and the resultant set of $\delta(H)$ values may then be compared with Eq. (13) for consistency to assure the correct choice of k_{11}.

C. Cholesteric–Nematic Transition: Twist Modulus

The most complete examination to date of an elastic constant of polymer liquid crystals is that of the twist modulus k_{22} of polypeptide solutions. Systematic studies have been made of the variation of k_{22} with concentration of polymer, molecular weight, and supporting solvent character (Duke and DuPré, 1974a,b; DuPré and Duke, 1975; Duke *et al.*, 1976; Guha-Sridhar

TABLE I

Solvent	P_0 (μm)	H_c (kG)	k_{22}/χ_a	χ_a (emu/cm^3)	k_{22} (dyne)
Dioxane	30	5.06	9.6	$0.64 \pm 0.07 \times 10^{-8}$	6.2×10^{-8}
Chloroform	42	6.87	33.7	$0.63 \pm 0.05 \times 10^{-8}$	2.1×10^{-7}
Dichloromethane	80	5.23	72.3	$0.74 \pm 0.05 \times 10^{-8}$	5.4×10^{-7}

et al., 1974; DuPré *et al.*, 1977a,b; Patel and DuPré, 1979a,b, 1980a,b). The influence of trace amounts of halogenated organic acids on viscoelastic properties has also been noted (DuPré *et al.*, 1977a). In this section we present some representative data from these studies and show how variations and trends correlate with theoretical predictions. Values for k_{22} were obtained by following the pitch dilation as the twisted macromolecular superstructure unwinds in the presence of a magnetic field. Bulk samples were examined with no special boundary orientations. The pitch of PBLG liquid crystals is large enough (5–200 μm) to be measured in an optical microscope equipped with a calibrated micrometer eyepiece. Because the natural pitch of PBLG solutions is relatively large for cholesterics, the field requirements for the transition to nematic order are modest ($H_c \lesssim 20$ kG). Identification of the critical magnetic field strength H_c along with independent measurements of χ_a yields k_{22} from Eq. (7). A typical plot of this transition is shown in Fig. 11. Table I lists selected parameters of a 310,000 molecular weight polymer (PBLG) in three solvents that support the cholesteric structure. The value of k_{22} for the 310,000 molecular weight sample in dioxane is somewhat lower than those obtained in thermotropic liquid crystals. The value of k_{22} for a sample of higher degree of polymerization (MW = 550,000) in a different solvent (CH$_2$Cl$_2$) has been measured to be 5.82×10^{-6} dyne (Guha-Sridhar *et al.*, 1974). A wide variability of this elastic modulus is clearly evident.

Values for k_{22} in the PBLG liquid crystal have been made to vary over a wide range by adjustment of solvent and polymer molecular weight. Only a slight concentration dependence of k_{22} throughout the liquid crystal range, however, has been noted.

1. Solvent Effects

Table I demonstrates that choice of solvent dramatically alters the pitch of the cholesteric texture. The magnetic anisotropy is essentially that of the polypeptide and is seen to be the same for PBLG in each of the solvents. The solvent molecule is only slightly ordered in the liquid crystal and principal values of the solvent susceptibility tensor are nearly the same. The ninefold change in k_{22} in going from dioxane to CH$_2$Cl$_2$ suggests that the solvent is attenuating intermacromolecular interactions. Since H_c and χ_a are almost insensitive to solvent, the nearly tenfold change in k_{22} (Table I) comes from the change in P_0.

2. Molecular Weight Dependence

In thermotropic liquid crystals, changes in molecular structure from one liquid crystal to another are rather subtle and k_{22} varies only slightly from compound to compound. As noted above, the value of k_{22} for liquid crystal PBLG of molecular weight 550,000 in CH_2Cl_2 is a factor of 90 larger than the value found for a 310,000 molecular weight sample in dioxane at the same temperature.

The unusual variability of k_{22} polypeptide liquid crystals can be partially explained (DuPré and Duke, 1975) by changing interparticle interactions induced by solvent and recent theoretical considerations on the elasticity of a suspension of long hard rods. Straley's (1973b) calculations of the twist elastic constant k_{22} (hard-rod liquid crystal using the Onsager approach) gives

$$k_{22}/\rho^2 L^4 DkT \simeq 0.02 \tag{14}$$

where ρ is the number density, L is the helix length, and D is the rod diameter. This relationship accounts for most of the difference in k_{22} for these two particular systems, although, as noted above, we do not find the strong concentration dependence that the ρ^2 factor in Eq. (14) would predict. Differences in molecular weight of the solute particles are substantial. The axial ratios L/D for the 550,000 and 310,000 MW PBLG are 150 and 85, respectively. According to Eq. (14) this alone would lead to a factor of ten larger k_{22} for the higher molecular weight polypeptide. The additional factor of nine may easily be the result of the different solvents involved (a ninefold increase was noted above in changing the solvent from dioxane to dichloromethane). Another factor that could influence the value of k_{22} is the effective diameter of the helical rod; D may be differentially modified by solvation of the macromolecule.

The theory cited contains no potential energy term other than hard-rod repulsions. The observed clearing point is, therefore, not in the physics of the model. However, at a given temperature in the mesophase regime it is valid to compare system to system once empirical solvent effects are introduced, since they apparently considerably influence the potential function.

3. Effect of Temperature

We have found P_0 to increase linearly with temperature in our polypeptide liquid crystals, a trend that is at variance with that found in thermotropic

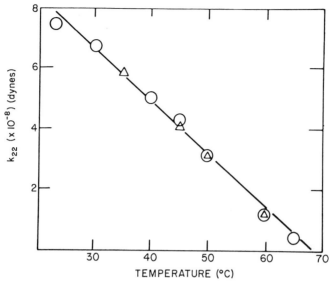

Fig. 14. Temperature dependence of k_{22} of a PBLG liquid crystal. (After DuPré, and Samulski, 1978.) ○, increasing temperature; △, decreasing temperature.

liquid crystals of small molecules that exhibit a negative temperature co-efficient: $dP/dT < 0$. The value of H_c for these solutions is inversely proportional to temperature, which allows the calculation of the temperature dependence of the ratio k_{22}/χ_a (DuPré and Duke, 1975). This quantity, along with our more recent determination (DuPré et al., 1977b) of $\chi_a(T)$, gives the temperature variation of the twist modulus itself (see Fig. 14). We have found that k_{22} decreases with increasing temperature, as do elastic moduli of more common materials. Magnetic susceptibility measurements have also given us the temperature dependence of the order parameter $S(T) = \bar{P}_2 \cos \theta$ throughout the mesomorphic range of this polymer (DuPré et al., 1977a,b). This variation is shown in Fig. 15 as a function of reduced temperature.

4. The Effect of Halogenated Organic Acids

The effect of halogenated organic acids on solution properties of polypeptides is both unusual and dramatic. A variety of experimental techniques has shown that in mixed solvent systems containing a helicogenic solvent and a denaturing solvent, dichloroacetic acid (DCA) or trifluoroacetic acid (TFA), synthetic polypeptides such as PBLG will undergo an intramolecular phase transition from a rigid rodlike. α-helical polymer to a flexible random coil (Fasman, 1967). TFA is more effective in that the transition occurs at 10–30% acid by volume (dependent on the nature of the helicogenic solvent

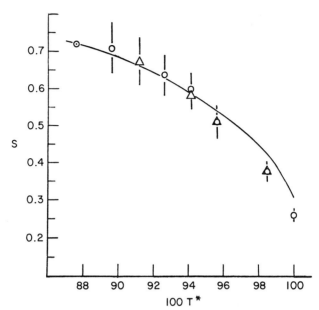

Fig. 15. Liquid crystal orientational parameter S as a function of reduced temperature, $T^* = T/T_c$. $T_c \simeq 65°C$ (see DuPré, *et al.*, 1977b). $\bigcirc$ increasing temperature; $\triangle$ decreasing temperature. $\bigcirc$ $S = 0.72$. (From Murthy *et al.*, 1976.)

and the polymer molecular weight), whereas $\sim 75\%$ of the weaker acid DCA is required. In addition, the literature contains many reports of more subtle gradual changes in the condition of the polypeptide macromolecule in the pretransition range (1–10% volume range for TFA). Such pretransition behavior at low acid concentrations has been accounted for with a variety of explanations and models including disruption of aggregated polymer helices, gradual unfreezing of side chains on the periphery of the macromolecule, and "melting" of low molecular weight helical moieties to random coils in polydisperse samples prior to the precipitous helix–coil transition itself.

It has been noted that small amounts of TFA have remarkable effects on liquid crystal formation and response at higher concentrations of the polymer. For example, a few percent TFA added to the helicogenic solvent promotes solubilization of the polymer and rapid formation and maturation of the cholesteric structure exhibited by this lyotropic liquid crystal. A large decrease in the response time of the mesophase to external perturbations such as electric or magnetic fields is also noted upon the addition of even 1% TFA. Indeed, the use of sparing quantities of this highly polar acid is a part of the art of handling in research on these macromolecular liquid crystals.

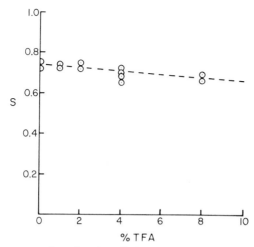

Fig. 16.   Order parameter S as a function of % TFA in PBLG:dioxane:TFA liquid crystals.

The effect of small amounts of TFA (0–10%) added to lyotropic solutions of PBLG has been examined (DuPré et al., 1977a) by measurements of the pitch, critical field, and diamagnetic susceptibility of the liquid crystal. We have found that although these quantities change dramatically with even slight additions of the acid, the order parameter is only slightly modified and the twist modulus is surprisingly invariant. The rotational viscosity coefficient does change abruptly, however (Duke et al., 1976).

In PBLG–dioxane liquid crystals, S decreases linearly from an initial value of 0.73 (in a liquid crystal without TFA) by about 10% when TFA is added in the range 0–10% TFA (see Fig. 16) (Murthy et al., 1976). The initial value of S is considerably less than that of a perfectly ordered system of close-packed rigid rods ($S = 1$). This reduced value of S is thought to be due to disorder in the liquid crystal in the form of correlated flexing of neighboring helices, i.e., the helices are not ideal rigid rods. In this context, addition of TFA to the liquid crystal would be expected to decrease S if the TFA perturbs the polymer so as to increase the flexibility of the helical conformation.

Utilizing the polarizing microscope to observe the retardation lines, the variation of the pitch for the PBLG–dioxane: TFA liquid crystal in zero magnetic field has been determined as a function of TFA concentration. In Fig. 17 the data has been normalized to the value of the pitch in the absence of TFA, $P_0 = 47 \ \mu m$. The effect of TFA is minimal below 2%. After this initial insensitivity, the pitch decreases more rapidly with TFA concentration in the remainder of the range 1–10% TFA (the helix–coil transition, determined by changes in the chemical shift of the $^{\alpha}C$—H proton, occurs at 25%

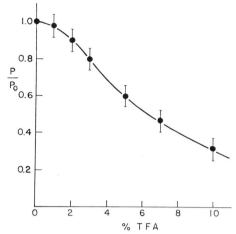

Fig. 17. The pitch P of the cholesteric structure normalized by the value for 0% TFA, P_0, as a function of % TFA in PBLG:dioxane:TFA liquid crystals (24 mg polymer/cm^3 dioxane).

TFA vol %). This pitch decrease is opposite to the trend observed in other mixed solvent systems when the concentration of a more polar moiety is increased.

The temperature dependence of the pitch over a range of TFA concentrations is shown in Fig. 18. Again, an unusual behavior is observed. At the higher acid concentration the pitch is noted to approach temperature independence. We know of only one other system exhibiting the latter behavior

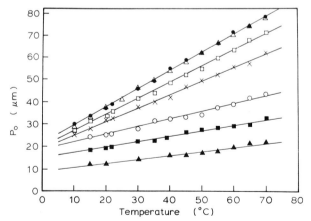

Fig. 18. Cholesteric pitch versus temperature for various TFA concentrations in PBLG:dioxane:TFA liquid crystals: ●, data for 0% TFA; △, 1% TFA; □, 2% TFA; ×, 3% TFA; ○, 5% TFA; ■, 7% TFA; and ▲, data points for 10% added TFA.

(Dolpin *et al.*, 1973), thermotropic mixtures of dextro-active and *racemic-p*-ethoxybenzal-*p*-(*β*-methylbutyl)aniline.

The variation of H_c with increasing TFA concentration has also been determined (DuPré and Duke, 1975). This, along with the acid dependence of P_0 and χ_a, is sufficient to determine the behavior of the twist modulus in the presence of TFA, which is shown in Fig. 19. The insensitivity of the modulus to TFA within experimental error appears to be one of the anomalies of the effect of the acid on this polymer. We have constructed an argument for this behavior based on published theory of the origin of cholesteric liquid crystals. The formation of a spontaneously twisted liquid crystal is apparently due to asymmetric terms that arise from molecular chirality in the angular dependent part of the potential of interaction between molecules (Goosens, 1971; Samulski and Samulski, 1977). Goosens has presented a general theory relating the origin of cholesteric twist to molecular chirality (Goosens, 1971). He derives an angular dependent intermolecular potential for planar molecules whose long axes lie in adjacent "twist planes" separated by a distance r_{ab}:

$$- V_{ab} = (A/r_{ab}^4) \cos 2\theta_{ab} + (B/r_{ab}^5) \sin 2\theta_{ab} \tag{15}$$

The planes are perpendicular to the cholesteric axis and θ_{ab} is the angle between the directions of the long axis alignment in respective planes a and b. The coefficient A of the symmetric part of this potential, is related to the anisotropy of the molecular polarizability. The cholesteric twist occurs because of the asymmetric part of the potential; B is related to the dispersion

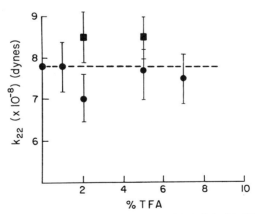

Fig. 19.   The twist elastic constant k_{22} as a function of % TFA (vol) in PBLG:dioxane:TFA liquid crystals: ●, PBLG mol wt = 300,000; ■, PBLG mol wt = 310,000.

energy determined by electric dipole–electric quadrupole interactions. The magnitude of the twist, θ_{ab}, is proportional to the ratio $B/r_{ab}A$. Physically the twist elastic constant is dependent on both short range and long range order in the liquid crystal. In principle k_{22} can be related to the intermolecular potential through the appropriate derivatives of a partition function that would involve a potential such as that above. Although TFA changes the equilibrium angular orientation of one PBLG molecule relative to its neighbor (θ_{ab}), k_{22} is not changed in the range 1–10% TFA. The value of P_0 is related to the ratio of the coefficients in Eq. (15) and k_{22}, presumably, is a function of the total V_{ab}. However, since $B \ll A$, a sizeable change in P_0 is not necessarily accompanied by a comparable change in k_{22}.

D. Light Scattering

At the time of this writing our light scattering results on this polymer liquid crystal are only preliminary and subject to a number of experimental difficulties. As mentioned above, these lyotropic preparations are remarkably clear, due to the low amplitude of the damped fluctuations of the director field in these viscous solutions. Measurements of the correlation time τ_c of depolarized light scattering from an unoriented bulk solution of 18% PBLG (MW = 296,000) in dioxane yield an extremely long τ_c of 38 sec. This should be compared with correlation times of the order of milliseconds usually found in studies of small molecule thermotropics. Because of the weak scattering processes involved here, the correlation function also takes several hours to build up. Our apparatus includes a 5 mW He–Ne laser (Spectral Physics Model 120), a 64-channel multibit realtime correlator, and an ITT FW 130 PMT with photon count amplification/discrimination circuitry mounted on a precision goniometer table.

Reduction of the solution viscosity by the addition of trace amounts of TFA reduces τ_c by a factor of two and considerably cleans up the autocorrelation function decay. Relaxation behavior is not perfectly exponential in the absence of TFA but becomes so with the addition of 2–6% of the acid. This use of sparing amounts of TFA should expedite the analysis of light scattering data on thin oriented samples in the geometries required to obtain the elastic constants. Our results will be published in the near near future (Fernandes and DuPré, 1982).

Acknowledgment

Portions of this work were supported by the National Science Foundation under Grant DMR-7903760 and by the National Institutes of Health under Grant HL24364.

References

Barrall, E. M., II, and Johnson, J. F. (1979). *J. Macromol. Sci., Rev. Macromol. Chem.* **C17**, 137.

Berne, B. J., and Pecora, R. (1976). "Dynamic Light Scattering." Wiley, New York.

Black, W. B., and Preston, J. (1979). "High Modulus Aromatic Polymers." Dekker, New York.

Blumstein, A., ed. (1978). "Liquid Crystalline Order in Polymers." Academic Press, New York.

Chandrasekhar, S. (1977). "Liquid Crystals." Cambridge Univ. Press, London and New York.

Chapoy, L. L., and DuPré, D. B. (1979). *J. Chem. Phys.* **70**, 2550.

Cheng, J., and Meyer, R. B. (1972). *Phys. Rev. Lett.* **29**, 1240.

Cheng, J., and Meyer, R. B. (1974). *Phys. Rev. A* **9**, 2744.

Chu, B. (1974). "Laser Light Scattering." Academic Press, New York.

Ciferri, A., and Ward, I. M. (1979). "Ultra High Modulus Polymers." Appl. Sci., London.

Cladis, P. E. (1972). *Phys. Rev. Lett.* **28**, 1626.

Cummins, H. Z., and Pike, E. R., eds. (1974). "Photon Correlation and Light Beating Spectroscopy." Plenum, New York.

de Gennes, P. G. (1968a). *C. R. Hebd. Seances Acad. Sci., Ser. B* **266**, 15.

de Gennes, P. G. (1968b). *Solid State Commun.* **6**, 163.

de Gennes, P. G. (1974). "The Physics of Liquid Crystals." Oxford Univ. Press (Clarendon), London and New York.

de Gennes, P. G. (1977). *Mol. Cryst. Liq. Cryst. (Lett.)* **34**, 177.

de Gennes, P. G., and Pincus, P. (1977). *Polym. Prepr., Am. Chem. Soc., Div. Polym. Chem.* **18**, 161.

de Jeu, W. H. (1980). "Physical Properties of Liquid Crystalline Materials." Gordon & Breach. New York.

de Jeu, W. H., and Claassen, W. A. P. (1977). *J. Chem. Phys.* **67**, 3705.

de Jeu, W. H., Claassen, W. A. P., and Spriujt, A. M. J. (1976). *Mol. Cryst. Liq. Cryst.* **37**, 269

Dolpin, D., Muljiani, Z., Cheng, J., and Meyer, R. B. (1973). *J. Chem. Phys.* **58**, 413.

Duke, R. W., and DuPré, D. B. (1974a). *J. Chem. Phys.* **60**, 2759.

Duke, R. W., and DuPré, D. B. (1974b). *Macromolecules* **7**, 374.

Duke, R. W., DuPré, D. B., Hines, W. A., and Samulski, E. T. (1976). *J. Am. Chem. Soc.* **98**, 3094.

DuPré, D. B., and Duke, R. W. (1975). *J. Chem. Phys.* **63**, 143.

DuPré, D. B., and Lin, F.-M. (1981). *Mol. Cryst. Liq. Cryst.* **75**, 217.

DuPré, D. B., and Patel, D. L. (1980). *Polym. Prepr., Am. Chem. Soc., Div. Polym. Chem.* **21**, 69.

DuPré, D. B., and Samulski, E. T. (1978). *In* "Liquid Crystals: The Fourth State of Matter" (F. Saeva, ed.), Chap. 5. Dekker, New York.

DuPré, D. B., Duke, R. W., and Samulski, E. T. (1977a). *Mol. Cryst. Liq. Cryst.* **40**, 247.

DuPré, D. B., Duke, R. W., and Samulski, E. T. (1977b). *J. Chem. Phys.* **66**, 2748.

Fasman, G. D. (1967). *In* "Poly-α-Amino Acids" (G. D. Fasman, ed.), p. 239. Dekker, Inc., New York.

Fernandes, J. R., and DuPré, D. B. (1981). *Mol. Cryst. Liq. Cryst. (Lett.)* **72**, 67.

Fernandes, J. R., and DuPré, D. B. (1982). "Ordered Fluids and Liquid Crystals," Vol. 4. Plenum, New York.

Flory, P. J. (1956). *Proc. R. Soc. London, Ser. A* **234**, 73.

Goosens, W. J. (1971). *Mol. Cryst. Liq. Cryst.* **12**, 237.

Gruler, H., Scheffer, T. J., and Meier, G. (1972). *Z. Naturforsch., A* **27A**, 966.

Guha-Sridhar, C., Hines, W. A., and Samulski, E. T. (1974). *J. Chem. Phys.* **61**, 947.

Harvey, T. B. (1977). *Mol. Cryst. Liq. Cryst. (Lett.)* **34**, 225.

Jen, S., Clark, N. A., and Pershan, P. S. (1977). *J. Chem. Phys.* **66**, 4653.

Krigbaum, W. R., Salaris, F., Ciferri, A., and Preston, J. (1979). *J. Polym. Sci., Polym. Lett. Ed.* **17**, 601.

Langevin, D. (1974). Ph.D. Thesis, Paris.

Langevin, D., and Bouchiat, M. A. (1975). *J. Phys. (Orsay, Fr.)* **36** (C1), 197.

Leenhouts, F., Roberts, H. J., Dekker, A. J., and Jonker, J. J. (1979). *J. Phys., Colloq. (Orsay, Fr.)* **40**, (C3), 291.

Martinand, J. L., and Durand, G. (1972). *Solid State Commun.* **10**, 815.

Meyer, R. B. (1968). *Appl. Phys. Lett.* **14**, 208.

Murthy, S., Knox, J. R., and Samulski, E. T. (1976). *J. Chem. Phys.* **65**, 4835.

Onsager, L. (1949). *Ann. N.Y. Acad. Sci.* **51**, 627.

Orsay Liquid Crystal Group (1969). *J. Chem. Phys.* **51**, 816.

Panar, M., and Beste, L. F. (1977). *Macromolecules* **10**, 1401.

Patel, D. L., and DuPré, D. B. (1979a). *J. Polym. Sci., Polym. Lett. Ed.* **17**, 299.

Patel, D. L., and DuPré, D. B. (1979b). *Mol. Cryst. Liq. Cryst.* **53**, 323.

Patel, D. L., and DuPré, D. B. (1980a). *J. Polym. Sci., Polym. Phys. Ed.* **18**, 1599.

Patel, D. L., and DuPré, D. B. (1980b). *J. Chem. Phys.* **72**, 2515.

Priest, R. G. (1973). *Phys. Rev.* **A 7**, 720.

Priestly, E. B., Wojtowicz, P. J., and Sheng, P., eds. (1975). "Introduction to Liquid Crystals." Plenum, New York.

Rapeni, A., and Papoular, M. (1969). *J. Phys. (Orsay, Fr.)* **30** (C4), 54.

Robinson, C. (1956). *Trans. Faraday Soc.* **52**, 571.

Robinson, C. (1961). *Tetrahedron* **13**, 219.

Robinson, C. (1966). *Mol. Cryst.* **1**, 467.

Robinson, C., and Ward, J. C. (1957). *Nature (London)* **180**, 1183.

Robinson, C., Ward, J. C., and Beevers, R. B. (1958). *Discuss. Faraday Soc.* **25**, 29.

Samulski, E. T., and DuPré, D. B. (1979). *Adv. Liq. Cryst.* **4**, 121.

Samulski, T. V., and Samulski, E. T. (1977). *J. Chem. Phys.* **67**, 824.

Saupe, A. (1960). *Z. Naturforsch.*, **A 15A**, 815.

Schad, H., and Osman, M. (1981). *J. Chem. Phys.* **72**, 880.

Schad, H., Baur, G., and Meier, G. (1979). *J. Chem. Phys.* **70**, 2770.

Schadt, M., and Muller, F. (1979). *Rev. Phys. Appl.* **14**, 265.

Straley, J. P. (1973a). *Mol. Cryst. Liq. Cryst.* **22**, 333.

Straley, J. P. (1973b). *Phys. Rev. A* **8**, 2181.

Uematsu, Y., and Uematsu, I. (1979). *Polym. Prepr., Am. Chem. Soc., Div. Polym. Chem.* **20**, 66.

VanEck, D. C., and Perdeck, M. (1978). *Mol. Cryst. Liq. Cryst. (Lett.)* **49**, 39.

8

Instabilities in Low Molecular Weight Nematic and Cholesteric Liquid Crystals

S. Chandrasekhar
U. D. Kini

Raman Research Institute
Bangalore, India

POLYMER LIQUID CRYSTALS

I. Introduction

The anisotropic properties of nematic liquid crystals result in certain novel instability mechanisms that are not encountered in the classical problems of hydrodynamic stability in ordinary liquids. Three types of convective instability in nematics have been investigated in considerable detail in recent years: electrohydrodynamic (EHD) instability brought about by an externally applied electric field, thermal instability in a temperature gradient, and hydrodynamic instability caused by flow. Of these, EHD instability had been noticed time and again almost since the beginning of the century, but a proper explanation of it was given only in 1969 (Helfrich, 1969a; Dubois-Violette et al., 1971). The aligning influence of a thermal gradient had also been observed many years ago (Stewart, 1936). However, it was the theoretical work on EHD that actually stimulated systematic studies on thermal as well as hydrodynamic instabilities (Dubois-Violette, 1971; Guyon and Pieranski, 1972; Pieranski and Guyon, 1973, 1974a) and it was soon established that there are close analogies in the mechanisms involved in all three cases. Excellent reviews on instabilities in nematics have been written by Guyon and Pieranski (1974) and by Dubois-Violette et al. (1977, 1978), while detailed accounts of the theoretical aspects of the subject have been given by Leslie (1979) and Jenkins (1978).

We present here a summary of the salient experimental and theoretical results in the field, emphasizing the physical principles underlying the observed phenomena. Some basic notions of the continuum theory of the nematic state are assumed, and the reader is referred to Chapter 3 of Chandrasekhar (1977), since we shall throughout adopt the same symbols and definitions as in that book. We shall confine our attention mainly to convective processes and not deal with effects like the Frederiks transition or flexoelectric distortion. Also, we shall restrict the discussion to instabilities in low molecular weight nematic and cholesteric liquid crystals, but, in principle, the ideas outlined here, with appropriate modifications of the material parameters, may perhaps be relevant to the studies initiated recently of similar phenomena in polymer liquid crystals (Krigbaum et al., 1980; Krigbaum, Chapter 10, this volume).

II. Electrohydrodynamic Instability in Nematics

A. The Carr–Helfrich Mechanism

EHD is a topic that has been covered quite thoroughly in a number of articles (see Chandrasekhar, 1977, Section 3.10; Chistyakov and Vistin, 1974; Durand, 1976; Dubois-Violette et al., 1978; Goossens, 1978; Blinov, 1978).

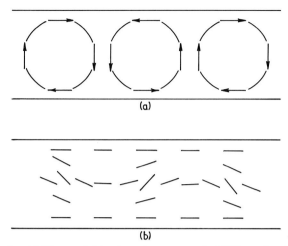

Fig. 1. (a) Flow and (b) orientation patterns in Williams domains.

We shall therefore content ourselves with giving a bare outline. The usual experimental arrangement for observing EHD instability consists of a nematic film of negative dielectric anisotropy (i.e., with $\varepsilon_a = \varepsilon_{||} - \varepsilon_{\perp} < 0$, where $\varepsilon_{||}$ and $\varepsilon_{\perp}$ are the principal dielectric constants parallel and perpendicular to the director) sandwiched between two glass plates with the director parallel to the glass surfaces. The surfaces are coated with a transparent electrically conducting material to enable optical observation. When a dc or low-frequency ac field is applied across the electrodes, there appear, above a threshold voltage, a regular set of parallel striations approximately perpendicular to the initial unperturbed orientation of the director, the spacing between the striations being roughly equal to the thickness of the film. The striations, often referred to as the *Williams* domains, are due to hydrodynamic motion of the fluid in the form of rolls as shown in Fig. 1. The threshold voltage is usually a few volts and is practically independent of the sample thickness. However, it increases rapidly with the frequency of the applied voltage (Fig. 2), and above a critical frequency ω_c, whose value depends on the conductivity of the sample, another type of domain pattern is obtained. Parallel striations, again approximately perpendicular to the initial orientation of the director but with a much shorter spacing (a few micrometers), are formed in the midplane of the sample. When the field is increased very slightly above the threshold, the striations bend and move to form a *chevron pattern*. The lower frequency range $\omega < \omega_c$ is called the *conduction* regime while $\omega > \omega_c$ is called the *dielectric* regime. In the dielectric regime the threshold is determined by a critical field (and not by a critical voltage as in the conduction regime). Both the threshold field strength and the spatial periodicity of the pattern are frequency dependent,

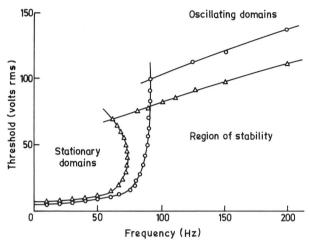

Fig. 2. Threshold voltage for EHD instability versus frequency for MBBA. Sample thickness 50 μm, $\bigcirc$, sinusoidal excitation; $\triangle$, square wave excitation. (From Orsay Liquid Crystal Group, 1972.)

the former increasing with frequency as $\omega^{1/2}$ and the latter diminishing with frequency. The relaxation time of the oscillating chevron pattern is a few milliseconds, while that of the stationary Williams domains is about 100 msec for a film of thickness 25 μm. The Williams domains as well as the chevron pattern give way to a turbulent mode when the voltage is increased to about twice the threshold value. The turbulence is accompanied by an intense scattering of light and it is therefore usually called the *dynamic scattering mode.*

The origin of this instability is now well understood and is referred to as the *Carr–Helfrich mechanism.* The current carriers in the nematic liquid crystal are ions whose mobility is greater along the director than perpendicular to it. The ratio of the principal conductivities $\sigma_{||}/\sigma_{\perp}$ is usually 1.3–1.5. Because of this anisotropy space charge can be formed by ion segregation in the liquid crystal itself as illustrated schematically in Fig. 3 for a bend fluctuation. The applied field E_z acts on the charges to give rise to material flow in alternating directions, which in turn exerts a torque on the molecules as indicated in Fig. 1. This is reinforced by the dielectric torque due to the transverse field E_y created by the space charge distribution. Under appropriate conditions, these torques may offset the stabilizing elastic and dielectric torques and the system may become unstable. Even a conductivity of the order of 10^{-9} Ω^{-1} cm^{-1} is enough to produce this type of fluid motion, and in practice, unless very special precautions are taken, the impurity conductivity is usually greater than this value.

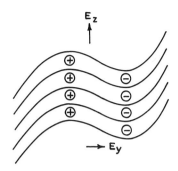

Fig. 3. Charge segregation is an applied electric field E_z caused by a bend fluctuation in a nematic of positive conductivity anisotropy ($\sigma_a > 0$). The resulting transverse field is E_y.

These experimental facts can be broadly explained on the basis of a one-dimensional theory (Helfrich, 1969a; Dubois-Violette *et al.*, 1971; Smith *et al.*, 1975). Let the initial unperturbed orientation of the director be along y and let the applied field E_z be along z. We consider a bend fluctuation of wavevector q_y such that the director is confined to the yz plane and makes an angle θ with y at any point. We choose as our variables the charge density q and the local director curvature $\psi = \partial\theta/\partial y$. Positive charges accumulate in the region of negative curvature and similarly a positive force qE_z increases the negative ψ. The behavior is described by two coupled equations:

$$\dot{q} + (q/\tau) + \sigma_H\psi E_z = 0 \tag{1}$$

$$\dot{\psi} + (\psi/\tau_\psi) + (qE_z/\eta) = 0 \tag{2}$$

where

$$\sigma_H = \sigma_{||}(\varepsilon_\perp/\varepsilon_{||} - \sigma_\perp/\sigma_{||}) \tag{3}$$

$\tau = \varepsilon_{||}/4\pi\sigma_{||}$ is the dielectric relaxation time and τ_ψ is the curvature relaxation time given by

$$\tau_\psi^{-1} = \left(\frac{1}{\lambda_1} + \frac{1}{\eta_0}\right)\left(-\frac{\varepsilon_a\varepsilon_\perp}{4\pi\varepsilon_{||}} E_z^2 + k_{33}q_y^2\right) \tag{4}$$

$$= \Lambda E_z^2 + \Lambda_0 \tag{5}$$

say, $q_y \simeq \pi/d$, d being the sample thickness, k_{33} the bend elastic constant, and λ_1, η_0, and η viscosity coefficients. It may be noted that the curvature relaxation time is strongly influenced by the applied field, whereas the dielectric relaxation time is independent of it.

For dc excitation the time dependence may be ignored in Eqs. (1) and (2), and the threshold condition becomes

$$1/\tau\tau_\psi = \sigma_H E_{th}^2/\eta \tag{6}$$

so that the threshold field may be expressed in the form

$$E_{th} = E_0/(\zeta^2 - 1)^{1/2} \tag{7}$$

where

$$E_0^2 = -(4\pi\varepsilon_{||}/\varepsilon_a\varepsilon_\perp)k_{33}q_y^2 \tag{8}$$

and ζ is a dimensionless quantity called the *Helfrich parameter* given by

$$\zeta^2 = \left(1 - \frac{\varepsilon_{||}\sigma_\perp}{\varepsilon_\perp\sigma_{||}}\right)\left[1 - \frac{\varepsilon_{||}}{\varepsilon_a}\frac{2\lambda_1^2}{(\lambda_1 - \lambda_2)(\lambda_1 + \eta_0)}\right] \tag{9}$$

the λs being viscosity coefficients. Typically ζ^2 is a small number; for example, for MBBA (4-methoxybenzylidene-4'-butylaniline) it is about three. For dc and low-frequency ac, $q_y \simeq \pi/d$, and therefore we have a voltage threshold independent of film thickness.

We have here treated the director distortion as a pure bend, but since the thickness of the sample is of the same order as the periodicity of the distortion, there will be a nonnegligible splay component. To allow for this it has been suggested that k_{33} should be replaced by $k_{33} + k_{11}(q_z/q_y)^2$, where q_z is also π/d. With this correction the dc threshold given by Eq. (7) is in good agreement with the experimental value.

In the ac case Eqs. (1) and (2) cannot in general be solved analytically since τ_ψ itself depends on E_z and t. For a square wave, E_z remains constant over any half period and exact solutions can be obtained (Smith *et al.*, 1975). The theoretical variation of q and ψ over a full period of the exciting wave is illustrated in Fig. 4 for three values of v ($= \omega/2\pi$).

For sinusoidal excitation the threshold field in the conduction regime turns out to be (Dubois-Violette *et al.*, 1971)

$$E_{th}^2 = \frac{E_0^2(1 + \omega^2\tau^2)}{\zeta^2 - 1 - \omega^2\tau^2} \tag{10}$$

In the dielectric regime simple analytical expressions cannot be derived for E_{th} and q_y for sinusoidal excitation, but for square-wave excitation they are given by (Dubois-Violette *et al.*, 1978)

$$E_{th}^2 = E_0^2(2.403)/\zeta^2\Lambda_0 \tag{11}$$

and

$$q_y^2 = \left(1.022 - \frac{2.403}{\zeta^2}\right)\frac{\eta_B\omega}{k_{33}} \tag{12}$$

where

$$1/\eta_B = (1/\lambda_1) + (1/\eta_0)$$

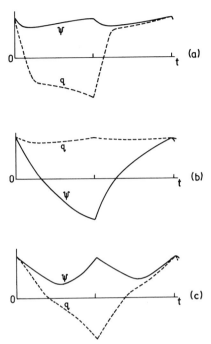

Fig. 4.   Time dependence of the charge q and the curvature ψ over one period of the square wave excitation. (a) Conduction regime ($\nu\tau \ll 1, \tau_\psi \gg \tau$). The charges oscillate but the domains are stationary. (b) Dielectric regime ($\nu\tau \gg 1, \tau_\psi \ll \tau$). The charges are stationary but the domains oscillate. (c) Conduction regime at a frequency close to the transition to the dielectric regime ($\nu\tau \lesssim 1, \tau_\psi \simeq 2\tau$). The turnaround of ψ follows the change of sign of q with a certain phase lag. (From Smith et al., 1975.)

A direct consequence of the above equations is that even for frequencies less than ω_c there can be quenching of the conductive instability at high fields. This is because τ_ψ given by Eq. (5) decreases with increasing E_z, and when it becomes equal to τ there can be restabilization. This gives a physical insight into the origin of the sigmoid shape of the experimental threshold curve (Fig. 2). Another conclusion that can be drawn is that the conduction regime can be suppressed altogether by using very thin samples. We have seen that for low frequencies $q_y \simeq \pi/d$. Now a decrease in sample thickness increases Λ_0 of Eq. (5). This in turn reduces the curvature relaxation time τ_ψ, and when $\tau_\psi < \tau$ the conduction induced instability is eliminated and only a dielectric instability is possible. A similar quenching can be achieved by applying a strong stabilizing magnetic field.

The two-dimensional problem, explicitly taking into account the variation of the deflection θ with z, has been treated by a number of authors (Pikin,

1971, 1972; Penz and Ford, 1972a,b; Penz, 1970, 1973, 1974; Pikin and Shtolberg, 1973; Sengupta and Saupe, 1974; Rigopoulos and Zenginoglou, 1976; Zenginoglou *et al.*, 1977; Zenginoglou and Kosmopoulos, 1977). These calculations essentially confirm the broad conclusions of the simpler theory. A number of other studies have been reported; these include the case of oblique alignment (Pikin and Indenbom, 1975; Pikin *et al.*, 1976a,b), the possibility of oscillatory instability (Laidlaw, 1979), nonlinear effects (Carroll, 1972; Hijikuro, 1975; Moritz and Franklin, 1976, 1977; Ben-Abraham, 1976, 1979; Akahoshi and Miyakawa, 1977), and the formation of defects (walls) and their role in material flow and molecular alignment (Carr and Kozlowski, 1980; Tarr and Carr, 1980), but we shall not discuss any of these problems here.

In principle, the Carr–Helfrich mechanism may also be expected to operate in a material having negative conductivity anisotropy ($\sigma_a < 0$) and positive dielectric anisotropy ($\varepsilon_a > 0$) (see Durand, 1976). This was demonstrated in a practical case recently by Moodithaya and Madhusudana (1980). The experiments were conducted on a 94:6 mol % mixture of 4,4′-diheptyloxyazoxybenzene (HOAB) and *trans-p-n*-ethoxy-α-methyl-cyanophenyl cinnamate (2OMCPC), which shows an induced smectic A phase. (The term "induced smectic phase" is used to mean the occurrence of a smectic phase in a mixture even though each component of the mixture does not by itself show a smectic phase.) At temperatures slightly above the smectic A–nematic transition point of the mixture, the nematic phase shows negative σ_a and positive ε_a because of the existence of cybotactic groups. EHD instability was observed in this temperature range in homeotropically aligned samples (i.e., with the director normal to the glass surfaces). From a study of the dependence of the threshold on frequency and temperature these authors have shown that the instability is due to the Carr–Helfrich mechanism, the charge segregation in this case being coupled with a splay fluctuation (Fig. 5).

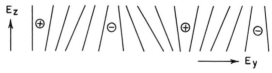

Fig. 5. Charge segregation in an applied electric field E_z caused by a splay fluctuation in a nematic of negative conductivity anisotropy ($\sigma_a < 0$). The resulting transverse field is E_y.

Other interesting combinations of parameters that can give rise to the Carr–Helfrich instability have been discussed by Dubois-Violette *et al.* (1978).

B. Some Unexplained Observations

EHD instability has also been observed in the isotropic phase of nematics, the threshold being comparable to that in the nematic phase. The precise origin of this instability does not seem to be fully understood. It is almost certainly related to the injection of charge carriers at the solid–liquid interface (Felici, 1969; Atten, 1975) though Barnik *et al.* (1976) have proposed a different mechanism that they call the electro-kinetic process. The latter authors have suggested that this isotropic type of instability may be the principal mechanism even in the nematic phase, but the question has been resolved by Ribotta and Durand (1979), who made a very careful study of the instability thresholds in the high frequency regime in nematic MBBA in the usual homogeneous configuration. Two distinct types of instability were found to occur, one resulting in the short-wavelength striations of the dielectric regime and the other in a slow convective hydrodynamic motion having a periodicity comparable to the sample thickness. The threshold for the former is higher than that for the latter, the difference being more pronounced the thicker the sample (Fig. 6). However, more significantly, the temperature dependence of the threshold shows a markedly different behavior in the two cases. For the dielectric instability, the threshold diverges as the temperature approaches the nematic–isotropic transition point and drops to zero in the isotropic phase, as is to be expected on the basis of the Carr–Helfrich mechanism (see Eq. (11)). On the other hand, for the convective instability the threshold shows no discontinuity at the transition indicating that it arises from a Felici-type or electrokinetic mechanism that is indifferent to the anisotropy of the medium. The isotropic mechanism would explain the occurrence of EHD instability in materials in which both ε_a and σ_a are negative (Blinov *et al.*, 1979; Goscianski and Leger, 1975) and in homeotropically aligned materials having positive ε_a (Barnik *et al.*, 1976; Sussman, 1976, 1978; Madhusudana and Moodithaya, 1979). In the low frequency (conduction) regime, it is expected that there will be a strong coupling between the Carr–Helfrich and isotropic mechanisms since both lead to identical flow patterns, and indeed these patterns persist in the isotropic phase without any apparent break at the transition (Koelmans and Van Boxtel, 1971).

Domain patterns with $q_y \simeq \pi/d$ have been observed in positive dielectric anisotropy materials at very high frequencies (10^4–10^6 Hz) around the dielectric relaxation frequency at which the sign of ε_a of the nematic changes from positive to negative (de Jeu *et al.*, 1972; de Jeu and Lathouwers, 1974; Goossens, 1972; Karat and Madhusudana, 1977). de Jeu and Lathouwers (1974) have identified this as being similar to the conduction regime because they were able to observe dynamic scattering at higher voltages and also chevrons above a second frequency limit. As emphasized by Smith *et al.*

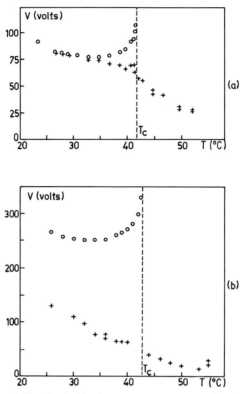

Fig. 6. EHD instability threshold voltage versus temperature for short wavelength stria-tions in the dielectric regime (○) and for convective rolls of spatial periodicity $\simeq$ sample thick-ness (+) for MBBA in the nematic and isotropic phases. Frequency of applied voltage, 200 Hz; sample thickness: (a) 26 μm and (b) 104 μm. (From Ribotta and Durand, 1979.)

(1975), ionic diffusion currents cannot be ignored at these frequencies. Theoretically, the effect of these diffusion currents is to give rise to a dielectric regime with $q_y \simeq \pi/d$ when $q_y \simeq q_D$, the Debye screening wave vector. Smith *et al.* have discussed other possibilities, but the precise origin of these observed instabilities does not appear to have been thoroughly investigated.

There is evidence that the flow patterns in the conduction regime may not be as simple in all cases as depicted in Fig. 1 (Madhusudana *et al.*, 1973; Karat and Madhusudana, 1973; Chang, 1973). In particular, it is more or less well established that within the conduction regime the nature of the pattern changes at well defined thresholds (Kai and Hirakawa, 1978; Hilsum and Saunders, 1980). The reason for this is still obscure.

Another observation that is not quite clear concerns the behavior of a freely suspended nematic film. Studies have been made on nematic MBBA with the initial orientation of the director normal to the free surfaces and the applied field parallel to them. Unlike the bounded film, the free film has no threshold for hydrodynamic motion (Meyerhofer *et al.*, 1972; Faetti *et al.*, 1979). However, above a certain critical field the apparently uncorrelated motion settles down to a stationary vortex motion having a much larger velocity (Faetti *et al.*, 1979). It has been suggested that the charges accumulating on the free surfaces in the presence of a field generate a stress that in turn causes this vortex motion.

Thus while it would be fair to say that the basic mechanism of EHD instability in nematics is reasonably well understood, a number of problems still remain to be explained in detail.

III. Two Illustrative Experiments on Nematic Flow

A. The Effect of Flow on the EHD Instability Threshold

Before proceeding to discuss flow-induced instabilities it is instructive to consider two experiments that bring out certain important aspects of nematic flow. The first one, which concerns the effect of flow on the EHD instability threshold, was suggested by de Gennes (1975) and performed by Guyon and Pieranski (1975a).

Nematic MBBA was taken in the usual sandwich geometry with the director aligned homogeneously along y and the fluid was made to flow along y by a pressure gradient $p_{,y}$. It was found that there is an enhancement of the threshold voltage for the formations of the Williams domains, the enhancement being greater, the greater the velocity of flow.

To appreciate the origin of this effect let us consider the case of simple shear flow of a nematic between two plane parallel plates. The viscous torque arising from fluid motion is given by the general expression

$$\Gamma_i = e_{ijk} t'_{jk} \tag{13}$$

where the viscous stress tensor

$$t'_{ji} = \mu_1 n_k n_m d_{km} n_i n_j + \mu_2 n_j N_i + \mu_3 n_i N_j + \mu_4 d_{ji}$$
$$+ \mu_5 n_j n_k d_{ki} + \mu_6 n_i n_k d_{kj} \tag{14}$$

$$N_i = \dot{n}_i - w_{ik} n_k \tag{15}$$

$$2d_{ij} = v_{i,j} + v_{j,i} \tag{16}$$

$$2w_{ij} = v_{i,j} - v_{j,i} \tag{17}$$

and μ_1–μ_6 are the coefficients of viscosity (see Section 3.1, Chandrasekhar, 1977). In the present geometry, $\mathbf{v} = (0, v_y, 0)$, $v_{i,j} = v_{y,z}$, and, if the director is confined to the yz plane and makes an angle θ with y, $n_y = \cos\theta$, $n_z = \sin\theta$, and $n_x = 0$. Then, in the steady state

$$\Gamma_x = v_{y,z}[\lambda_1 + \lambda_2 \cos 2\theta]/2 \qquad (18)$$

where $\lambda_1 = \mu_2 - \mu_3$ and $\lambda_2 = \mu_5 - \mu_6$, or making use of Parodi's relation $\mu_2 + \mu_3 = \mu_6 - \mu_5$,

$$\Gamma_x = v_{y,z}(\mu_2 \sin^2\theta - \mu_3 \cos^2\theta) \qquad (19)$$

Γ_x vanishes when the director orientation assumes an equilibrium value θ_0, given by

$$\cos 2\theta_0 = -(\lambda_1/\lambda_2) \qquad \text{or} \qquad \tan^2\theta_0 = \mu_3/\mu_2 \qquad (20)$$

Now both μ_2 and μ_3 are negative in ordinary nematics, and θ_0 is usually a small angle. For example, $\theta_0 \simeq 7°$ for MBBA (see the appendix). This equilibrium orientation θ_0 is attained in practice in relatively thick samples at high flow rates so that the aligning effect of the walls has a negligible influence. In thin samples and at moderate flow rates, the elastic terms cannot be neglected (as we have done in the above discussion) and the velocity and orientation profiles can be evaluated only by numerical techniques taking explicit account of the director orientation at the boundaries.

In certain nematics, e.g., HBAB (*p-n*-hexyloxybenzylidene-*p*-amino-benzonitrile) it is found that $\mu_3 > 0$ at temperatures close to the nematic–smectic A transition point (see the appendix). Under these circumstances the contributions of μ_2 and μ_3 to Γ_x given by Eq. (19) cannot annul each other and there is no equilibrium value of θ_0 as defined by Eq. (20). Thus in the absence of an orienting effect due to the walls or a strong external field, the flow becomes unstable (see Section IV.H).

Returning to the experiment in question, it is clear that the flow produces an additional stabilizing torque that reinforces the normal elastic and dielectric torques. Thus the Williams threshold is increased, and instead of Eq. (6), we now have (de Gennes, 1975)

$$\eta^{-1}\sigma_H\tau E_{\text{th}}^2 = \tau_\psi^{-1} + (q_y\alpha D)^{1/2} \qquad (21)$$

where

$$\alpha = p_{,y}/(\mu_3 + \mu_4 + \mu_6)$$

and

$$D = k_{11} \left/ \left[\frac{2\mu_2^2}{\mu_4 + \mu_5 - \mu_2} + \lambda_1 \right] \right.$$

Fig. 7. Dependence of the Williams threshold
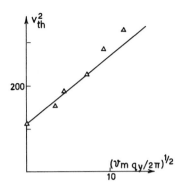

Fig. 7. Dependence of the Williams threshold (V_{th}) on the velocity of flow (v_m): V_{th}^2 versus $(v_m q_y/2\pi)^{1/2}$ for MBBA shows a linear relationship in agreement with the theory (full line). (From Guyon and Pieranski, 1975a.)

the simplifying assumption being that $\theta_0 = 0$. The second term on the right hand side of Eq. (21) adds to the relaxation rate of the director. Taking α as proportional to v_m, the maximum velocity in the midplane of the sample, a plot of E_{th}^2 versus $(v_m q_y)^{1/2}$ should yield a straight line, as was indeed found to be the case experimentally (Fig. 7).

B. Transverse Pressure and Secondary Flow

In the well-known Miesowicz experiment the nematic is oriented by applying a strong magnetic field and the viscosity coefficients are measured in the three principal geometries (i) $\mathbf{n} = (0,1,0)$, parallel to the flow; (ii) $\mathbf{n} = (0, 0, 1)$, parallel to the velocity gradient; and (iii) $\mathbf{n} = (1, 0, 0)$, perpendicular to the flow and to the velocity gradient. The corresponding viscosity coefficients are, respectively (see Section 3.6.1, Chandrasekhar, 1977)

$$\eta_1 = \tfrac{1}{2}(\mu_3 + \mu_4 + \mu_6), \qquad \eta_2 = \tfrac{1}{2}(-\mu_2 + \mu_4 + \mu_5), \qquad \eta_3 = \tfrac{1}{2}\mu_4 \quad (22)$$

We shall now examine the case of oblique orientation of the director. As before, the flow is along y (in a plane Poiseuille geometry) and the velocity gradient along z, but the director is oriented by a strong magnetic field in the xy plane at an angle ϕ with respect to x. (The velocity is assumed to be small enough not to give rise to any distortion in the director orientation.) Under such circumstances there develops a transverse pressure gradient and flow along x.

Let the sample thickness (measured along z) be d, and its lateral width (along x) be l. Strictly speaking, the finite lateral width l of the sample and the appropriate boundary conditions have to be taken into account in treating this problem exactly, but since l is small we shall make the simplifying assumption that there is no secondary flow v_x but that a pressure gradient $p_{,x}$ builds up. We take $\mathbf{n} = (\cos\phi, \sin\phi, 0)$ throughout the sample, and thus, in the absence of director gradients, the torque equation (Eq. 3.3.2, Chandra-sekhar, 1977) may be ignored. We now make use of the equation for momentum transport

$$\rho \dot{v}_i = f_i + t_{ji,j} \tag{23}$$

where ρ is the density, f_i is the body force per unit volume, and t_{ji} is the stress tensor. In the steady state, the x component of Eq. (23) yields

$$g_1 v_{y,z} = p_{,x} z \tag{24}$$

and the y component yields

$$g_2 v_{y,z} = p_{,y} z \tag{25}$$

where

$$g_1 = (\eta_1 - \eta_3) \sin \phi \cos \phi \qquad \text{and} \qquad g_2 = \eta_3 \cos^2 \phi + \eta_1 \sin^2 \phi$$

Imposing the boundary conditions $v_y(\pm \frac{1}{2}d) = 0$, we have from Eq. (25)

$$v_y = p_{,y}(x^2 - \tfrac{1}{4}d^2)/2g_2$$

and from Eq. (24) the transverse pressure gradient

$$p_{,x} = \left(\frac{g_1}{g_2}\right) p_{,y} = \left[\frac{(\eta_1 - \eta_3)\sin \phi \cos \phi}{(\eta_1 \sin^2 \phi + \eta_3 \cos^2 \phi)}\right] p_{,y} \tag{26}$$

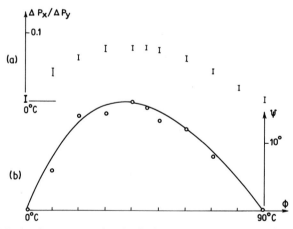

Fig. 8. (a) Ratio of transverse to longitudinal pressure gradient versus director orientation ϕ in plane Poiseuille flow of nematic MBBA. Sample thickness $d = 200 \ \mu m$. Length of cell $L = 4$ cm and lateral width $l = 4$ cm. (b) Deflection ψ of flow lines with respect to the y axis versus ϕ in wide cells ($L/l \sim 1/10$). Full line represents theoretical variation. (From Pieranski and Guyon, 1974b.)

The transverse pressure difference was directly demonstrated by Pieranski and Guyon (1974b) by measuring the liquid level difference in tubes connected to holes facing each other across the width l of the cell. Its dependence on ϕ predicted by Eq. (26) was verified (Fig. 8a). It vanishes for $\phi = 0$ and $\pi/2$. Further, it was confirmed that it changes sign when the flow is reversed or when the field is rotated so that $\phi \to -\phi$.

For wide cells ($l \gg L$, the length of the cell measured along y) it was found by observing the motion of dust particles that the flow lines are not along y (in other words, the assumption that $v_x = 0$ is no longer valid). In the central region of the cell the flow lines were deflected by an angle $\psi(\phi)$ with respect to y, which also shows the expected dependence on ϕ (Fig. 8b).

Though this experiment is related to a particularly simple situation, the basic principle that it illustrates is relevant to the discussions that follow in the next few sections on hydrodynamic instability.

IV. Hydrodynamic Instabilities in Nematics

A. Homogeneous Instability in Shear Flow

We consider first what is called the *homogeneous instability* (HI), which is the hydrodynamic analog of the Frederiks transition (Pieranski and Guyon, 1973, 1974a). Consider simple shear flow between two plane parallel plates (caused by moving the plates parallel to each other at a constant relative velocity v). Let the director be initially along x, the flow along y, and the velocity gradient along z. At low values of v, the sample retains its planar configuration, but above a critical velocity v_c—or a critical shear rate $S_c = v_c/d$, where d is the gap width—the director tilts towards the yz plane, S_c varying inversely as the square of the sample thickness. In the presence of a strong stabilizing magnetic field H_x along x, S_c increases as H_x^2.

The physical interpretation of this distortion is as follows. (Unless otherwise stated we shall assume throughout these discussions that the material is an ordinary nematic like MBBA with $\mu_3 < 0$.) Let us suppose that there is a fluctuation $\theta > 0$ in the xy plane that rotates the director from its initial orientation $(1, 0, 0)$. The flow now exerts a viscous torque, which from Eq. (13) is given by $\Gamma_z = -\mu_2 S\theta$ (Fig. 9). Since $\mu_2 < 0$, $\Gamma_z > 0$ and gives rise to a small twist $\phi > 0$. Now, a deflection ϕ results in a viscous torque $\Gamma_y = \mu_3 S\phi$, and since $\mu_3 < 0$, the sign of Γ_y is such as to increase θ further. Thus the viscous torques have a destabilizing effect, and above a critical shear rate they overcome the elastic and magnetic torques and the system becomes unstable.

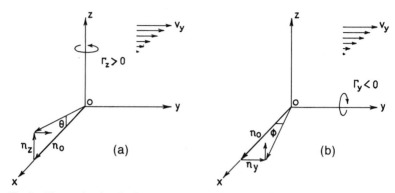

Fig. 9. The mechanism for homogeneous instability: the flow is along y and the velocity gradient along z. (a) An angular fluctuation $\theta = n_z > 0$ results in a viscous torque $\Gamma_z > 0$ such that a small twist $\phi = n_y > 0$ is produced, (b) a deformation $\phi > 0$ results in a torque $\Gamma_y < 0$ such that the initial deformation θ is enhanced.

There is in fact another torque that comes into effect. As we have seen earlier, a deflection of ϕ in the director orientation gives rise to a secondary flow v_x. The secondary velocity gradient $v_{x,z}$ causes a torque Γ_y that makes an additional destabilizing contribution (except close to the boundaries).

The rigorous theory of HI threshold has been developed by Manneville and Dubois-Violette (1976a) and by Leslie (1976a) but for the present purpose it is enough to discuss the approximate treatment given by Pieranski and Guyon (1973) neglecting secondary flow. Taking $\mathbf{n} = [1, \phi(z), \theta(z)]$, $\mathbf{v} = (0, Sz, 0)$, and balancing the viscous torque against the elastic and magnetic torques, one obtains the coupled equations

$$k_{11}(d^2\theta/dz^2) - \chi_a H_x^2 \theta = \mu_3 S\phi \tag{27}$$

$$k_{22}(d^2\phi/dz^2) - \chi_a H_x^2 \phi = \mu_2 S\theta \tag{28}$$

where k_{11} and k_{22} are the elastic constants for splay and twist, respectively; $\chi_a = \chi_{||} - \chi_\perp$ is the diamagnetic anisotropy per unit volume of the nematic. Putting $\theta = \theta_1 \cos(\pi z/d)$ and $\phi = \phi_1 \cos(\pi z/d)$, we get from the condition for the compatibility of Eqs. (27) and (28) the threshold shear rate for $H_x = 0$:

$$S_c = \frac{\pi^2}{d^2}\left(\frac{k_{11}k_{22}}{\mu_2\mu_3}\right)^{1/2} \tag{29}$$

and in the presence of the magnetic field

$$S_c(H_x) = \frac{\chi_a H_x^2}{(\mu_2\mu_3)^{1/2}}\left[\left(\frac{H_{c1}}{H_x}\right)^2 + 1\right]^{1/2}\left[\left(\frac{H_{c2}}{H_x}\right)^2 + 1\right]^{1/2} \tag{30}$$

H_{c1} and H_{c2} are the Frederiks threshold fields for the splay and twist geometries. It is seen that $S_c \propto d^{-2}$, and for large H_x, $S_c \propto H_x^2$; these predictions are in agreement with observations. For a sample thickness d of 200 μm, the zero field critical threshold velocity $v_c = 11.5$ μm/sec for MBBA (Pieranski and Guyon, 1974a).

It is possible to define a dimensionless quantity called the Ericksen number

$$E_r = (Sd^2/4)(\mu_2\mu_3/k_{11}k_{22})^{1/2}$$

which from Eq. (29) is seen to be $\pi^2/4$ at the threshold (de Gennes, 1976). The more rigorous calculation, taking into account secondary flow, yields a critical value of $E_r = 2.309$ for MBBA at the threshold (Manneville and Dubois-Violette, 1976a). It follows from the exact solutions that E_r does not involve all the material parameters and therefore changes from one substance to another.

B. Roll Instability in Shear Flow

We have noted that under steady (dc) shear, the HI threshold increases with increasing strength of the stabilizing magnetic field H_x. However, if the field becomes large enough, the instability does not appear as a homogeneous distortion but as a regular series of bright and dark lines parallel to y, the direction of primary flow (Pieranski and Guyon, 1974a). This is referred to as the *roll instability* (RI). The spacing between the lines is of the order of the sample thickness and the pattern is reminiscent of the Williams domains (see Fig. 1). The lines arise from a cellular motion of the fluid in the xz plane superposed on the primary flow along y. A laser beam produces a diffraction pattern, and as in the case of the Williams domains, the diffraction pattern vanishes when the light is polarized perpendicular to the long axes of the molecules.

Another way of producing this pattern is to apply an ac shear. In this case the RI appears even in the absence of a magnetic field (except, of course, at very low frequencies, in which case it reduces in effect to dc shear and HI is regained). Two distinct regimes can be identified by optical observations. These are referred to as the Y and Z regimes. In the Y regime the y component of the director n_y changes sign at each half period of the ac shear but the z component n_z does not, and vice versa in the Z regime. Nematic MBBA was used in the experiment with a stabilizing magnetic field H_x as well as a high frequency stabilizing electric field E_z (the latter is stabilizing because $\varepsilon_a < 0$ for MBBA). Sinusoidal shears of up to about 2 Hz were applied. The threshold curves for different values of the effective velocity v_{eff} of the upper plate (= total plate displacement per half period) as functions of the

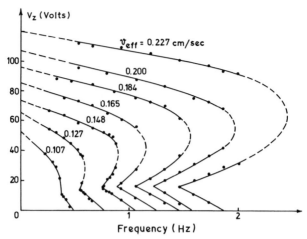

Fig. 10. Experimental roll instability threshold curves for MBBA as functions of the applied voltage and frequency of shear for different values of the effective plate velocity v_{eff}; $d = 240\ \mu\text{m}$, $H_x = 3200$ G. (From Pieranski and Guyon, 1974a.)

frequency of shear and of the applied voltage V_z are shown in Fig. 10. To the left of each curve is the domain of existence of the RI. At larger frequencies the rolls do not develop.

To understand these curves let us examine the curve for a given v_{eff} (Fig. 11). Let us suppose that the frequency of shear is kept constant at 0.8 Hz and the voltage V_z is increased. At $V = 0$, the instability is of the Y type; at V_1 the instability disappears; at $V_2 > V_1$ the instability reappears

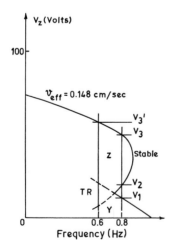

Fig. 11. Roll instability threshold curve for $v_{\text{eff}} = 0.148$ cm/sec from Fig. 10 showing the regimes Y, Z, and TR (see text). (From Pieranski and Guyon, 1974a.)

but is now of the Z type; and finally at $V_3 > V_2$ it disappears again. The spatial periodicity of the rolls remains of the order of the sample thickness, though it is somewhat larger in the branch just above the cusp. At a frequency slightly less than that at the cusp, 0.6 Hz say, the instability is of the Y type at $V = 0$. At much larger voltages it is of the Z type, but for $V = V'_3$ the rolls disappear. The changeover from Y to Z takes place in a transition regime TR that is defined by extrapolating the two branches to lower frequencies. In TR the rolls do not extend over large regions but seem to correspond to an interchange between the Y and Z types along a given roll. The laser diffraction pattern now has satellite spots in the y direction and also shows that the spacing between the rolls has doubled.

The principal mechanism for the onset of RI is shown schematically in Fig. 12. A spatially periodic ϕ fluctuation (or n_y fluctuation) of the form $\cos q_x x$ results in a secondary velocity v_z that is also periodic in x. This effect has been referred to as *hydrodynamic focusing*. Under appropriate conditions, v_z can have a destabilizing effect and generate convective rolls.

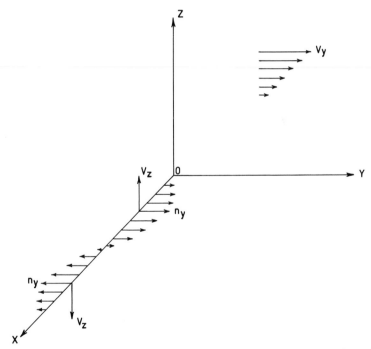

Fig. 12. A fluctuation $\phi = n_y$ that is spatially periodic in x results in a secondary velocity v_z that is also spatially periodic. This can have a destabilizing effect and generate convective rolls.

For a given ϕ there is the torque $\mu_3 S\phi$ due to the primary velocity (as in HI) as well as a torque due to the secondary velocity gradient $v_{z,x}$ and together the total viscous torque Γ_y brings about a distortion θ. In turn, a θ distortion gives rise to a torque Γ_z that increases ϕ. (We have already seen in the case of HI that $\Gamma_z = -\mu_2 S\theta$, but this value is now modified slightly by a torque due to $v_{y,z}$.)

The complete calculation of the threshold for RI involves a solution of the form $\exp i(q_x x + q_z z)$ with boundary conditions for θ, ϕ, v_x, v_y, and v_z. Exact solutions have been obtained by Manneville and Dubois-Violette (1976a) for dc shear. The ac case has been discussed with some approximations by Pieranski and Guyon (1974a; see also Dubois-Violette et al., 1977). We shall present only the salient results of the theory. Assuming $k_{11} = k_{22} = k_{33} = k$, and ignoring material derivatives in Eq. (23), the behavior under ac shear can be expressed in terms of two coupled equations with $n_y (=\phi)$ and $n_z (=\theta)$ as the variables:

$$\dot{n}_y + (n_y/\tau_y) + ASn_z = 0 \tag{31}$$

$$\dot{n}_z + (n_z/\tau_z) + BSn_y = 0 \tag{32}$$

where the relaxation rates are

$$\tau_y^{-1} = (kq^2 + \chi_a H_x^2)/\gamma_y, \qquad \tau_z^{-1} = \left(kq^2 + \chi_a H_x^2 - \frac{\varepsilon_a}{4\pi} E_z^2 \right)\Big/\gamma_z$$

γ_y and γ_z are effective viscosity coefficients, A and B are functions of the viscosity coefficients and of the wave vectors q_x and q_z, and $q^2 = q_x^2 + q_z^2$. The value of τ_z is influenced by the external electric field whereas τ_y is not; $\tau_z > \tau_y$ gives the Y regime and $\tau_z < \tau_y$ the Z regime.

It is seen at once that Eqs. (31) and (32) bear a close similarity to Eqs. (1) and (2) for EHD; n_y, n_z and S correspond to the charge, curvature, and electric field, and the Y and Z regimes correspond to the conduction and dielectric regimes, respectively.

Let the shear rate $S = S_0 \cos \omega t$ ($\omega = 2\pi/T$). At steady state n_y and n_z are periodic functions of time. Suppose that $\tau_z > \tau_y$, $\tau_z > T$, and that n_z is constant in time. Then from Eq. (32)

$$n_z \simeq -\frac{B\tau_z S_0}{T} \int_0^T n_y(t) \cos \omega t \, dt \tag{33}$$

Using Eq. (33) in Eq. (31) and integrating

$$n_y \simeq -\frac{AS_0 n_z[(\cos \omega t)/\tau_y + \omega \sin \omega t]\tau_y^2}{(1 + \omega^2 \tau_y^2)} \tag{34}$$

Compatibility between Eqs. (33) and (34) leads to

$$ABS_c^2\tau_z\tau_y/[2(1 + \omega^2\tau_y^2)] = 1 \qquad (35)$$

which is the threshold condition for the Y regime. On the other hand, with $\tau_y > \tau_z$, $\tau_y > \tau$ one gets in a similar fashion

$$ABS_c^2\tau_y\tau_z/[2(1 + \omega^2\tau_z^2)] = 1 \qquad (36)$$

as the threshold condition for the Z regime. Eqs. (35) and (36) can be cast in a parametric form by defining $x = \omega\tau_y$, $y = (\tau_y/\tau_z)^{1/2}$, $c = ABS_c^2\tau_y^2/2$. Then

$$y^2 = c/(1 + x^2), \qquad x \gg y^2, \quad y > 1 \qquad (37)$$

describes the Y regime, while

$$x^2 + y^4 - cy^2 = 0, \qquad x \gg 1, \quad y < 1 \qquad (38)$$

describes the Z regime. The theoretical diagram for threshold is shown in Fig. 13. These expressions do not, of course, describe the complex behavior in the transition region between the two regimes.

Calculations confirm that for dc shear in the presence of a sufficiently strong stabilizing magnetic field H_x the critical shear for RI is less than that for HI. For fields greater than $\simeq 1$ kG, $S_c^2(\text{RI})/S_c^2(\text{HI}) \simeq 1/3.5$ for MBBA. To understand qualitatively the onset of RI under ac shear even without a

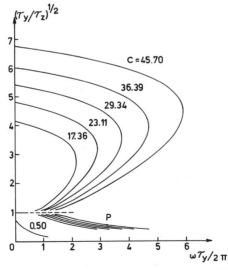

Fig. 13. Calculated roll instability curves fitted to the results of Fig. 11 by adjusting the values at zero frequency, at the cusp and at a point P to define the frequency scale. (From Pieranski and Guyon, 1974a.)

stabilizing magnetic field (except, of course, at very low frequencies) we note that, in writing down Γ_y and Γ_z, the viscous torques arising from $\dot{\theta}$ and $\dot{\phi}$ (the angular velocities of the director) play the same role as the magnetic torques $\chi_a H^2 \theta$ and $\chi_a H^2 \phi$. The former increase with frequency, and thus the field has to be large for very low frequencies, and vice versa, in order that rolls should set in. In practice the frequency has to be less than about 0.3 Hz for HI to occur.

C. The Effect of a Low-Frequency Electric Field on the RI Threshold

We shall next describe an elegant experiment that illustrates in a direct manner the similarity between the rolls produced in shear flow and the Williams domains (Dubois-Violette et al., 1977). In the former case a perturbation n_y, periodic along x, induces a flow $v_z(x)$ along z; in the latter, $n_z(x)$ gives rise to space charge segregation in an electric field, which in turn results in a flow $v_z(x)$. In both cases it is the velocity $v_z(z)$ that is primarily responsible for generating the rolls. Thus if a shear and a low-frequency electric field were applied simultaneously to a material like MBBA, one would expect that the critical shear for RI should diminish with increasing voltage and drop to zero at the Williams threshold. This has been verified experimentally (Fig. 14a).

Shear-induced RI involves both n_y and n_z distortions, while EHD requires only the n_z component. Qualitatively, therefore, the amplitude of n_y may be expected to decrease with increasing voltage, and this is borne out by experiment (Fig. 14b). Other conclusions that can be drawn from this experiment have been discussed in detail by Dubois-Violette et al. (1977).

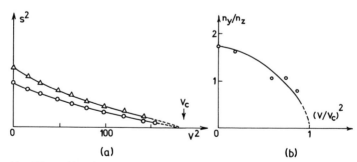

Fig. 14. Effect of low frequency electric field (40 Hz) on roll instability in shear flow of MBBA. Sample thickness, 200 μm. (a) Decrease of threshold shear rate with increase of applied voltage. Frequency of shear 0.5 Hz ($\triangle$) and 1.0 Hz ($\bigcirc$). Magnetic field $H = 1500$ G. V_c is the Williams threshold. (b) Decrease of the amplitude of the n_y fluctuations with increase of applied voltage. Frequency of shear 0.6 Hz, $H = 2500$ G. (From Dubois-Violette et al., *Journal de Mêchanique* **16**, 733–767 (1977), Fig. 13.15. Gauthier-Villars, Publisher, Paris.)

D. Elliptical Shear Flow

Pieranski and Guyon (1977) have investigated the effect of elliptical shear flow in a homeotropically aligned nematic. Relatively high frequency ac shear (a few hundred Hz) was applied simultaneously along two mutually perpendicular directions. At any given frequency the director precesses on the surface of a cone whose axis is normal to the plates. (This observation was made by stroboscopic conoscopy.) If the amplitude of one of the displacements, say U_{x0}, is fixed and the other, U_{y0}, is increased, then at a certain threshold value of U_y periodic rolls appear that give a row of diffraction spots making an angle ψ with the y axis, the value of ψ depending on the ratio U_{y0}/U_{x0}. The threshold is practically independent of the sample thickness. For circular excitation the threshold amplitude ($= U_{x0} = U_{y0}$) decreases as $\omega^{-1/2}$. For a given frequency the threshold curve is roughly hyperbolic. Pieranski and Guyon (1977) give an explanation of these effects by showing that the time average of shear stresses t_{ji} contain certain purely elliptic terms. Consider for simplicity,

$$\mathbf{n} = \left[n_{x0}^a \sin \omega t + n_x^s(r), \, n_{y0}^a \cos \omega t + n_y^s(r), \, 1 \right]$$

and

$$\mathbf{S} = \left[S_{x0}^a \cos \omega t + S_x^s, \, S_{y0}^a \cos \omega t + S_y^s, \, S_z^s \right]$$

as the expressions for the director and shear rate, under the assumption that every quantity can be written as the sum of a time dependent part and a stationary part (superscripts a and s, respectively) and that the time dependent part of n is small. By linearizing the torque equation one can write $n_{x0}^a \approx \mu_2 x_0/\lambda_1 d \approx x_0/d$ and $n_{y0}^a \approx -\mu_2 y_0/\lambda_1 d \approx -y_0/d$, where d is the sample thickness. If one takes a typical term such as $\mu_5 n_x (n_x v_{x,z} + n_y v_{y,z} + n_z v_{z,z})$ of the viscous tensor component t'_{xz} (Eq. (14)) and obtains the time average, it reduces to

$$\langle t'_{xz} \rangle_{\mu 5} = \tfrac{1}{4}(\mu_5 n_x^a S_{y0}^a n_y^s) \approx \tfrac{1}{4}\mu_5(x_0/d)(y_0\omega/d)n_y^s \tag{39}$$

which will not vanish as long as x_0, $y_0 \neq 0$. The proportionality factor for stress and orientation is $x_0 y_0 \omega/d^2$ (while in the case of HI the factor is S multiplied by some viscosity coefficient). Just as in HI the Ericksen number was defined as $\propto Sd^2$, we can define a dimensionless number $N \propto (x_0 y_0 \omega/d^2)d^2$, which attains a certain critical value N_c at threshold. Thus $x_0 y_0 \omega = N_c$ is the threshold condition. This expression qualitatively accounts for all the observations.

A more rigorous approach has been adopted by Dubois-Violette and Rothen (1978) whose calculation accounts for most of the experimental observations and also predicts that the angle made by the rolls with x and y

should be independent of sample thickness and frequency. These points are yet to be checked.

Recently Dreyfus and Guyon (1981) have found that when the amplitude of the shear is increased beyond the threshold, a square grid pattern appears above a second threshold. On increasing S even further, the stationary two-dimensional pattern loses its long range order and becomes nonstationary.

E. Poiseuille Flow

Qualitatively, instabilities in plane Poiseuille flow are similar to those in simple shear flow. Experiments have been conducted for steady and alternating flows, and both HI and RI are found to occur (Guyon and Pieranski, 1975b; Janossy *et al.*, 1976). Some results are presented in Figs. 15 and 16,

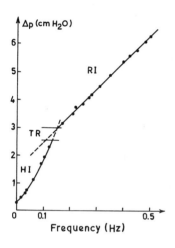

Fig. 15. Experimental threshold curves for instabilities in alternating plane Poiseuille flow of MBBA. Sample thickness, 200 μm and length of cell, 10.8 cm. (a) Threshold pressure difference Δp versus frequency (for $H = 0$) showing the regimes Y, Z and TR. (From Janossy *et al.*, 1976.)

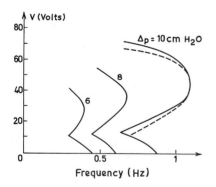

Fig. 16. Threshold curves in plane Poiseuille flow of MBBA as functions of the applied voltage and frequency of ac flow for different values of Δp. The broken curve was obtained for decreasing voltage and indicates a hysteresis effect. Dimensions of cell same as in Fig. 15. (From Janossy *et al.*, 1976.) See Fig. 10 for corresponding curves in shear flow.

and, as can be seen, the trends are practically identical to what has been discussed previously for shear flow. There are, however, some differences that arise from the fact that under a pressure gradient the shear rate varies with z and, in particular, reverses sign at the midplane of the sample. The profiles of the velocity and director fluctuations are modified accordingly (Manneville and Dubois-Violette, 1976b; Manneville, 1979). These differences are summarized diagrammatically in Fig. 17. All the curves refer to MBBA.

F. Couette Flow

Dubois-Violette and Manneville (1978) have given a theoretical description of instabilities in Couette flow with the initial orientation of the director parallel to the axis of the cylinders ($n_r = 0 = n_\psi, n_z = 1$). It turns out that the results are similar to those of shear flow except for additional terms arising out of the coupling with the Coriolis force.

However, if the director is initially oriented along the direction of flow ($n_r = 0 = n_z, n_\psi = 1$) or radially ($n_r = 1$, $n_\psi = 0 = n_z$), a Taylor instability may set in at a threshold much lower than that for an isotropic liquid (Pieranski and Guyon, 1975). In the case of an isotropic liquid, the steady state flow is a function of r alone and has only a ψ component. If perturbations v_r and v_ψ are imposed on this flow such that (in a one-dimensional model) they depend sinusoidally on z, one gets a positive feedback instability mechanism whereby v_ψ produces an additional centrifugal force F_r, which gives rise to a fluctuation v_r, which in turn produces an additional Coriolis force F_r and further enhances v_ψ. For a nematic ($\mu_3 < 0$) originally aligned homogenously along the flow one finds by simple calculation that in addition to the usual isotropic term F_r, which arises from v_ψ, one gets a contribution $F'_r \approx \eta n_{z,z}/r^2$ due to a fluctuation n_z. At the same time, from the torque equation one finds that v_ψ can give rise to a fluctuation $n_z \approx \mu_3 v_\psi/q_z k$. The comparatively small value of k is responsible for bringing down the Taylor instability threshold to low values. If one starts with homeotropic alignment ($n_r = 1$, $n_\psi = 0 = n_z$), one is led to a similar conclusion except that the instability will occur at higher shear rates when flow alignment has taken place (i.e., when $n_r = \sin\theta_0$, $n_\psi = \cos\theta_0$, $n_z = 0$ over almost the entire gap).

G. Other Studies

Leslie (1976b) has considered the effect of a magnetic field H_x on the shear flow of a nematic with the director initially aligned in the yz plane at an angle θ_0 (Eq. (20)) with y the direction of flow. A twist fluctuation that rotates the director away from the yz plane gives rise to a transverse flow

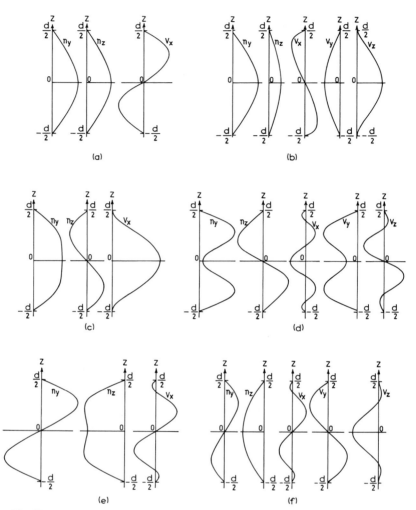

Fig. 17. Profiles of fluctuations of director orientation and of velocity in homogeneous instability (HI) and roll instability (RI). Calculations are for MBBA. Throughout v_y represents a fluctuation superposed on the steady flow along y. (a) HI in dc shear flow; n_y and n_z are symmetric with respect to the midplane and v_x is antisymmetric. $H_x = 0$. (From Manneville and Dubois-Violette, 1976a.) (b) RI in dc shear flow; n_y, n_z, v_y, v_z are symmetric and v_x is antisymmetric. $H_x = 1400$ G. (From Manneville and Dubois-Violette, 1976a.) For HI and RI in shear flow there is another mode that is theoretically possible, but this mode involves antisymmetric n_y and n_z and is energetically not favorable. Stabilizing fields enhance the RI threshold and also the critical wave vector at the threshold. (c) HI in dc Poiseuille flow. Twist mode: n_y and v_x are symmetric but n_z is antisymmetric. $H_x = 0$. (From Manneville and Dubois-Violette, 1976b.) This mode, which involves a net secondary flow, is observed in the field-free case and also with a stabilizing field H_x. In ac Poiseuille flow this mode is observed for very low frequencies just above the threshold (Guyon and Pieranski, 1975b). (d) RI in dc Poiseuille flow. Twist mode:

v_x. It is readily verified that the viscous torque arising from the secondary velocity gradient $v_{x,z}$ tends to reduce ϕ, i.e., it has a stabilizing effect. Thus the threshold magnetic field for the Frederiks deformation is enhanced because of the shear S, and can now be written in the form $H_{xc} = (A + BS)^{1/2}$, where A and B are constants depending on the material parameters. For zero shear one recovers the normal Frederiks threshold field.

The stability of planar solutions of the type $\mathbf{n} = (0, \cos\theta, \sin\theta)$ against homogeneous perturbations in and perpendicular to the plane of shear has been investigated by Currie (1978; Currie and Macsithigh, 1979). Some consequences of weak anchoring of the director at the boundaries, in particular the possibility of a new type of instability, have been discussed by Kleman and Pikin (1979). Approximate nonlinear analyses have been carried out on the threshold of HI by Chaure (1980) and on the distortion above the threshold by Manneville (1978).

Instabilities have been observed even at ultrasonic frequencies (Kessler and Sawyer, 1970; Mailer *et al.*, 1971; Bertolotti *et al.*, 1978; Scudieri *et al.*, 1976, 1978). Scudieri *et al.* (1976, 1978) have shown very clearly that above a threshold intensity of the ultrasound the instability appears as convective rolls, precisely of the type illustrated schematically in Fig. 1. The mechanism of this acoustohydrodynamic instability has not yet been discussed theoretically.

H. Nematics with $\mu_3 > 0$

In a nematic that undergoes a transformation to the smectic A phase at a lower temperature, μ_3 changes sign from negative to positive as the temperature approaches the nematic–smectic A transition point. This reversal of sign is related to the formation of smectic-like clusters in the nematic prior to the transition to the smectic phase. As remarked in Section III.A, when

---

Fig. 17. (*continued*)

n_y, v_x, and v_y are symmetric but n_z and v_z are antisymmetric. $H_x = 0$. (From Manneville, 1979.) In ac Poiseuille flow this mode is observed just above the threshold (Guyon and Pieranski, 1975b). (e) HI in dc Poiseuille flow. Splay mode: n_z is symmetric but n_y and v_x are antisymmetric. $H_x = 0$. (From Manneville and Dubois-Violette, 1976b.) This mode involves no net secondary flow and has a higher threshold than the twist mode for the field-free case as also in a stabilizing field. H_x. However it has been shown theoretically (Kini, 1978) that in a destabilizing field H_z the splay mode can become more favorable than the twist mode in the vicinity of the Frederiks transition. In ac Poiseuille flow this mode is favorable at very low frequency well above the threshold (Guyon and Pieranski, 1975b). (f) RI in dc Poiseuille flow. Splay mode: n_z and v_z are symmetric but n_y, v_x, and v_y are antisymmetric. $H_x = 0$. (From Manneville, 1979.) In ac Poiseuille flow this mode is found when the shear rate is increased well above the threshold for the RI twist mode (Guyon and Pieranski, 1975b).

$\mu_3 > 0$ there is no equilibrium flow alignment angle θ_0 defined by Eq. (20), and consequently such materials have two distinct flow regimes, as was discovered by Gahwiller (1972) and later confirmed by Pieranski and Guyon (1974c, 1976; see also Gasparoux *et al.*, 1975). Two compounds that have been very thoroughly studied are HBAB and CBOOA (*p*-cyanobenzylidene-*p'*-octyloxyaniline).

Pieranski and Guyon (1974c, 1976; Pieranski *et al.*, 1976) conducted shear flow experiments on these two compounds and observed that when $\mu_3 > 0$ there is a cascade of instabilities. Starting with the director along the flow (y axis), the first instability involves a large deformation ($-\theta$) in the yz plane. On increasing S there occurs a second instability that gives rise to a twist that rotates the director out of the yz plane. The twist angle increases with increasing S till finally, when the director is aligned along x, RI sets in with the roll axis along y.

To understand this sequence, let us start with an orientation $n_x = n_z = 0$, $n_y = 1$. On applying a (positive) shear $v_{y,z}$, the torque $\Gamma_x = -v_{z,y}\mu_3 < 0$ causes a negative value of θ to appear. If now there is a fluctuation in θ (or n_z), the viscous torque due to primary flow $\Gamma_x = 2\lambda_2 \sin\theta(\cos\theta)v_{y,z}n_z < 0$ enhances the distortion. There is also a small contribution due to backflow but we ignore it in this qualitative discussion. Beyond a critical shear rate S_{c1}, Γ_x overcomes the elastic torque and the fluctuation n_z grows to give a large negative deformation θ. Approximate analysis shows (Pieranski *et al.*, 1976) that $S_{c1} = 4.8\, k_{11}/(\mu_2\mu_3)^{1/2}d^2$. Now if perturbations n_x and v_x are imposed on the sample having a large negative θ, there arise torques Γ_z and Γ_y. For instance, the torque $\Gamma_z = -\mu_2(\sin\theta)n_x v_{y,z} < 0$ due to the primary flow is destabilizing. The other torques due to primary and secondary flows can be worked out similarly. At $S_{c2} = 9.5(k_{11}k_{22}/\mu_2\mu_3)^{1/2}d^2$ a second instability occurs. On increasing S further, the twist angle increases until the director is aligned along x in the bulk of the sample. From the arguments put forward in Section IV.A it will be clear that when $\mu_3 > 0$ and the director is initially oriented along x, the system is stable against homogeneous perturbations, but as long as μ_3 is not too large, RI can set in with the roll axis along y.

In the case of CBOOA, μ_3 becomes very large and positive close to the nematic–smectic A point, and under such circumstances even RI cannot be observed because of the strong stabilizing role of μ_3. However, weakly damped oscillations of the director around the equilibrium position are seen (Pieranski and Guyon, 1976). This can be explained by using Eqs. (31) and (32) for n_y and n_z, which can be combined to yield

$$\ddot{n}_z + \dot{n}_z(1/\tau_y + 1/\tau_z) + n_z((1/\tau_y\tau_z) - ABS^2) = 0 \qquad (40)$$

The condition for oscillations, viz., $(\tau_y^{-1} - \tau_z^{-1})^2 + 4ABS^2 < 0$, is found to be satisfied if $\mu_3/\mu_2 > (\mu_5 + \mu_2)/2\eta_2$. For large μ_3/μ_2, the ABS^2 term dominates so that the oscillation frequency varies linearly with shear rate.

Experiments on the Couette flow of HBAB and CBOOA with the initial alignment $n_r = 1$, $n_\psi = n_z = 0$ were reported earlier by Cladis and Torza (1975). They observed that at a critical shear rate S_{c1} the director tumbles in the $r\psi$ plane, and when $S > S_{c1}$ there is stationary flow with the director in this new configuration. Optical observations revealed two "tumble lines" parallel to the cylinder axis indicating a change in the director orientation by π and 2π. The possibility of the director tumbling for $\mu_3 > 0$ had been theoretically anticipated by de Gennes (1972). However, on further increase of S, additional tumble lines did not appear—rather the two lines moved apart. There was then a tilt of the director out of the plane of shear giving rise to a nonzero n_z component and, at a higher threshold S_{c2}, a secondary cellular flow developed similar to the Taylor vortex flow in isotropic liquids. In the case of CBOOA, S_{c2} increased rapidly as the temperature was lowered to the nematic–smectic A transition point and diverged as μ_3 became large and positive. Cladis and Torza have discussed these observations in terms of the elastic and viscous torques.

Dubois-Violette and Manneville (1978) have theoretically discussed Couette flow instabilities when the director is initially parallel to the axis z of the cylinders. Apart from the additional terms arising out of a coupling between inertial forces and the director orientation, they obtain results qualitatively similar to those of shear flow.

A recent calculation (Kini, 1980a) on the shear flow normal to the helical axis of twisted HBAB (the twist being assumed to be produced by the addition of chiralic impurities) suggests that the threshold of RI may be enhanced due to the presence of the twist. This is because the shear must first act against the stabilizing effect of the intrinsic twist in order to align the director along x and it is only then that RI can come into existence. Another calculation on the action of destabilizing magnetic fields (H_y and H_z) on the RI of HBAB (Kini, 1980b) indicates that as the field strength increases, the threshold and wave vector of deformation at threshold decrease. It is also shown that by using the stabilizing influence of shear it is possible to excite HI in HBAB by the application of destabilizing fields larger than the Frederiks threshold.

Cohen et al. (1977) have studied the instability in shear flow of CBOOA with the director initially aligned in the yz plane at a positive angle θ with y and under the influence of a high frequency electric field E_z along z. (The positive θ is produced by first applying a negative shear on a homogeneously aligned sample and then applying E_z.) When θ is positive the electric field plays a stabilizing role. On increasing the shear rate θ can become negative in the middle of the sample, and in these circumstances both the electric field and shear have a destablizing influence. It is interesting to note here that, to the extent that the elastic forces oppose the electric field in producing a positive θ initially, they actually play a destabilizing role.

V. Thermal Instabilities in Nematics

A. Thermal Convection in Isotropic Liquids

We shall first give a brief outline of the classical Benard–Rayleigh problem of thermal convection in isotropic liquids since it illustrates many of the features concerning thermal instabilities in nematics. When a horizontal layer of isotropic liquid bounded between two plane parallel plates spaced d apart is heated from below a steady convective flow is observed when the temperature difference between the plates exceeds a critical value ΔT_c. The flow has a stationary cellular character with a spatial periodicity of about $2d$. The mechanism for the onset of convection may be looked upon as follows. A fluctuation T' in temperature creates warmer and cooler regions, the former tending to move upward and the latter downward. When $\Delta T < \Delta T_c$ the production of the internal energy by v_z is dissipated by viscous effects and heat losses due to conductivity and the fluctuations die out in time. At the threshold the energy losses are balanced exactly, and beyond the threshold the instability develops. Assuming a one-dimensional model in which T' and v_z vary as $\exp(iq_y y)$ with $q_y \simeq \pi/d$, and linearizing the heat conduction and Navier–Stokes equations, one gets under the Boussinesq approximation

$$\dot{T}'_0 + (T'_0/\tau_T) - \beta v_{z0} = 0 \tag{41}$$

$$\dot{v}_{z0} + (v_{z0}/\tau_v) - \alpha g T'_0 = 0 \tag{42}$$

where T'_0 and v_{z0} are the amplitudes of the temperature and velocity fluctuations, $\tau_T = (kq_y^2)^{-1}$ is the relaxation time for the temperature fluctuation in the absence of flow, $k = K/\rho C$ the thermal diffusivity, K the thermal conductivity, ρ the density, C the specific heat, $-\beta$ the temperature gradient $(T_u - T_l)/d$ (u and l standing for the upper and lower plates), α the coefficient of expansion, g the acceleration due to gravity, $\tau_v = (vq_y^2)^{-1}$ the time constant for velocity diffusion in the absence of T', $v = \eta/\rho$ the kinematic viscosity, and η the viscosity coefficient. Ignoring time dependence, the compatibility of Eqs. (41) and (42) yields at the threshold

$$\frac{\tau_v \tau_T}{(1/\alpha g \beta_c)} = \frac{\tau_v \tau_T}{\tau^2} = 1 \tag{43}$$

or

$$R_c = \alpha g d^4 \beta_c / k v = \pi^4 \tag{44}$$

where R is a dimensionless quantity called the Rayleigh number and $\tau = (\alpha g \beta)^{-1/2}$ may be regarded as the time associated with the destabilized motion of a hot spot from the bottom to the top plate. Rigorous calculation shows that π^4 of Eq. (44) should be replaced by 120 for stress-free insulating boundaries and 1704 for rigid conducting boundaries. Naively one may say that convection takes place when the diffusion times τ_v and τ_T are of the same order as τ.

Equation (44) gives a dependence of d^{-3} for ΔT_c. For a liquid having material parameters comparable to the average value for MBBA ($k \sim 10^{-3}$, $\eta \sim 1$, $\alpha \sim 10^{-3}$ cgs) the corrected form of Eq. (44) gives $\Delta T_c \approx 2°C$ for $d = 1$ cm and $\Delta T_c \approx 2 \times 10^{3}°C$ for $d = 1$ mm.

B. Stationary Convection in Nematics

The situation is profoundly modified in the case of a nematic because of its anisotropic transport properties. Dubois-Violette (1971) was the first to give an approximate theoretical treatment of thermal convection in a planar (homogeneously aligned) nematic and to show by considerations of torques that such a system will be unstable against cellular flow when the film is heated from below if $K_a > 0$, or when it is heated from above if $K_a < 0$, where $K_a = K_{\parallel} - K_{\perp}$ is the anisotropy of thermal conductivity. [It may be mentioned here that $K_a > 0$ for all known nematics (see, e.g., Hervet et al., 1980).] Dubois-Violette also showed that the critical temperature gradient β_c should be much less than that for an isotropic liquid. Experimental observations of convective rolls in a homogeneously aligned film of MBBA heated from below were reported by Guyon and Pieranski (1972) and later by Dubois-Violette et al. (1973) who also gave a somewhat more detailed theoretical analysis. As expected, the threshold value of β_c was about 10^{-3} times that for an isotropic liquid of comparable average physical properties. Optical observations confirmed that the rolls are essentially as depicted in Fig. 1 for the Williams domains. When the bottom plate was heated to a temperature greater than the nematic–isotropic transition point the convection disappeared, showing that it is the anisotropy that is responsible for the low threshold.

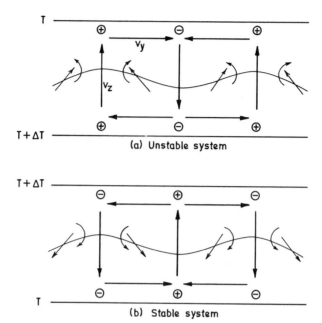

Fig. 18. (a) Thermal instability in a nematic heated from below. Initial orientation of the director is horizontal (along y). A bend fluctuation causes "heat focusing" because of the anisotropy of thermal conductivity ($K_{\parallel} > K_{\perp}$) and gives rise to warmer ($+$) and colder ($-$) regions. The warmer regions move up and the colder regions down due to buoyancy effects and this in turn results in a velocity v_y (the fluid being assumed to be incompressible). The transverse velocity gradient $v_{z,y}$ induces a major destabilizing viscous torque, while the vertical gradient $v_{y,z}$ induces only a very weak stabilizing torque. The resultant torques, shown by the curved arrows, are destabilizing. Here the long straight arrows represent the translational velocities v_z and v_y, the short straight arrows the heat fluxes. (b) The same geometry as in (a) but with the top plate at a higher temperature. The system is stable.

Figures 18 and 19 illustrate the destabilizing mechanisms involved. Because of the anisotropy of thermal conductivity, a thermal fluctuation of the director along y creates warmer ($+$) and cooler ($-$) regions. This is referred to as *heat focusing*. Because of the buoyancy effect, these regions move up and down creating a velocity fluctuation v_z, which in turn gives rise to a torque Γ_x that stabilizes or destabilizes the orientation depending on the sign of K_a. We seek solutions of the form $\mathbf{n} = [0, 1, \theta(y, t)]$, $\mathbf{v} = [0, 0, v_z(y, t)]$, $T = -\beta z + T'(y, t)$, and obtain (using Eqs. 3.1.6–3.1.9 of Chandrasekhar, 1977) for a variation $\exp(iq_y y)$ of the fluctuation (Dubois-Violette *et al.*, 1978)

$$\exp(iq_y y)[\dot{v}_{z0} + (v_{z0}/\tau_v) - \alpha g T_0'] - (\mu_2/\rho)\dot{\psi} = 0 \qquad (45)$$

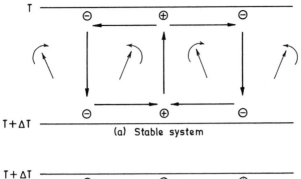

(a) Stable system

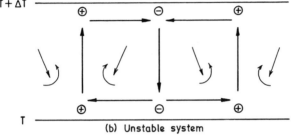

(b) Unstable system

Fig. 19. (a) The initial orientation of the director is along z. A splay fluctuation gives rise to warmer and colder regions. In this case, $v_{y,z}$ causes a dominant stabilizing viscous torque while $v_{z,y}$ causes only a very weak destabilizing torque. The system is therefore stable against stationary convection (see, however, Section V.C). (b) The same geometry as in (a) but with the top plate at a higher temperature. The system is unstable (see legend of Fig. 18a).

$$\exp(iq_y y)[\dot{T}'_0 + (T'_0/\tau_T) - \beta v_{z0}] + K_a\beta(\partial\psi/\partial y) = 0 \tag{46}$$

$$\exp(iq_y y)[\dot{\theta}_1 + (\theta_1/\tau_\psi)] - (\mu_2/\lambda_1)(\partial v_z/\partial y) = 0 \tag{47}$$

where $\psi = \partial\theta/\partial y$ is the director curvature and θ_1, v_{z0}, T'_0 are amplitudes of fluctuations. In Eq. (47) $\tau_\psi = -\lambda_1/k_{33}q_y^2$ is the relaxation time for the director in the absence of any coupling. Ignoring $\dot{v}_{z0}$ in Eq. (45) and eliminating v_{z0} in Eqs. (46) and (47) one finds the condition for threshold to be

$$(\tau_v\tau_T/\tau^2)[1 + (\tau_\psi/\tau_a)] = 1$$

where $\tau_a = (K_a q_y^2 \lambda_1/\mu_2)^{-1}$. In an isotropic liquid $K_a = 0$, $\tau_a = \infty$, and we recover Eq. (44). For a nematic film of thickness $d \sim 1$ mm, $\tau_\psi \sim 10^3$ sec, $\tau_a \sim 1$ sec, and ΔT_c reduces to 10^{-3} times the isotropic value. It is the large ratio τ_ψ/τ_a that leads to a very low threshold. A magnetic field H_x (stabilizing) or H_z (destabilizing) can be used to decrease or increase τ_ψ, or, equivalently, increase or decrease the threshold temperature difference ΔT_c. Dubois-Violette et al. (1973) verified a linear relationship between ΔT_c and H_x^2 (or H_z^2).

Pieranski *et al.* (1973) then showed that when a homeotropically aligned film of MBBA is heated from above, the orientation becomes unstable above a critical value of ΔT_c that is of the same order as that for the planar orientation. They also verified that the film is stable when heated from below. A model of the destabilizing mechanism is no longer purely one-dimensional involving only v_z (Fig. 19b). A thermal fluctuation causes a fluctuation v_z as before, but because of the incompressibility of the fluid this in turn causes a velocity fluctuation v_y that contributes the major destabilizing torque $\Gamma_x \sim \mu_2 v_{y,z}$ for $K_a > 0$. Again with a stabilizing magnetic field H_z there is a linear relationship between ΔT_c and H_z^2. A field H_y favors rolls with axes normal to y, but in the field-free case the rolls degenerate into a square pattern that may be thought of as a linear superposition of crossed convection rolls. When ΔT is increased well beyond ΔT_c a complex hexagonal structure is found with a nematic–isotropic interface if the temperature of the upper plate is large enough.

An exact solution of the threshold in the above problems involves θ, T', v_y, and v_z fluctuations that are supposed to depend on y and z in the form $g(z)\exp(iq_y y)$ and that satisfy the boundary conditions of vanishing at $z = \pm d/2$. Equations 3.1.6–3.1.9 of Chandrasekhar (1977) reduce to four linear differential equations, which lead to a compatibility condition for the boundary conditions to be satisfied. This condition yields a value for the threshold T for any given value of q_y. At a particular value q_{yc} of q_y, T has a minimum value T_c that is taken as the true threshold. Exact numerical and semianalytical calculations have been performed by Dubois-Violette (1974) and by Barratt and Sloan (1976).

The studies described above were the most important ones that established the fundamental principles of thermal instability in nematics. Other geometries have since been investigated both theoretically and experimentally: (i) the director initially in the yz plane at an angle θ with y, and the temperature gradient along z (Guyon *et al.*, 1978a; Barratt and Bramley, 1980); (ii) the sample tilted with respect to the gravitational field (Horn *et al.*, 1976; Fitzjarrald and Owen, 1979); (iii) a twisted nematic heated from below (Barratt and Sloan, 1978); and (iv) the sample subjected to a temperature gradient in the annular space between two coaxial cylinders executing a solid body rotation about the common axis (Carrigan and Guyon, 1975).

Probably the earliest observation of the effect of a temperature gradient on a nematic was by Stewart (1936), who found by x-ray diffraction that the molecular alignment in PAA is vertical when heated from below and horizontal when heated from above. Currie (1973) has shown that this observation can be explained from stability considerations taking $K_a > 0$.

Attempts have been made recently to understand thermal instability thresholds, taking into account nonlinear fluctuations (Dubois-Violette and

Rothen, 1979, 1980). An interesting consequence of nonlinearity is that the induced flow fluctuation can reduce the thermal gradient. The director fluctuation that is the slowest relaxing variable controls the critical slowing down of the fluctuations close to the threshold. Calculations have been made for the free–free model.

Miyakawa (1975) has shown theoretically that in a homeotropic nematic, which is in a temperature gradient directed downward, the power spectrum of the fluctuations breaks up into three Lorentzians, one of which corresponds to the nonpropagating soft mode whose frequency goes to zero at the instability threshold.

Recent observations by Urbach *et al.* (1977) indicate that surface tension gradients may give rise to an instability in nematic droplets. These authors also give an estimate of the threshold. More complete calculations have since been presented by Guyon and Velarde (1978) and by Velarde and Zuniga (1979).

C. Oscillatory Convective Instability in Nematics

Lekkerkerker (1977) predicted that a homeotropic nematic heated from below (which, it will be recalled, is stable against stationary convection) should become unstable with respect to oscillatory convection. The essential idea is that if there are oscillatory perturbations in director, velocity, and temperature with a frequency ω such that $\tau_\psi^{-1} \ll \omega \ll \tau_T^{-1}$, where τ_ψ and τ_T are the director and thermal relaxation times, it follows from Eqs. (45)–(47) that the director and velocity perturbations will be out of phase by 90°, whereas the velocity and temperature fluctuations will be in phase. Thus the stabilizing effect of heat focusing, which prevents stationary convection, is practically eliminated and the possibility of an oscillatory instability (or overstability) exists. Detailed calculations by Lekkerkerker (1977) show that such an instability will be favorable only for thick samples.

The phenomenon was demonstrated experimentally by Guyon *et al.* (1978b, 1979). A thick sample (~ 5 mm) was homeotropically aligned by the combined effect of surface treatment and a magnetic field. Thin chromel–alumel wires (diameter ~ 80 μm) were stretched across the midplane of the sample and were connected as differential thermocouples. A short (~ 30 sec) heat pulse was applied by means of a resistive wire parallel to the thermocouples. When the pulse was switched off, thermocouples detected oscillations, characteristic of overstability (Fig. 20). Guyon *et al.* also studied experimentally the effect of a stabilizing magnetic field and presented a simple one-dimensional model that explained all the observed effects. A more general thermodynamic analysis of these effects has been developed by Lekkerkerker (1979) and by Velarde and Zuniga (1979).

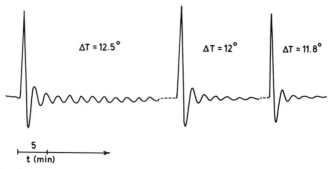

Fig. 20. Response of a thermocouple to a short (30 sec) heat pulse applied from below in a homeotropically aligned sample of MBBA. The decay rate of the oscillation decreases as the threshold $\Delta T_c \sim 12.6°C$ is approached. Magnetic field $= 465$ G. (From Guyon *et al.*, 1979, with permission of Cambridge University Press.)

VI. Instabilities in Cholesteric Liquid Crystals

A. Flow Properties

The flow properties of a cholesteric liquid crystal are not as well understood as those of a nematic. Its behavior is highly nonNewtonian, the apparent viscosity (η_{app}) increasing by about a million times as the shear rate drops to a very low value (Porter *et al.*, 1966). Because of experimental difficulties, most of the earlier rheological studies were carried out without special regard to the alignment of the sample and the data are therefore difficult to interpret. Of late, however, some measure of understanding has been achieved for two special geometries, viz., flow normal to and along the helical axis.

Leslie (1969) has discussed the basic theory of flow normal to the helical axis and Kini (1979, 1980c) has presented detailed calculations of η_{app}, orientation, and velocity profiles as functions of pitch and sample thickness. The results of Kini are in agreement with trends observed by Candau *et al.* (1973) and Bhattacharya *et al.* (1978) on samples aligned with the helix normal to the bounding surfaces. The important fact that emerges from these studies is that η_{app} is of the same order of magnitude as for a nematic even at the lowest shear rates.

The very high η_{app} observed by Porter *et al.* (1966) at low shear rates may therefore be attributed to flow along the helical axis. Helfrich (1969b) proposed a simple physical mechanism to account for such high η_{app}. At very low shears the structure remains undistorted but the director executes rapid rotational motion about the helical axis and dissipates the energy gained by the translational motion of the fluid. A simple calculation shows that $\eta_{app} \cong 10^6 \lambda_1$. A somewhat more detailed theory based on the Ericksen–Leslie equations was given by Kini *et al.* (1975; see Chandrasekhar, 1977, Section 4.5.1) that confirms the Helfrich model. Flow along the helical axis is often referred to as *permeation*, though it should be emphasized that permeation is not a radically new mechanism but falls conceptually within the framework of the continuum theory.

Another special case that has been studied is a planar cholesteric sample sandwiched between two disks, one of which is rotated with respect to the other (Janossy, 1980a,b, 1981). Janossy observed that for one sense of rotation, dust particles drifted toward the center of the disks, while for the opposite sense of rotation they drifted away from the center. Chandrasekhar *et al.* (1980) have shown that this effect is due to a transverse (radial) flow under the action of a torsional shear about the helical axis.

However, the general flow behavior of a cholesteric appears to be much more complex. Pochan and coworkers (see Pochan, 1979) have examined the textural changes brought about by flow and have inferred from optical observations that the helical axis itself gets distorted and even rotated under the action of shear. There is no quantitative explanation of these effects.

It is not surprising therefore that comparatively little has been done on instabilities in cholesterics. Only field-induced instability has been studied in any detail. Some theoretical work has been done on thermal instabilities.

B. *Field-Induced Instability: the Square Grid Pattern*

When a magnetic field H_z is applied along the helical axis of a planar cholesteric film composed of molecules of positive diamagnetic anisotropy $\chi_a > 0$, and the boundary constraints are such as to maintain the orientation of the helix, the structure becomes distorted, as illustrated schematically in Fig. 21. The layers become corrugated along two orthogonal directions so that there results what is called a *square grid pattern*. If the field is increased to a much higher value, a possible deformation is for the director to tilt toward the field over the entire sample, i.e., a conical distortion (Meyer, 1968; Leslie, 1970).

The theory of the square grid pattern was proposed by Helfrich (1970, 1971) and subsequently elaborated by Hurault (1973). (For a full discussion of the theory see Chandrasekhar, 1977, Sections 4.6.2–3). If $q_0 = 2\pi/P_0$, P_0 being the pitch of the cholesteric, q_y is the wave vector of the distortion, and $q_z = \pi/d$, we have in practice $q_z \ll q_y \ll q_0$. Under these conditions, the

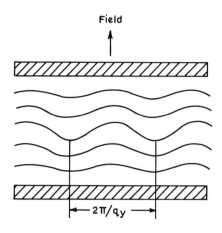

Fig. 21. Deformation of a planar structure by a magnetic or electric field acting along the helical axis of a cholesteric liquid crystal. A similar deformation in an orthogonal direction results in a square grid pattern.

threshold field is given by

$$H_{th}^2 = \chi_a^{-1}(6k_{22}k_{33})^{1/2}q_0q_z \tag{48}$$

or

$$H_{th} \propto (P_0 d)^{-1/2} \tag{49}$$

and the wavevector of the distortion

$$q_y^4 = (8k_{22}/3k_{33})q_0^2 q_z^2 \tag{50}$$

or

$$q_y \propto (P_0 d)^{-1/2} \tag{51}$$

An analogous situation arises with an electric field except that now conduction must be taken into account. The Carr–Helfrich mechanism comes into play but with some modifications. First, the twist and bend distortions are now coupled; second, fluid motion along z can take place only by the process of "permeation," which, as we have seen, is associated with a very high apparent viscosity, and therefore in practical situations material flow and the consequent viscous torques play a negligible part. The distortion wavevector q_y is the same as in the magnetic case (Eq. (50)), while the threshold for dc field is given by

$$E_{th}^2 = \frac{4\pi}{\varepsilon_\perp}\frac{\sigma_\parallel + \sigma_\perp}{\sigma_a}(\tfrac{3}{2}k_{22}k_{33})^{1/2}q_0q_z \tag{52}$$

or

$$E_{th} \propto (P_0 d)^{1/2} \tag{53}$$

where $\sigma_{||}$ and $\sigma_{\perp}$ are the conductivities parallel and perpendicular to the local director axis, and $\varepsilon_{||}$ and $\varepsilon_{\perp}$ are similarly defined. For ac field in the conduction regime

$$\overline{E_{\mathrm{th}}^2} = \frac{8\pi^3}{\varepsilon_{\perp}} \frac{\varepsilon_{||} + \varepsilon_{\perp}}{\varepsilon_{\mathrm{a}}} \frac{1 + \omega^2\tau^2}{1 - \zeta + \omega^2\tau^2} (\tfrac{3}{2}k_{22}k_{33})^{1/2}(P_0 d)^{-1} \tag{54}$$

where

$$1 - \zeta = \frac{\sigma_{\mathrm{a}}}{(\sigma_{||} + \sigma_{\perp})} \frac{(\varepsilon_{||} + \varepsilon_{\perp})}{\varepsilon_{\mathrm{a}}} \tag{55}$$

and τ is the dielectric relaxation time given by

$$\frac{1}{\tau} = 4\pi \frac{\sigma_{||} + \sigma_{\perp}}{\varepsilon_{||} + \varepsilon_{\perp}} \tag{56}$$

For negative dielectric anisotropy ($\varepsilon_{\mathrm{a}} < 0$) the conduction regime occurs when $\omega < \omega_{\mathrm{c}}$, where

$$\omega_{\mathrm{c}} = 4\pi(\zeta - 1)^{1/2} \frac{\sigma_{||} + \sigma_{\perp}}{\varepsilon_{||} + \varepsilon_{\perp}} \tag{57}$$

All these results are in quantitative agreement with observations (Gerritsma and Van Zanten, 1971; Rondelez and Hulin, 1972; Rondelez et al., 1972; Arnould-Netillard and Rondelez, 1974). In particular, the dependence of the spatial periodicity of the pattern and of the threshold field on d and P_0 have been verified for both magnetic and electric fields. Also for $\varepsilon_{\mathrm{a}} < 0$ the electric field threshold has the frequency dependence in the conduction regime as described by Eq. (54) and above a critical frequency ω_{c} there is a transition to the dielectric regime. The behavior in the dielectric regime does not appear to be fully explained. As long as ε_{a} is not too large, an increase of the field well above the threshold value in the conduction regime leads to dynamic scattering.

   Studies have been made using wedge-shaped samples that form the Grandjean–Cano lines; the threshold voltage shows an abrupt change on crossing a line and this has been interpreted in terms of the Helfrich–Hurault theory (Belyaev and Blinov, 1976; De Zwart, 1978; De Zwart and Van Doorn, 1979; Eber and Janossy, 1979). For cholesterics of large negative dielectric anisotropy it has been noticed that at higher voltages there is a sudden increase in the effective number of pitches in the sample at definite values of the applied voltage, and in consequence there is an alternation between planar texture and square grid pattern with increasing voltage. Mechanisms have been proposed for this change of pitch via discontinuities (de Zwart and Lathouwers, 1975; de Zwart, 1978; de Zwart and Van Doorn, 1979; Kohno et al., 1978). There is evidence that the isotropic mechanism for EHD instability (see Section II.B) may operate in cholesterics also (Belyaev, 1978).

The square grid pattern can also be produced by increasing the separation between the glass plates by mechanical means. A quantitative study of this effect appears to have been reported only for smectics and not for cholesterics (see Chandrasekhar, 1977, Section 5.3.8).

C. Thermal Instability

The theory of thermal instability of a planar cholesteric film in a thermal gradient along the helical axis z is rather similar to the electric field case (Dubois-Violette, 1973). Heat focusing takes place due to the anisotropy of thermal conductivity. Considering, as before, a distortion of wavelength much longer than the pitch but much shorter than the sample thickness (i.e., $q_z \ll q_y \ll q_0$), one again gets $q_y \propto (q_0 q_z)^{1/2} \propto (P_0 d)^{-1/2}$, and the threshold thermal gradient

$$\beta_c = 2 \frac{K_{\parallel} + K_{\perp}}{K_a} \frac{k_{22} q_0^2 q_z^2}{g\alpha} \tag{58}$$

Using MBBA parameters, $d \sim 1$ cm, $P_0 \sim 10\ \mu m$, $\Delta T_c \sim 10°C$, which is about 10^3–10^4 times greater than that for a nematic. Physically the reason for this is that in the nematic, the material flow along z gives rise to a viscous torque that plays a major destabilizing role, whereas in the cholesteric this is suppressed because of the very high viscosity associated with permeation.

The thermally induced convective instability should also lead to a square grid pattern. Thus a magnetic field along z should in principle reduce the thermal threshold, which should drop to zero when H equals the value given by Eq. (48). Parsons (1975) has also discussed the same problem on the basis of a continuum theory of cholesterics developed by Lubensky (1973), and his results are in qualitative agreement with those of Dubois-Violette (1973). However, he finds that if the pitch is assumed to decrease with increasing temperature and $K_a > 0$, then oscillatory convection is possible when the sample is heated from below.

Recently Pleiner and Brand (1981) have considered modifications of the general theory of Martin et al. (1972) by adding higher order terms, and have discussed in some detail both stationary and oscillatory convective instabilities in cholesterics.

APPENDIX

Material Constants for MBBA and HBAB[a]

	MBBA[b] at 25°C	HBAB[c] at 70°C
μ_1 (poise)	0.065	0.06
μ_2 (poise)	-0.775	-0.33
	(-0.93^d)	
μ_3 (poise)	-0.012	0.0034
	(-0.04^d)	
μ_4 (poise)	0.832	0.275
μ_5 (poise)	0.463	0.222
μ_6 (poise)	-0.324	-0.102
$\lambda_1 = \mu_2 - \mu_3$ (poise)	-0.763	-0.333
$\lambda_2 = \mu_5 - \mu_6$ (poise)	0.787	0.324
θ_0	7.0°	
	$(11.75^{\circ d})$	
k_{11} (dyne)	6×10^{-7}	8×10^{-7}
k_{22} (dyne)	3×10^{-7}	5×10^{-7}
k_{33} (dyne)	7×10^{-7}	11×10^{-7}
$\varepsilon_\parallel$ (cgs)	4.71	23.5
$\varepsilon_\perp$ (cgs)	5.26	7.7
ε_a (cgs)	-0.55	$+15.8$
$\chi_\parallel$ (cgs)	5.8×10^{-7}	
$\chi_\perp$ (cgs)	4.66×10^{-7}	
χ_a (cgs)	1.14×10^{-7}	0.75×10^{-7}
$\sigma_\parallel / \sigma_\perp$	1.5	
$K_\parallel$ (cgs)	6.65×10^{-4}	
$K_\perp$ (cgs)	4×10^{-4}	
$K_\parallel / K_\perp$	1.65	

[a] MBBA is 4-methoxybenzylidene-4′-butylaniline and HBAB is p-n-hexyloxybenzylidene-p-aminobenzonitrile.

[b] $\mu_1 - \mu_6$ (Gahwiller, 1971, 1973), k_{ii} (Haller, 1972), ε and σ (Diguet et al., 1970), χ (Gasparoux et al., 1971), and K (Vilanove et al., 1974).

[c] $\mu_1 - \mu_6$, χ (Gahwiller, 1973), k_{ii} (Cladis, 1975; Cladis and Torza, 1975) and ε (Schadt, 1972).

[d] A recent determination by Skarp and Carlsson (1978).

Acknowledgments

We are grateful to Dr. G. V. Vani and Mr. A. N. Kalkura for their help in the preparation of the diagrams.

References

Akahoshi, S., and Miyakawa, K. (1977). *J. Phys. Soc. Jpn.* **42**, 1997–2005.

Arnould-Netillard, H., and Rondelez, F. (1974). *Mol. Cryst. Liq. Cryst.* **26**, 11–31.

Atten, P. (1975). *J. Mec.* **14**, 461–495.

Barnik, M. I., Blinov, L. M., Grebenkin, M. F., and Trufanov, A. N. (1976). *Mol. Cryst. Liq. Cryst.* **37**, 47–56.

Barratt, P. J., and Bramley, J. S. (1980). *J. Phys. C* **13**, 535–548.

Barratt, P. J., and Sloan, D. M. (1976). *J. Phys. A: Math. Gen.* **9**, 1987–1998.

Barratt, P. J., and Sloan, D. M. (1978). *J. Phys. D* **11**, 2131–2138.

Belyaev, S. V. (1978). *Zh. Eksp. Teor. Fiz.* **75**, 705–711.

Belyaev, S. V., and Blinov, L. M. (1976). *Zh. Eksp. Teor. Fiz.* **70**, 184–190.

Ben-Abraham, S. I. (1976). *Phys. Rev. A* **14**, 1251–1257.

Ben-Abraham, S. I. (1979). *J. Phys. (Orsay, Fr.)* **40** (C3), 259–262.

Bertolotti, M., Scudieri, F., and Sturla, E. (1978). *J. Appl. Phys.* **49**, 3922–3926.

Bhattacharya, S., Hong, C. E., and Letcher, S. (1978). *Phys. Rev. Lett.* **41**, 1736–1739.

Blinov, L. M. (1978). "Electro and Magneto-Optics of Liquid Crystals." Nauka, Moscow.

Blinov, L. M., Barnik, M. I., Lazareva, V. T., and Trufanov, A. N. (1979). *J. Phys. (Orsay, Fr.)* **40** (C3), 263–268.

Candau, S., Martinoty, P., and Debeauvais, F. (1973). *C. R. Hebd. Seances Acad. Sci., Ser. B* **277**, 769–772.

Carr, E. F., and Kozlowski, R. W. H. (1980). *Liq. Cryst., Proc. Int. Conf., Bangalore, 1979* (S. Chandrasekhar, ed.), pp. 287–295. Heyden, London.

Carrigan, C. R., and Guyon, E. (1975). *J. Phys., Lett. (Orsay, Fr.)* **36**, 145–147.

Carroll, T. O. (1972). *J. Appl. Phys.* **43**, 1342–1346.

Chandrasekhar, S. (1977). "Liquid Crystals." Cambridge Univ. Press, London and New York.

Chandrasekhar, S., Kini, U. D., and Ranganath, G. S. (1980). *Liq. Cryst., Proc. Int. Conf., Bangalore, 1979* (S. Chandrasekhar, ed.), pp. 247–253. Heyden, London.

Chang, R. (1973). *Mol. Cryst. Liq. Cryst.* **20**, 267–278.

Chaure, A. (1980). *Int. J. Eng. Sci.* **18**, 761–774.

Chistyakov, I. G., and Vistin, L. K. (1974). *Kristallografiya* **19**, 195–215.

Cladis, P. E. (1975). *Phys. Rev. Lett.* **35**, 48–51.

Cladis, P. E., and Torza, S. (1975). *Phys. Rev. Lett.* **35**, 1283–1286.

Cohen, M., Pieranski, P., Guyon, E., and Mitescu, C. D. (1977). *Mol. Cryst. Liq. Cryst.* **38**, 97–108.

Currie, P. K. (1973) *Rheol. Acta* **12**, 165–169.

Currie, P. K. (1978). *Iran. J. Sci. Technol.* **7**, 71–77.

Currie, P. K., and Macsithigh, G. P. (1979). *Q. J. Mech. Appl. Math.* **32**, 499–511.

de Gennes, P. G. (1972). *Phys. Lett. A* **41A**, 479.

de Gennes, P. G. (1975). *C. R. Hebd. Seances Acad. Sci., Ser. B* **280**, 9–11.

de Gennes, P. G. (1976). *In* "Molecular Fluids" (R. Balian and G. Weill, eds.), pp. 373–400. Gordon & Breach, New York.

de Jeu, W. H., and Lathouwers, T. W. (1974). *Mol. Cryst. Liq. Cryst.* **26**, 235–243.

de Jeu, W. H., Gerritsma, C. J., Van Zanten, P., and Goossens, W. J. A. (1972). *Phys. Lett. A* **39A**, 355–356.

de Zwart, M. (1978). *J. Phys. (Orsay, Fr.)* **39**, 423–431.

de Zwart, M., and Lathouwers, T. W. (1975). *Phys. Lett. A* **55A**, 41–42.

de Zwart, M., and Van Doorn, C. Z. (1979). *J. Phys. (Orsay, Fr.)* **40** (C3), 278–284.

Diguet, D., Rondelez, F., and Durand, G. (1970). *C. R. Hebd. Seances Acad. Sci., Ser. B* **271**, 954–957.

Dreyfus, J. M., and Guyon, E. (1981). *J. Phys. (Orsay, Fr.)* **42**, 283–292.
Dubois-Violette, E. (1971). *C. R. Hebd. Seances Acad. Sci., Ser. B* **273**, 923–926.
Dubois-Violette, E. (1973). *J. Phys. (Orsay, Fr.)* **34**, 107–113.
Dubois-Violette, E. (1974). *Solid State Commun.* **14**, 767–771.
Dubois-Violette, E., and Manneville, P. (1978). *J. Fluid Mech.* **89**, 273–303.
Dubois-Violette, E., and Rothen, F. (1978). *J. Phys. (Orsay, Fr.)* **39**, 1039–1047.
Dubois-Violette, E., and Rothen, F. (1979). *J. Phys. (Orsay, Fr.)* **40**, 1013–1023.
Dubois-Violette, E., and Rothen, F. (1980). *Liq. Cryst., Proc. Int. Conf., Bangalore, 1979*
   (S. Chandrasekhar, ed.), pp. 239–245. Heyden, London.
Dubois-Violette, E., de Gennes, P. G., and Parodi, O. (1971). *J. Phys. (Orsay, Fr.)* **32**, 305–317.
Dubois-Violette, E., Guyon, E., and Pieranski, P. (1973). *Mol. Cryst. Liq. Cryst.* **26**, 193–212.
Dubois-Violette, E., Guyon, E., Janossy, I., Pieranski, P., and Manneville, P. (1977). *J. Mec.*
   16, 733–767.
Dubois-Violette, E., Durand, G., Guyon, E., Manneville, P., and Pieranski, P. (1978). *Solid
   State Phys., Suppl.* **14**, 147–208.
Durand, G. (1976). *In* "Molecular Fluids" (R. Balian and G. Weill, eds.), pp. 405–440. Gordon
   & Breach, New York.
Eber, N., and Janossy, I. (1979). *Mol. Cryst. Liq. Cryst. (Lett.)* **49**, 137–142.
Faetti, S., Fronzoni, L., and Rolla, P.A. (1979). *J. Phys. (Orsay, Fr.)* **40** (C3), 197–201.
Felici, N. (1969). *RGE, Rev. Gen. Electr.* **78**, 717–734.
Fitzjarrald, D. E., and Owen, E. W. (1979). *J. Phys. E* **12**, 1075–1082.
Gahwiller, C. (1971). *Phys. Lett. A* **36A**, 311–312.
Gahwiller, C. (1972). *Phys. Rev. Lett.* **28**, 1554–1556.
Gahwiller, C. (1973). *Mol. Cryst. Liq. Cryst.* **20**, 301–318.
Gasparoux, H., Regaya, B., and Prost, J. (1971). *C. R. Hebd. Seances Acad. Sci., Ser. B* **272**,
   1168–1171.
Gasparoux, H., Hardouin, F., Achard, M. F., and Sigaud, G. (1975). *J. Phys. (Orsay, Fr.)* **36**
   (C1), 107–111.
Gerritsma, C. J., and Van Zanten, P. (1971). *Phys. Lett. A* **37A**, 47–48.
Goossens, W. J. A. (1972). *Phys. Lett. A* **40A**, 95–96.
Goossens, W. J. A. (1978). *Adv. Liq. Cryst.* **3**, 1–39.
Goscianski, M., and Leger, L. (1975). *J. Phys. (Orsay, Fr.)* **36** (C1), 231–235.
Guyon, E., and Pieranski, P. (1972). *C. R. Hebd. Seances Acad. Sci., Ser. B* **274**, 656–658.
Guyon, E., and Pieranski, P. (1974). *Physica (Amsterdam)* **73**, 184–194.
Guyon, E., and Pieranski, P. (1975a). *C. R. Hebd. Seances Acad. Sci., Ser. B* **280**, 187–188.
Guyon, E., and Pieranski, P. (1975b). *J. Phys. (Orsay, Fr.)* **36** (C1), 203–208.
Guyon, E., and Velarde, M. G. (1978). *J. Phys., Lett. (Orsay, Fr.)* **39**, 205–208.
Guyon, E., Pieranski, P., and Boix, M. (1978a). *J. Phys. (Orsay. Fr.)* **39**, 99–103.
Guyon, E., Pieranski, P., and Salan, J. (1978b). *C. R. Acad. Sci., Ser. B* **287**, 41–43.
Guyon, E., Pieranski, P., and Salan, J. (1979). *J. Fluid. Mech.* **93**, 65–81.
Haller, I. (1972). *J. Chem. Phys.* **57**, 1400–1405.
Helfrich, W. (1969a). *J. Chem. Phys.* **51**, 4092–4105.
Helfrich, W. (1969b). *Phys. Rev. Lett.* **23**, 373–374.
Helfrich, W. (1970). *Appl. Phys. Lett.* **17**, 531–532.
Helfrich, W. (1971). *J. Chem. Phys.* **55**, 839–842.
Hervet, H., Rondelz, F., and Urbach, W. (1980). *Liq. Cryst., Proc. Int. Conf., Bangalore, 1979*
   (S. Chandrasekhar, ed.), pp. 263–266. Heyden, London.
Hijikuro, N. (1975). *Prog. Theor. Phys.* **54**, 592–593.
Hilsum, C., and Saunders, F. C. (1980). *Mol. Cryst. Liq. Cryst. (Lett.)* **64**, 25–31.
Horn, D., Guyon, E., and Pieranski, P. (1976). *Rev. Phys. Appl.* **11**, 139–142.

Hurault, J. P. (1973). *J. Chem. Phys.* **59**, 2068–2075.

Janossy, I. (1980a), *Proc. Liq. Cryst. Conf. Soc. Countries, 3rd, Budapest, 1979* (L. Bata, ed.), pp. 625–631. Pergamon Press, Oxford and New York.

Janossy, I. (1980b). *J. Phys. (Orsay, Fr.)* **41**, 437–442.

Janossy, I. (1981). *J. Phys., Lett. (Orsay, Fr.)* **42**, 41–43.

Janossy, I., Pieranski, P., and Guyon, E. (1976). *J. Phys. (Orsay, Fr.)* **37**, 1105–1113.

Jenkins, J. T. (1978). *Annu. Rev. Fluid Mech.* **10**, 197–219.

Kai, S., and Hirakawa, K. (1978). *Prog. Theor. Phys.* **64**, 212–243.

Karat, P. P., and Madhusudana, N. V. (1975). *Liq. Cryst., Proc. Int. Conf., Bangalore, 1973* (S. Chandrasekhar, ed.), pp. 285–288.

Karat, P. P., and Madhusudana, N. V. (1977). *Mol. Cryst. Liq. Cryst.* **42**, 57–65.

Kessler, L. W., and Sawyer, S. P. (1970). *Appl. Phys. Lett.* **17**, 440–441.

Kini, U. D. (1978). *Pramana* **10**, 143–153.

Kini, U. D. (1979). *J. Phys. (Orsay, Fr.)* **40** (C3), 62–66.

Kini, U. D. (1980a). *Liq. Cryst., Proc. Int. Conf. Bangalore, 1979* (S. Chandrasekhar, ed.), pp. 255–261. Heyden, London.

Kini, U. D. (1980b). *Pramana* **15**, 231–244.

Kini, U. D. (1980c). *Pramana* **14**, 463–475.

Kini, U. D., Ranganath, G. S., and Chandrasekhar, S. (1975). *Pramana* **5**, 101–106.

Kleman, M., and Pikin, S. A. (1979). *J. Mec.* **18**, 661–672.

Koelmans, H., and Van Boxtel, A. M. (1971). *Mol. Cryst. Liquid Cryst.* **12**, 185–191.

Kohno, T., Miike, H., and Ebina, Y. (1978). *J. Phys. Soc. Jpn.* **44**, 1678–1684.

Krigbaum, W. R., Lader, H. J., and Ciferri, A. (1980). *Macromolecules* **13**, 554–559.

Laidlaw, W. G. (1979). *Phys. Rev. A* **20**, 2188–2192.

Lekkerkerker, H. N. W. (1977). *J. Phys., Lett. (Orsay, Fr.)* **38**, 277–281.

Lekkerkerker, H. N. W. (1979). *J. Phys. (Orsay, Fr.)* **40** (C3), 67–72.

Leslie, F. M. (1969). *Mol. Cryst. Liq. Cryst.* **7**, 407–420.

Lesile, F. M. (1970). *Mol. Cryst. Liq. Cryst.* **12**, 57–72.

Leslie, F. M. (1976a). *J. Phys. D* **9**, 925–937.

Leslie, F. M. (1976b). *Mol. Cryst. Liq. Cryst.* **37**, 335–352.

Leslie, F. M. (1979). *Adv. Liq. Cryst.* **4**, 1–81.

Lubensky, T. C. (1973). *Mol. Cryst. Liq. Cryst.* **23**, 99–109.

Madhusudana, N. V., and Moodithaya, K. P. L. (1979). *Mol. Cryst. Liq. Cryst.* **51**, 137–148.

Madhusudana, N. V., Karat, P. P., and Chandrasekhar, S. (1973). *Curr. Sci.* **42**, 147–149.

Mailer, H., Likins, K. L., Taylor, T. R., and Fergason, J. L. (1971). *Appl. Phys. Lett.* **18**, 105–107.

Manneville, P. (1978). *J. Phys. (Orsay, Fr.)* **39**, 911–925.

Manneville, P. (1979). *J. Phys. (Orsay, Fr.)* **40**, 713–724.

Manneville, P., and Dubois-Violette, E. (1976a). *J. Phys. (Orsay, Fr.)* **37**, 285–296.

Manneville, P., and Dubois-Violette, E. (1976b). *J. Phys. (Orsay, Fr.)* **37**, 1115–1124.

Martin, P. C., Parodi, O., and Pershan, P. S. (1972). *Phys. Rev. A* **6**, 2401–2420.

Meyer, R. B. (1968). *Appl. Phys. Lett.* **12**, 281–282.

Meyerhofer, D., Sussmann, A., and Williams, R. (1972). *J. Appl. Phys.* **43**, 3685–3689.

Miyakawa, K. (1975). *J. Phys. Soc. Jpn.* **39**, 628–633.

Moodithaya, K. P., and Madhusudana, N. V. (1980). *Liq. Cryst., Proc. Int. Conf., Bangalore, 1979* (S. Chandrasekhar, ed.), pp. 297–302. Heyden, London.

Moritz, E., and Franklin, W. (1976). *Phys. Rev. A* **14**, 2334–2337.

Moritz, E., and Franklin, W. (1977). *Mol. Cryst. Liq. Cryst.* **40**, 229–237.

Orsay Liquid Crystal Group (1972). *Phys. Lett. A* **39A**, 181–182.

Parsons, J. D. (1975). *J. Phys. (Orsay, Fr.)* **36**, 1363–1370.

Penz, P. A. (1970). *Phys. Rev. Lett.* **24**, 1405–1409.

Penz, P. A. (1973). *Mol. Cryst. Liq. Cryst.* **23**, 1–11.
Penz, P. A. (1974). *Phys. Rev. A* **10**, 1300–1310.
Penz, P. A., and Ford, G. W. (1972a). *Phys. Rev. A* **6**, 414–425.
Penz, P. A., and Ford, G. W. (1972b). *Phys. Rev. A* **6**, 1676–1683.
Pieranski, P., and Guyon, E. (1973). *Solid State Commun.* **13**, 435–437.
Pieranski, P., and Guyon, E. (1974a). *Phys. Rev. A* **9**, 404–417.
Pieranski, P., and Guyon, E. (1974b). *Phys. Lett. A* **49A**, 237–238.
Pieranski, P., and Guyon, E. (1974c). *Phys. Rev. Lett.* **32**, 924–926.
Pieranski, P., and Guyon, E. (1975). *Adv. Chem. Phys.* **32**, 151–161.
Pieranski, P., and Guyon, E. (1976). *Commun. Phys.* **1**, 45–49.
Pieranski, P., and Guyon, E. (1977). *Phys. Rev. Lett.* **39**, 1280–1282.
Pieranski, P., Dubois-Violette, E., and Guyon, E. (1973). *Phys. Rev. Lett.* **30**, 736–739.
Pieranski, P., Guyon, E., and Pikin, S. A. (1976). *J. Phys. (Orsay, Fr.)* **37** (C1), 3–6.
Pikin, S. A. (1971). *Zh. Eksp. Teor. Fiz.* **60**, 1185–1190.
Pikin, S. A. (1972). *Zh. Eksp. Teor. Fiz.* **63**, 1115–1119.
Pikin, S. A., and Indenbom, V. L. (1975). *Kristallografiya* **20**, 1127–1129.
Pikin, S. A., and Shtolberg, A. A. (1973). *Kristallografiya* **18**, 445–453.
Pikin, S. A., Chigrinov, V. G., and Indenbom, V. L. (1976a). *Mol. Cryst. Liq. Cryst.* **37**, 313–320.
Pikin, S., Ryschenkow, G., and Urbach, W. (1976b). *J. Phys. (Orsay, Fr.)* **37**, 241–244.
Pleiner, H., and Brand, H. (1981). *Phys. Rev. A* **23**, 933–958.
Pochan, J. M. (1979). *In* "Liquid Crystals, The Fourth State of Matter" (F. D. Saeva, ed.), pp. 275–304. Dekker, New York.
Porter, R. S., Barrall, E. M., and Johnson, J. F. (1966). *J. Chem. Phys.* **45**, 1452–1456.
Ribotta, R., and Durand, G. (1979). *J. Phys. (Orsay, Fr.)* **40** (C3), 334–337.
Rigopoulos, R. A., and Zenginoglou, H. M. (1976). *Mol. Cryst. Liq. Cryst.* **35**, 307–318.
Rondelez, F., and Hulin, J. P. (1972). *Solid State Commun.* **10**, 1009–1012.
Rondelez, F., Arnould, H., and Gerritsma, C. J. (1972). *Phys. Rev. Lett.* **28**, 735–737.
Schadt, M. (1972). *J. Chem. Phys.* **56**, 1494–1497.
Scudieri, F., Bertolotti, M., Melone, S., and Albertini, G. (1976). *J. Appl. Phys.* **47**, 3781–3783.
Scudieri, F., Ferrari, A., and Fedtchouk, A. (1978). *J. Appl. Phys.* **49**, 1289–1290.
Sengupta, P., and Saupe, A. (1974). *Phys. Rev. A* **9**, 2698–2706.
Skarp, K., and Carlsson, T. (1978). *Mol. Cryst. Liq. Cryst. (Lett.)* **49**, 75–82.
Smith, I. W., Galerne, Y., Lagerwall, S. T., Dubois-Violette, E., and Durand, G. (1975). *J. Phys. (Orsay, Fr.)* **36** (C1), 237–259.
Stewart, G. W. (1936). *J. Chem. Phys.* **4**, 231–236.
Sussman, A. (1976). *Appl. Phys. Lett.* **29**, 633–635.
Sussman, A. (1978). *J. Appl. Phys.* **49**, 1131–1138.
Tarr, C. E., and Carr, E. F. (1980). *Solid State Commun.* **33**, 459–462.
Urbach, W., Rondelez, F., Pieranski, P., and Rothen, F. (1977). *J. Phys. (Orsay, Fr.)* **38**, 1275–1284.
Velarde, M. G., and Zuniga, I. (1979). *J. Phys. (Orsay, Fr.)* **40**, 725–731.
Vilanove, R., Guyon, E., Mitescu, C., and Pieranski, P. (1974). *J. Phys. (Orsay, Fr.)* **35**, 153–162.
Zenginoglou, H. M., and Kosmopoulos, I. A. (1977). *Mol. Cryst. Liq. Cryst.* **43**, 265–277.
Zenginoglou, H. M., Rigopoulos, R. A., and Kosmopoulos, I. A. (1977). *Mol. Cryst. Liq. Cryst.* **39**, 27–32.

9

Rheo-Optical Studies of Polymer Liquid Crystalline Solutions

Tadahiro Asada

Department of Polymer Chemistry
Kyoto University
Kyoto, Japan

247

I. Introduction

Study of the deformation and flow mechanisms of polymer liquid crystals should provide fundamental knowledge clarifying technologically significant phenomena, including the mechanism of fiber formation during liquid crystal spinning. Rheological properties alone cannot fully elucidate the deformation mechanisms of polymer liquid crystalline systems. The latter usually form complicated superstructures or textures, the instabilities of which have prevented us from studying the relationship between the physical properties and structures of polymer liquid crystals. The necessity for rheo-optical studies to clarify the relation between rheological properties and structure has recently become apparent. Some investigators (Horio, 1978; Kiss *et al.*, 1979; Aoki *et al.*, 1980) observed flow process directly with a microscope, while others (Asada *et al.*, 1975, 1978a,b, 1980; Asada, 1976; Asada and Onogi, 1980; Baird *et al.*, 1979) measured optical quantities simultaneously with rheological properties.

The author and co-workers (Asada *et al.*, 1975, 1978a,b, 1980; Asada, 1976; Asada and Onogi; 1980) also developed new rheo-optical techniques to investigate the relation between structure and the rheological properties of various liquid crystals. Two new types of apparatus have been designed for this purpose. Some interesting information about the mechanisms of mesophase transition and orientation of lyotropic polymer liquid crystals subjected to shear deformation have been obtained by use of the equipments for studying the rheo-optical properties. In this chapter these techniques will be described, as well as some recent experimental results obtained for lyotropic polymeric liquid crystals of poly(γ-benzyl-L-glutamate) (PBLG), racemic poly(γ-benzyl glutamate) (PBG), hydroxypropylcellulose (HPC), etc.

II. Structure of Solutions of Liquid Crystalline Polymers

A. Solution Structure

Investigation of the rheological properties of polymeric liquid crystals have been concerned with concentrated solutions of (a) rigid and semi rigid molecules, (b) rodlike helical molecules, and (c) other molecules. Polymers of group (a) are mostly aromatic polyamides such as poly(p-phenyleneterephthalamide) and polyterephthalamide of p-aminobenzhydrazide (X-500). Great attention has been devoted to these polymers recently, due to their importance in the production of synthetic fibers (Kwolek, 1972; Papkov *et al.*, 1974; Ciferri, 1975; Marrucci and Ciferri, 1977; Valenti and Ciferri, 1978; Baird, 1978; Aoki *et al.*, 1978, 1979a,b; White and Fellers, 1978; Wong *et al.*, 1978; Baird *et al.*, 1979). Those of group (b) are mostly polypeptides (Yang, 1958; Hermans, 1962; Robinson, 1956; Iizuka, 1974; Asada *et al.*, 1978a,b, 1980; Kiss and Porter, 1978; Aoki *et al.*, 1979a,b; Patel and DuPré, 1979). Group (c) includes cellulose derivatives (Asada *et al.*, 1981; Asada and Onogi, 1980; Aoki *et al.*, 1980) and biological polymers (Duke and Chapoy, 1976).

Solutions of these liquid crystalline polymers are usually isotropic below a limiting concentration *A*, which is a function of the molecular weight of the polymer. Above this concentration two phases, an isotropic phase and a birefringent phase coexist in equilibrium. The birefringent phase dispersed in an isotropic phase easily forms spherulites due to the interfacial tension between the two phases. Above a still higher concentration *B* only the birefringent phase can exist, and it forms a liquid crystalline texture.

B. *Textures of Liquid Crystalline Solutions above the B Point*

The texture of a liquid crystal varies considerably with the preparation procedure and history of the sample. For example, polarized light photomicrographs of the cholesteric phase of a low molecular weight thermotropic mesogen prepared under different conditions are shown in Fig. 1. The textures of the samples obtained by slow cooling or by quenching from the isotropic melt between thin glass plates are very different. A polarized light photomicrograph of a high density polyethylene film prepared by slow

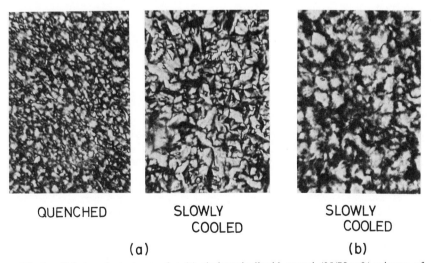

QUENCHED	SLOWLY COOLED	SLOWLY COOLED
(a)		(b)

Fig. 1. Polydomain textures of a (a) cholesteric liquid crystal (25/75 wt% mixture of cholesteryl chloride and cholesteryl oleyl carbonate) and (b) a high density polyethylene film, between crossed polarizers. (From Asada, 1976.)

cooling from the melt is also shown for comparison. As can be seen from the figure, the polydomain texture of a liquid crystal resembles the superstructure of polyethylene, a typical crystalline polymer, though the liquid crystal is of low molecular weight. Thus the texture is a determining factor for the rheological properties of liquid crystals. Disregard of the texture often causes misunderstanding or disagreement among investigators. This is why we must utilize rheo-optical techniques to study liquid crystalline systems. In general, the texture of a polymeric liquid crystal is much more stable to external forces compared to liquid crystals of low molecular weight. Therefore, careful consideration of the bulk structure may be required to understand rheological properties. However, the textures of single phase polymeric liquid crystalline solutions in general cannot be characterized simply.

When a thin sample of a *nematic* liquid crystal, for example, a racemic mixture of PBLG and PBDG ($\bar{M}_w = 15 \times 10^4$), is held between glass plates, a typical threaded Schlieren texture is observed, which changes to the Schlieren texture with disclination points resembling those seen in low molecular weight thermotropic nematic phases. Polymer solutions require much longer to form typical Schlieren textures. However, Schlieren textures cannot be observed in thick samples, but polydomains appear, due to disclination points or lines.

When a *cholesteric* liquid crystalline solution is held between glass plates, polydomains are observed that may consist of complete and incomplete spherulitic superstructures, and in which regions of parallel equidistant lines are seen sporadically.

III. Experimental Methods

A. Polarized Light Method (Transmission)

1. Apparatus I

A block diagram of one of the rheo-optical instruments (apparatus I) is shown in Fig. 2. A cone and plate rheometer equipped with a transparent cone and plate made of quartz is made a part of the optical system. A monochromatic laser light beam ($\lambda = 6328$ Å) passes through a polarizer (P), quartz plate (E), sample (F), quartz cone (G), and analyzer (A). The transmitted light is finally detected by a photomultiplier tube (PM_1). Thus this apparatus enables us simultaneously to measure the rheological properties and transmission of polarized light through a sheared sample. The diameter of the cone and plate is kept constant (8 cm), but cones of different cone angles are available.

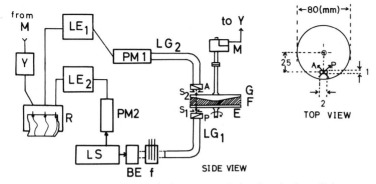

Fig. 2. Block diagram of rheo-optical apparatus I showing the laser light source (LS), beam expander (BE), filter (F), input glass fiber tube (light guide, 10 mm i.d. × 30 μm) (LG$_1$), polarizer (P), entrance slit (S$_1$), quartz plate (E), sample (F), quartz cone (G), exit slit (S$_2$), analyzer (A), output glass fiber tube (LG$_2$), photomultiplier tube (PM$_1$), amplifier (LE$_1$), photomultiplier tube (for monitor use) (PM$_2$), amplifier (for monitor use) (LE$_2$), recorder (R), transducer for detecting torque (M), and amplifier (for torque) (Y). (Reprinted with permission from Asada *et al.*, *Macromolecules* **13**, 867 (1980). Copyright 1980 American Chemical Society.)

2. Measurements

The shear stress and intensity of the transmitted light ($I_\times$, $I_\parallel, I_E$) are measured simultaneously as functions of shear rate, where $I_\times$, $I_\parallel$, and I_E are fractional transmitted light intensities with crossed polarizers (at $\psi = 45°$ in Fig. 3), parallel polarizers (at $\psi = 45°$), and crossed polarizers at the extinction position (at $\psi = 0°$), respectively. The geometries are shown in Fig. 3.

B. Spectrophotometric Method

1. Apparatus II

A cone and plate rheometer equipped with a transparent quartz cone and plate is combined with a spectrophotometer (Jasco UVIDEC-1), which was modified to measure the wavelength dependence of light transmitted by the sample between the cone and plate. The block diagram of apparatus II is shown in Fig. 4. The standard path of the sample and reference beams is replaced by a newly designed light guide box (F). The original standard light source was replaced by a 500-W xenon lamp (A). The right half (I-K-P–V) of the block diagram serves as a cone and plate rheometer. Rotation of the plate at a constant speed applies a shear to the sample at a definite shear rate, and the torque is detected by a transducer (Q), which provides the shear stress and the viscosity. The left half of the diagram is the optical

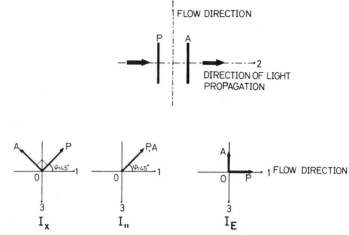

Fig. 3. Geometry of the optical system. Here 1 is the flow direction, 2 the direction of light propagation, and 3 the radial direction of the cone and plate. OP is the transmission axis of the polarizer and OA is the transmission axis of the analyzer. (Reprinted with permission from Asada *et al.*, *Macromolecules* **13**, 867 (1980). Copyright 1980 American Chemical Society.)

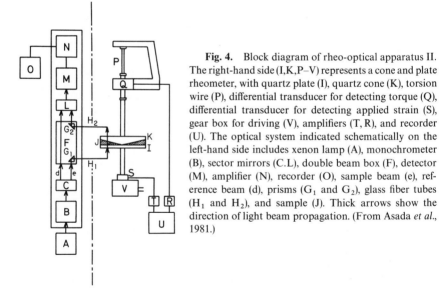

Fig. 4. Block diagram of rheo-optical apparatus II. The right-hand side (I,K,P–V) represents a cone and plate rheometer, with quartz plate (I), quartz cone (K), torsion wire (P), differential transducer for detecting torque (Q), differential transducer for detecting applied strain (S), gear box for driving (V), amplifiers (T, R), and recorder (U). The optical system indicated schematically on the left-hand side includes xenon lamp (A), monochrometer (B), sector mirrors (C.L), double beam box (F), detector (M), amplifier (N), recorder (O), sample beam (e), reference beam (d), prisms (G_1 and G_2), glass fiber tubes (H_1 and H_2), and sample (J). Thick arrows show the direction of light beam propagation. (From Asada *et al.*, 1981.)

system. The standard sample beam (e) is introduced normally to the shear plate (I) by the prism (G_1) and the glass fiber tube (H_1), and passes through the sample (J) between the plate (I) and the cone (K). The transmitted light is introduced to the receiving sector mirror (L) through the glass fiber tube (H_2) and the prism (G_2).

Thus the apparatus enables us to measure simultaneously rheological properties and wavelength dependence of the transmitted light through a sheared sample.

2.   Measurements

The wavelength dependence of the light transmitted through a sample subjected to shear has been measured quantitatively at various shear rates. The absorbance spectra after sudden cessation of steady shear flow is also observed as a function of time.

The slit of the spectrophotometer is triangular, with a half-width of less than 0.5 nm over the experimental range of wavelengths (400–760 nm). When experimental data are compared with those generated by computer fitting, the slit function is taken into account.

Various cones having different cone angles are available. The diameters of cone and plate are 8 cm. The direction of light propagation is always normal to the plate. The thickness of the sample at the position where the light beam passes through can be varied to suit the experimental purpose. Arc-shaped slits having 2-mm width and 10-mm length are used in both the entrance and exit paths. The circle of arc of a slit is coaxial with the rotational axis of the plate, and the diameter can be set to suit a desired thickness. To keep the experimental conditions constant the cone and plate may be rubbed beforehand along a certain definite direction with a sheet of gauze.

IV.   General Optical Considerations for the Polarized Light Method

A.   Uniform System

When the retardation Γ of a uniaxially oriented phase undergoes continuous changes, $I_\times$ and $I_{||}$ vary in a quasi-periodic manner, as expressed by

$$I_\times = K \sin^2 (\pi\Gamma/\lambda), \qquad I_{||} = K[1 - \sin^2 (\pi\Gamma/\lambda)] \qquad (1)$$

and shown schematically in Fig. 5. In Eq. (1), λ is the wavelength of the light

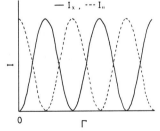

Fig. 5.   Schematic representation of the variation of the transmitted light intensities $I_\times$, $I_{||}$ with retardation Γ according to Eq. (1).

and K is a coefficient that is unity when there is no absorption and scattering. In such a simple case this equation enables us to evaluate the birefringence from $I_\times$ and $I_{||}$ (the transmission method). The same concept can be applied to a liquid crystal system, so long as the system is optically inactive and behaves as a monodomain [indicated schematically in Fig. 6(3)].

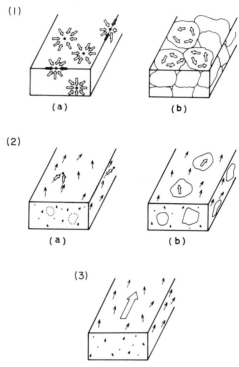

Fig. 6. Schematic representation of the bulk structure of a polymeric liquid crystal. (1) Polydomain system, (2) degraded polydomain system, and (3) monodomain.

If there is optical rotation due to molecular or structural optical activity, this effect may be superimposed on the birefringence. In such a case it is helpful to know whether or not optical rotation is involved, because a change in optical rotatory power causes a wavy variation of I_E. On the other hand, for a uniaxially birefringent system without optical rotation, I_E should be zero. When the optical rotation is very small, the birefringence can still be evaluated by the polarized light method. In practice, we have successfully obtained the streaming birefringence of isotropic solutions of poly(γ-benzyl-L-glutamate) using this polarized light method (Asada *et al.*, 1978a).

B. Polydomain System of Liquid Crystals

Generally, liquid crystals appear in polydomain textures (for example, see Fig. 1). Lyotropic polymeric liquid crystals also appear in polydomain textures as discussed in Section II. For such polydomain textures, many domains exist in the field, as indicated schematically in Fig. 6(1). The polarized incident light will be depolarized to various degrees owing to the depolarization by the various domains. No difference between I_x and $I_{||}$ will be observed, since the transmitted light consists of variously depolarized beams. Thus for such polydomain systems it is difficult to determine the birefringence arising from molecular orientation by the polarized light method.

As mentioned above, I_E is useful for studying the deformation process in such polydomain systems. Suppose a uniaxially birefringent body is placed between crossed polarizers, as shown in Fig. 7. The transmission axis of the polarizer (OP) coincides with the flow direction 1. The angle between the optic axis and the transmission axis of the polarizer is denoted by ψ. When the uniaxially birefringent body is oriented so that its optic axis is parallel to the flow direction, the angle ψ becomes zero, and hence the intensity of the transmitted light, I_E, tends to zero. Thus for an ideal system without light scattering or optical activity, $I_E = 0$ indicates either that all domains are oriented so that their optic axes are parallel to the flow direction or that all domains coalesce to a monodomain having its optic axis parallel to the flow direction. Here a volume element that acts as a birefringent body is referred to as a domain.

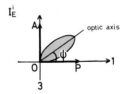

Fig. 7. A uniaxially birefringent domain between crossed polarizers. The transmission axis of the polarizer (OP) coincides with flow direction 1, and ψ is the angle between the optical axis and the polarizer axis. The transmitted light intensity I_E^i vanishes when $\psi \to 0$: $I_E^i = k \sin^2 (2\psi) \sin^2(\pi\Gamma/\lambda)$ when $\psi \to 0$, $I_E \to 0$.

Now consider the case of many domains existing in the field [Fig. 6(1)]. The thickness and the direction of the optic axes of the domains are assumed to be randomly distributed. Owing to depolarization of the many domains, the polarized incident light is depolarized, and I_E, I_x and $I_{||}$ become as high as 50%. Such a system consisting of many domains, as illustrated in Fig. 6(1), is referred to as a "polydomain system." The value of I_E tends to zero when the optic axes of all domains are oriented parallel to the flow direction to form a monodomain, as shown in Fig. 6(3). When domains suspended in a continuous phase become partially organized, as shown in Fig. 6(2), they weaken the wavy change in $I_\times$ and $I_{||}$ by depolarizing, and I_E takes an intermediate value.

C. Liquid Crystals with Strong Optical Rotatory Power

The preceding discussion of optical properties has been limited to a simple case, such as a nematic phase. Now we consider the case of a liquid crystal having strong optical rotatory power, such as a cholesteric phase. Except for the ideal case in which all the planes of the molecular layers are completely parallel to the plane of the plate, that is, a perfect monodomain, domains having various retardations and different rotatory powers will be randomly distributed in a field. Consequently, polarized incident light will be depolarized during its passage. Thus a high value will again be expected for all three quantities I_E, $I_\times$, and $I_{||}$.

So long as the system retains its cholesteric structure under shear, one does not expect I_E to vanish.

V. Processes of Structure Formation in Cholestric Liquid Crystals Studied by Spectrophotometric Methods

A. A Model for Calculating the Absorbance Spectrum

The optical data shown below may be interpreted by assuming disorientation of Bragg sites during steady flow and their reorientation after cessation of the flow, as suggested for low molecular weight cholesteric materials by Pochan and Marsh (1972). Certain types of orientation distribution functions of the Bragg reflection elements will be introduced, based on current concepts of cholesteric reflection as developed by Chandrasekhar and Prasad (1971).

Let us consider small orientation elements that act as Bragg sites and satisfy the following assumptions:

(i) The distribution of intensity versus wavelength of the reflected light from a single "Bragg element" is Lorentzian with respect to $\lambda_0 = \bar{n}P_0$, where P_0 is the cholesteric pitch and $\bar{n}$ is the mean refractive index of the matrix.

(ii) The distribution of the orientation of Bragg elements is represented approximately by a rotationally symmetric distribution of the Gaussian type with respect to a certain orientation reference axis (OZ).

As is shown later in Fig. 14, the experimental result for a Grandjean-like texture of an HPC liquid crystal 40 min after the cessation of steady flow can be fitted by computer by assuming that the reflection curve from each Bragg element is Lorentzian. Comparison of computer fitted and theoretical results reported by Chandrasekhar and Prasad (1971) for Bragg elements of finite thickness supports the adequacy of assumption (i). Assumption (ii) is partly based on experimental evidence obtained by microscopic observations indicating that HPC molecules are easily aligned by a glass wall; because of the wall effect, a cholesteric structure always grows so that its optic axis is normal to the glass wall.

Now we consider the case in which the orientation axis (reference axis) is normal to the plate and parallel to the direction of light propagation. From assumption (i), the wavelength dependence curve of absorbance $f(\lambda)$ for a Grandjean texture with optic axis parallel to the direction of light propagation (cf. Fig. 8a) can be represented by

$$f(\lambda) = \frac{C \cdot B}{(\lambda - \lambda_0)^2 + B^2} \tag{2}$$

with

$$\lambda_0 = \bar{n}P_0 \tag{3}$$

where λ, P_0, and $\bar{n}$ are, respectively, wavelength, the cholesteric pitch, and the mean refractive index. The term B is the half-value width of the Lorentzian curve, which depends on the thickness of the cholesteric element and also the perfection of the cholesteric structure. For reference, comparison with the theoretical curves based on the Chandrasekhar–Prasad theory (1971) shows that $B \cong 15$ for 5-pitch thicknesses and $B \cong 8$ for 10-pitch thicknesses of

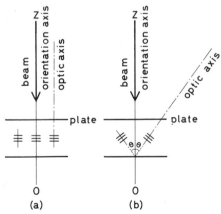

Fig. 8. Geometry of cholesteric Bragg elements related to the orientation axis OZ. The direction of the incident light is shown by a thick arrow. In (a) the helical screw axis of a cholesteric Bragg element is normal to the plate, while in (b) it makes an angle θ to the normal.

perfect cholesteric structure. The term C will be proportional to the number of Bragg elements for simple systems.

When the axis of the element is tilted from the orientation axis by angle θ (as is shown schematically in Fig. 8b), the wavelength dependence curve of absorbance $f(\lambda)$ is given by

$$f(\lambda) = \frac{C \cdot B \cos \theta}{(\lambda - \lambda_0 \cos \theta)^2 + B^2 \cos^2 \theta} \tag{4}$$

with

$$\lambda_M = \lambda_0 \cos \theta \tag{5}$$

where λ_M is the wavelength at which the spectrum has a maximum. When $\theta = 0$, Eq. (4) reduces to Eq. (2), and $\lambda_M = \lambda_0$.

From assumption (ii) the orientation distribution function or probability $F(\theta)$ for elements with optic axes at angle θ to the orientation axis OZ, parellel to the direction of light propagation, is given by

$$F(\theta) = \sin \theta \exp(-D^2\theta^2) \bigg/ \int_0^{\pi/2} \sin \theta \exp(-D^2\theta^2)\, d\theta \tag{6}$$

Here D is a parameter characterizing the sharpness of the orientation distribution. The reciprocal of D is analogous to the standard deviation of a Gaussian distribution. In Fig. 9, the variation of the orientation distribution function $F(\theta)$ with D is shown.

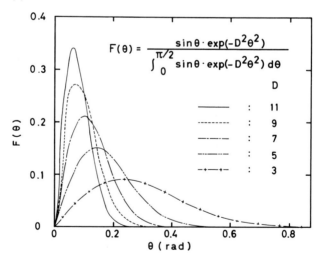

Fig. 9. Variation of the shape of the orientation distribution function $F(\theta)$ with D.

Fig. 10. Wavelength dependence of absorbance calculated according to Eq. (7). In the example shown, only D varies. (From Asada et al., 1981.)

The wavelength dependence of the absorbance can be calculated by combining Eqs. (4) and (6):

$$\log\left(\frac{I_0}{I}\right) = \int_0^{\pi/2} \frac{C \cdot B \cdot \cos\theta \cdot F(\theta)}{(\lambda - \lambda_0 \cos\theta)^2 + B^2 \cos^2\theta}\, d\theta + \frac{E}{\lambda^4} \qquad (7)$$

The second term on the right-hand side of Eq. (7) represents the base line of the spectrum.

Figure 10 illustrates computer calculations based on Eq. (7). In this particular case, λ_0, B, and C were kept constant, as are indicated in the figure, and only D was changed from 0 to 4. One sees that the shape of the calculated absorption spectrum drastically changes with D, becoming sharper as D increases.

When the angle between the orientation axis and the direction of light propagation is θ_0, as shown schematically in Fig. 11, the angle χ between the optic axis (which is tilted by θ with respect to the orientation axis) and the direction of light propagation is represented by the following equation

Fig. 11. Calculated wavelength dependence for a geometry in which orientation axis OZ is at angle θ_0 to the normal. The direction of incident light is shown by a thick arrow and the direction of the optic axis of a Bragg element is given by (θ_0, θ, ψ).

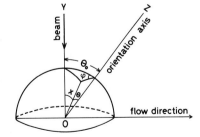

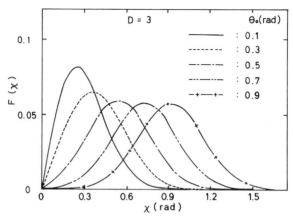

Fig. 12. Variation of the shape of the orientation distribution function $F(\chi)$ with χ for various θ_0 values (radians).

in terms of the cosine theorem of spherical trigonometry:

$$\cos \chi = \cos \theta_0 \cos \theta + \sin \theta_0 \sin \theta \cos \psi \tag{8}$$

The variation of the shape of the orientation distribution function $F(\chi)$ with θ_0 is shown in Fig. 12. The maximum of the distribution changes with θ_0 and its height depends upon D.

Thus when the orientation axis tilts by an angle of θ_0, as schematically shown in Fig. 11, Eq. (7) becomes

$$\log\left(\frac{I_0}{I}\right) = \int_0^{\pi/2} d\theta \int_0^{2\pi} \frac{C \cdot B \cdot \cos \chi \cdot F(\theta)}{(\lambda - \lambda_0 \cos \chi)^2 + B^2 \cos^2 \chi} \, d\psi + \frac{E}{\lambda^4} \tag{9}$$

where

$$F(\theta) = \sin \theta \cdot \exp(-D^2\theta^2) \bigg/ \int_0^{\pi/2} \sin \theta \cdot \exp(-D^2\theta^2) \, d\theta$$

B. Experimental Results for Aqueous HPC (Hydroxypropylcellulose)

The HPC sample ($\bar{M}_w \cong 1.04 \times 10^5$, $\bar{M}_w/\bar{M}_n \cong 2.5$, $MS = 3.5$, $DS = 2.3$) was obtained from Nihonsoda Co. Ltd. Concentrated aqueous solutions (43–68 wt%) of this polymer at 25°C, form cholesteric liquid crystals which exhibit beautiful iridescent color. A 61-wt % aqueous solution studied at 25°C was ascertained to be a single phase liquid crystalline solution in the quiescent state by microscopic observation.

Unless otherwise indicated, all data were obtained under the following conditions. A cone angle of 1.01° was used, and the direction of light propaga-

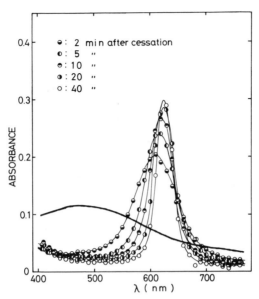

Fig. 13. Recovery of the absorbance spectrum after cessation of steady shear flow ($\dot{\gamma} = 1.19 \, \text{sec}^{-1}$). The solid line shows the spectrum under steady state shear flow. (From Asada *et al.*, 1981.)

tion was always normal to the plate. The sample thickness where the light beam passed was 0.44 mm. The circle of arc of the slit was 25 mm from the rotational axis of the plate. To keep the experimental conditions fixed, cone and plate were always rubbed tangentially beforehand with a sheet of gauze.

During shear at constant rate, the spectrum of transmitted light changes with time until it reaches a steady state. (Figure 26 shows the steady-state spectrum obtained at various shear rates.) The shape of the steady-state spectrum depends upon the shear rate.

When the shear is suddenly stopped the spectrum recovers quickly. Figure 13 shows the recovery of the spectrum for $\dot{\gamma} = 1.19 \, \text{sec}^{-1}$. The solid curve, which represents the steady state spectrum in shear flow, is broad in shape. When shear is stopped, however, the spectrum becomes very sharp within 2 min, and continues to become sharper with time. After a certain time λ_M no longer changes and only the height of the spectrum grows with time. Eventually the sample becomes transparent and seems to achieve a Grandjean-like texture. The experimental result for the Grandjean-like texture of HPC liquid crystals formed 40 min after the cessation of steady flow can be computer-fitted by assuming that the reflection curve from each Bragg element is Lorentzian, as shown in Fig. 14.

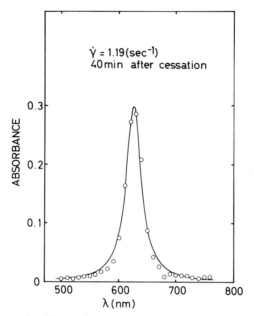

Fig. 14. Computer fitted Lorentzian curve based on assumption (i). Open circles represent experimental results for a Grandjean-like texture. (From Asada *et al.*, 1981.)

C. Structural Formation Processes in Polymeric Cholesteric Liquid Crystals

The equation introduced above is helpful for discussing a process of structure formation of cholesteric liquid crystals. Figure 15 shows a computer fit to the experimental results shown in Fig. 13 for the structure formation process after the cessation of steady shear flow. The calculated values (lines) coincide with experimental data (circles) when suitable values are assumed for the parameters. The change in shape of the spectrum with time will be discussed below in terms of the change in the parameters.

As shown in Fig. 16, parameter B decreases rapidly with time after the cessation of steady shear flow. It is noteworthy that for low shear rates (0.119 and 0.238 sec^{-1}), B remains large. Furthermore, the higher the shear rate, the more rapidly B decreases. From this result it follows that the higher the shear rate, the more rapidly and completely the cholesteric structure reforms after the cessation of steady flow. Figure 17 illustrates the time dependence of D, the sharpness of the orientation distribution. At shear rates below 0.594 sec^{-1} the orientation distribution does not change with time, while at higher shear rates D increases rapidly. It is interesting to note

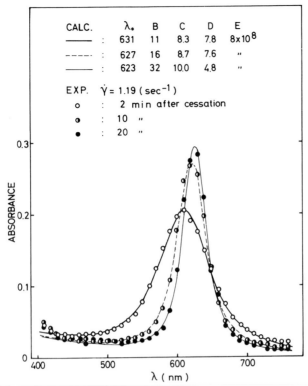

Fig. 15. Computer fitted curve using Eq. (7) to represent the behavior following cessation of steady shear flow ($\dot{\gamma} = 1.19$ sec^{-1}) using parameters listed in the figure.

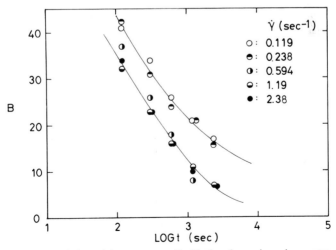

Fig. 16. Variation of the parameter B with time for various shear rates.

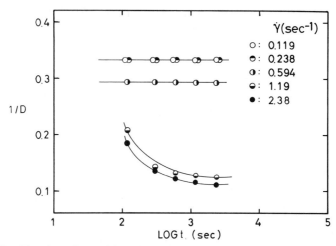

Fig. 17. Time dependence of the reciprocal of D, the broadness of orientation distribution, for various shear rates.

that the shear rate region in which D is independent of time corresponds to the yielding region of the flow curve, in which the apparent viscosity increases with decreasing rate of shear.

The apparent viscosity is plotted against rate of shear in Fig. 18. This viscosity curve can be divided into two regions, the yielding region at lower rates and the Newtonian region above a certain critical rate of shear $\dot{\gamma}_c$ (neglecting the shear-thinning region observed at the high shear-rate end). Below the critical shear rate $\dot{\gamma}_c$ the sample probably flows with retention of some superstructure in which many elements are packed in a regular manner.

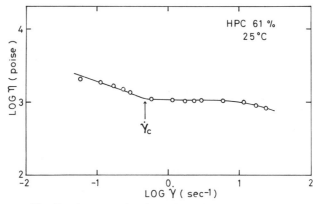

Fig. 18. Apparent viscosity as a function of shear rate ($\dot{\gamma}$).

Under such circumstances, Bragg elements can hardly be reoriented, even after cessation of steady shear flow. On the other hand, above the critical shear rate the superstructure is broken down into smaller flow units, which may be as small as the Bragg elements. Thus when the shear perturbation ceases, these elements are reoriented very rapidly. Consequently, the uniformity of the orientation of Bragg elements can be improved, as reflected by a substantial increase in D with time.

VI. Deformation and Flow Mechanisms for Some Typical Polymeric Liquid Crystalline Solutions

A. PBG [Racemic Poly(γ-benzyl glutamate)]

The rheo-optical properties of nematic liquid crystals of racemic poly(γ-benzyl glutamate) ($\bar{M}_w = 15 \times 10^4$) in m-cresol depend markedly on sample thickness. A typical example is shown in Fig. 19, where d denotes the sample thickness. Measurements of optical intensities were carried out by means of 2 parallel plate viscometer instead of a cone and plate. The shear rate $\dot{\gamma}_i$, marking the onset of the wavy change of $I_\times$ and $I_{\parallel}$ as well as the decrease in I_E, is considered the shear rate at which the continuous phase becomes predominant in the sample. The value of $\dot{\gamma}_i$ is smaller, the thinner the sample.

The effective region of a disclination may be reduced by shear, or some disclinations may be squeezed out by continuous shearing. The squeezing of disclinations may be done more effectively in a thinner sample than in a thicker one. Therefore, a monodomain [Fig. 6(3)] is more easily achieved in thinner samples.

The rheo-optical properties measured for 20 and 40 wt% solutions by means of a cone and plate with a cone angle of 0.865 are shown in Figs. 20 and 21. The value of I_E is very large in the undisturbed state for all the solutions examined.

For a 20% solution $\dot{\gamma}_i$ is as low as 8×10^{-2} sec^{-1}. This indicates that the polydomain structure of the 20% solution in the undisturbed state changes easily to the degraded polydomain structure [Fig. 6(2)] even at low shear rates. The continuous phase grows with increasing shear rate, the residual disclinations are gradually removed, and a monodomain is formed at high shear rates. The rheo-optical behavior of a 30% solution is almost the same as that of a 20% solution.

In the case of a 40% solution, I_E, $I_\times$, and $I_{\parallel}$ begin to decrease at lower shear rate ($\dot{\gamma} = 7 \times 10^{-2}$ sec^{-1}). This decrease in I_E does not indicate the disappearance or orientation of the polydomain structure, but arises from dynamic light scattering due to polydomain flow. Since $\dot{\gamma}_i$ is as high as

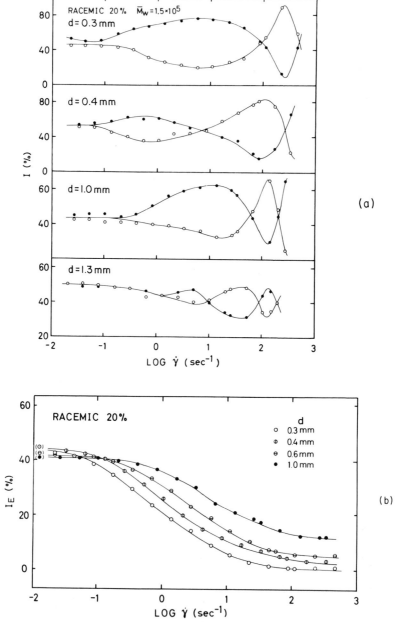

Fig. 19. Effect of sample thickness d for a 20% solution of PBG in m-cresol. (a) $I_\times$ (○) and $I_{||}$ (●) plotted against shear rate $\dot{\gamma}$. (b) I_E plotted against shear rate $\dot{\gamma}$. (Reprinted with permission from Asada *et al.*, *Macromolecules* **13**, 867 (1980). Copyright 1980 American Chemical Society.)

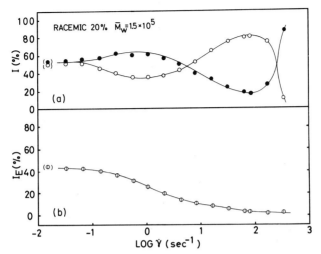

Fig. 20. Variation of (a) $I_{\times}$ (○), $I_{\parallel}$ (●), and (b) I_E (⊕) with shear rate $\dot{\gamma}$ for 20% PBG solution. (From Asada and Onogi, 1980.)

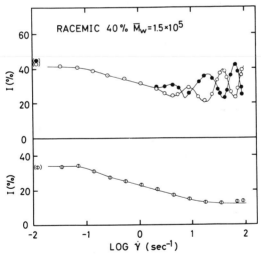

Fig. 21. Variation of $I_{\times}$ (○), $I_{\parallel}$ (●), and I_E (⊕) with shear rate $\dot{\gamma}$ for 40% PBG solution. (From Asada and Onogi, 1980.)

2.2 sec^{-1}, high shear rates are required for the system to form a monodomain. These results indicate that textures in a 40% solution are more stable than those in less concentrated solutions.

Figure 22 shows viscosity data for solutions at three different concentrations. The apparent viscosities η of the 20 and 30% solutions do not depend

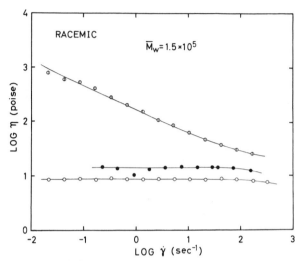

Fig. 22. Shear rate dependence of the apparent viscosity of three PBG solution: 20% (○); 30% (●); 40% (①). (Reprinted with permission from Asada *et al.*, *Macromolecules* **13**, 867 (1980). Copyright 1980 American Chemical Society.)

on shear rates below 100 sec^{-1}, while that of the 40% solution exhibits a strong shear rate dependence in the entire region of $\dot{\gamma}$.

Presumably, at very low rates a shear rate dependence of the apparent viscosity would also be observed for the 20 and 30% solutions. The decrease in the apparent viscosity for the 40% solution indicates yielding, and may arise from plastic flow of domains. These results for the 40% solution suggest that domains in concentrated solutions are more stable than those in less concentrated solutions. The higher mechanical stability of the domains in the 40% solution is also responsible for the decrease in light transmission with increasing shear rate due to the dynamic scattering.

B. PBLG[Poly(γ-benzyl-L-glutamate)]

In Figure 23 I_E measured for 10, 15, 20, and 40% m-cresol solutions of PBLG ($\bar{M}_w \cong 5.1 \times 10^5$) in the steady state is plotted against log $\dot{\gamma}$. These solutions are anisotropic in the unsheared state at 26°C. The results for I_E, $I_\times$ and $I_\parallel$ for the 10% solution indicate that the polydomain cholesteric structure of the 10% solution in the unsheared state transforms easily to a continuous nematic phase, even at low shear rates. For the 40% solution I_E remains high, even at the highest shear rate, and the system maintains its polydomain cholesteric structure under flow. In this case no long range orientation of the molecules occurs at high shear rates. Again, the organiza-

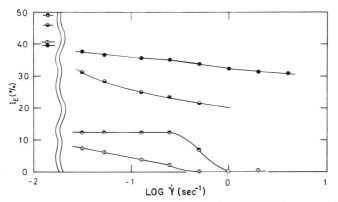

Fig. 23.   The variation of I_E with shear rate $\dot\gamma$ for solutions of PBLG in m-cresol: 10% (◐); 15% (◓) v 20% (◒); 40% (●); (26°C). (From Asada *et al.*, 1978b.)

tion of domains in concentrated solutions is found to be more stable than that in less concentrated solutions, just as in PBG solutions. This conclusion is also supported by the recovery of I_E after the cessation of steady flow, as is shown in Fig. 24. The value of I_{ER} (%) in this figure represents the percent recovery of I_E to its value in the unsheared state. A comparison with Fig. 23 clearly shows that I_{ER} is approximately zero when the steady state I_E is very small. On the contrary, I_{ER} becomes higher than 80% when the steady state value of I_E is high, as can be seen for the 15% solution in the low shear-rate region, and for the 20 and 30% solutions over the entire region. When the system flows as an oriented nematic phase, as in the 10 and 15% solutions in the high shear rate region, the system retains the same texture after the cessation of steady flow, and I_E remains near zero.

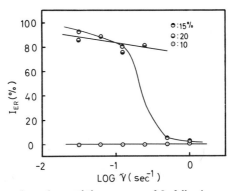

Fig. 24.   Shear rate dependence of the recovery of I_E following cessation of steady flow for solutions of PBLG in m-cresol. I_{ER}(%) represents the percentage of recovery to the value in the unsheared state. (From Asada *et al.*, 1978b.)

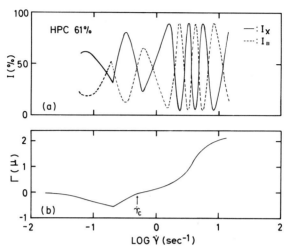

Fig. 25. Shear rate dependence of I_x and $I_{||}$ for a 61% HPC solution (top). Shown below is the shear rate dependence of the retardation (Γ) obtained from the data above using Eq. (1).

C. Hydroxypropylcellulose (HPC)

Some aspects of the deformation and flow mechanisms of aqueous cholesteric liquid crystalline solutions of HPC have been discussed in Section V. The behavior in steady shear has been shown in Fig. 18 [Section V]. In Fig. 25a, the steady state I_x and $I_{||}$ are plotted against shear rate. The wavy behaviors of I_x and $I_{||}$ can be converted by means of Eq. (1) to the retardation behavior versus shear rate shown in Fig. 25b. One sees that the birefringence is negative during the yielding region (below $\dot\gamma_c$). This negative birefringence can be explained by tilting of the rather thick cholesteric blocks, the optic axes of which were initially normal to the shear plate. Above $\dot\gamma_c$ orientation of the individual molecules may occur and positive birefringence appears.

Steady-State spectra obtained at various shear rates are shown in Fig. 26b. In Fig. 26a are shown calculated curves obtained by computer fitting, using Eq. (9) and the values of parameters listed in the figure. The experimental curve obtained at the shear rate in the yielding region shows a maximum around 580 nm. The height of the maximum decreases with increasing shear rate. In the Newtonian region the λ_M (the wavelength at the peak) shifts to the blue with increasing shear rate. The values of the parameters indicate that the structural change during yielding arises from a change in B; the cholesteric element becomes thinner and the perfection of the structure decreases with increasing shear rate. The structural change in the Newtonian

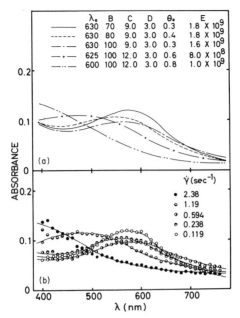

Fig. 26. Computer generated values using Eq. (9) for steady shear flow. (a) Calculated curves using the parameters listed, where θ_0 is the tilt angle of the orientation axis in radians. (b) Experimental results. (From Asada *et al.*, 1981.)

region arises from the change in θ_0; the tilting angle θ_0 in radians increases considerably with increasing shear rate, while θ_0 is almost unchanged at lower shear rates. As described above, the superstructure is disintegrated in this high shear-rate region into smaller flow units. These units can easily be oriented so that their optic axes are somewhat tilted from the normal, because of their small size. When shear flow ceases, the units are oriented so that their optic axes are normal to the surface of the plate. This is the structure formation process discussed above.

Highly oriented Grandjean-like textures can be attained only by reorientation of the smaller flow units after cessation of shear flow, and for this to occur, liquid crystals must flow at rates exceeding the critical shear rate.

V. Concluding Remarks

Both the rheological and rheo-optical properties of polymeric liquid crystals may be interpreted by the structural model shown in Fig. 6. Under shear or other external excitation, polydomains can be transformed to a monodomain texture. However, no long-range orientation is expected so long as polydomain flow takes place. To achieve uniform molecular orientation throughout the sample, one must find conditions under which the monodomain texture appears. To obtain an oriented monodomain texture, the disclinations must be forced out of the nematic liquid crystal, or the many domains in the polydomain system must coalesce and grow larger. For this purpose the applied stress (or strain) must be delivered effectively to individual domains. Slippage at the interface of domains may prevent the effective delivery of the stress (or strain). Consequently, in order to obtain a highly oriented sample, it is necessary to take account of the mode of deformation, including the geometry of the apparatus as well as the size and stability of the domains or superstructures, which may depend on molecular weight, concentration, temperature, pressure, and other factors.

The fact that extensional deformation can give a highly oriented sample much more easily than shear deformation can may be related to the problem of strain propagation. Further investigation of the mode of propagation of strain (or stress) in polydomain systems appears to be important in connection with fiber spinning in the liquid crystalline state.

The above results for PBG and PBLG show that the superstructures of domains at higher concentrations are smaller and more stable to mechanical excitation, and that higher shear rates (or larger stresses) are required to obtain an oriented continuous phase. Application of both rheo-optical methods (the polarized light and the spectrophotometric) makes it possible to clarify the deformation and flow mechanisms of complicated cholesteric liquid crystalline aqueous solutions of HPC in more detail. One conclusion from this study is that a highly oriented Grandjean-like texture can only be achieved when a system has been deformed at rates exceeding the critical shear rate.

References

Aoki, H., Coffin, D. R., Hancock, Harwood, D., Lenk, R. S., Fellers, J. F., and White, J. L. (1978). *J. Polym. Sci., Polym. Symp.* **65**, 29.
Aoki, H., Harwood, D., Lee, Y., Fellers, J. F., and White, J. L. (1979a). *J. Appl. Polym. Sci.* **23**, 2155.
Aoki, H., White, J. L., and Fellers, J. F. (1979b). *J. Appl. Polym. Sci.* **23**, 2293.
Aoki, H., Onogi, Y., White, J. L., and Fellers, J. F. (1980). *Polym. Eng.* **20**, 221.
Asada, T. (1976). *Nippon Reoroji Gakkaishi* **4**, 102.
Asada, T., and Onogi, S. (1980). *In* "Rheology [I]" (G. Astarita, G. Marrucci, and L. Nicolais eds.), pp. 127–147. Plenum Press, New York.
Asada, T., Maruhashi, Y., and Onogi, S. (1975). *Nippon Reoroji Gakkaishi* **3**, 129.
Asada, T., Maruhashi, Y., Kuroki, Y., and Onogi, S. (1978a). *Nippon Reoroji Gakkaishi* **6**, 14.
Asada, T., Muramatsu, H., and Onogi, S. (1978b). *Nippon Reoroji Gakkaishi* **6**, 130.
Asada, T., Muramatsu, H., Watanabe, R., and Onogi, S. (1980). *Macromolecules* **13**, 867.
Asada, T., Toda, K., and Onogi, S. (1981). *J. Mol. Cryst. Liq. Cryst.* **68**, 231.

Baird, D. G. (1978). *In* "Liquid Crystalline Order in Polymers" (A. Blumstein, ed.), pp. 237–259. Academic Press, New York.

Baird, D. G., Ciferri, A., Krigbaum, W. R., and Salaris, F. (1979). *J. Polym. Sci., Polym. Phys. Ed.* **17**, 1649.

Chandrasekhar, S., and Prasad, J. S. (1971). *J. Mol. Cryst. Liq. Cryst.* **14**, 115.

Ciferri, A. (1975). *Polym. Eng. Sci.* **15**, 191.

Duke, R. W., and Chapoy, L. L. (1976). *Rheol. Acta* **15**, 548.

Hermans, J. (1962). *J. Colloid Sci.* **17**, 638.

Horio, M. (1978). *Koenshu—Kyoto Daigaku Nippon Kagaku Sen'i Kenkyusho* **35**, 87.

Iizuka, E. (1974). *Mol. Cryst. Liq. Cryst.* **25**, 287.

Kiss, G., and Porter, R. S. (1978). *J. Polym. Sci., Polym. Symp.* **65**, 193.

Kiss, G., Orrell, T. S., and Porter, R. S. (1979). *Rheol. Acta* **18**, 657.

Kwolek, S. L. (1972). U.S. Patent 3,671,542.

Marrucci, G., and Ciferri, A. (1977). *Polym. Lett.* **15**, 643.

Papkov, S. P., Kulichikin, V. G., Kalmykova, V. P., and Malkin, A. Y. (1974). *J. Polym. Sci., Polym. Phys. Ed.* **12**, 1753.

Patel, D. L., and Dupré, D. B. (1979). *Rheol. Acta* **18**, 543.

Pochan, J. M., and Marsh, D. G. (1972). *J. Chem. Phys.* **57**, 1193.

Robinson, C. (1956). *Trans. Faraday Soc.* **52**, 571.

Valenti, B., and Ciferri, A. (1978). *Polym. Lett.* **16**, 657.

White, J. L., and Fellers, J. F. (1978). *J. Appl. Polym. Sci.: Appl. Polym. Symp.* No. 33, 137.

Wong, C. P., Ohnuma, H., and Berry, G. C. (1978). *J. Polym. Sci., Polym. Symp.* **65**, 173.

Yang, J. T. (1958). *J. Am. Chem. Soc.* **80**, 1783.

10

The Effects of External Fields on Polymeric Nematic and Cholesteric Mesophases

W. R. Krigbaum

Paul M. Gross Chemical Laboratory
Duke University
Durham, North Carolina

I. Introduction

The observation by Reinitzer (1888) of what appeared to be two melting transitions in cholesteryl benzoate is usually cited as the event that initiated the study of mesomorphism or liquid crystalline behavior. His findings were confirmed by Lehmann (1889), who coined the term "liquid crystal" to denote the partially ordered liquid phase that forms upon melting

275

the crystalline solid. By the time Friedel (1922) proposed a new nomenclature, which included the terms "mesomorphic," "nematic," and "smectic," a number of low molecular weight compounds had been synthesized that exhibited liquid crystalline or mesomorphic behavior. Turning to macro-molecules, Bawden and Pirie (1937) reported that, when its concentration was increased a solution of tobacco mosaic virus separated into two phases, one of which was birefringent. Later Elliott and Ambrose (1950) found that a chloroform solution of γ-benzyl-L-glutamate (PBLG) spontaneously forms a birefringent phase under similar conditions. Thus when Flory (1956) pre-sented his first lattice model treatments of anisotropic phases in solutions of rodlike and semiflexible polymers, only biopolymers could be cited as examples. In those cases the requisite rodlike shape is achieved by virtue of the rigidity of a helical structure. Although the block copolymers such as the Shell Kratons can be considered to be mesomorphic, the current interest in polymers exhibiting liquid crystalline behavior was certainly sparked by the Kwolek (1971) patent, and by its subsequent commercialization as Kevlar fiber. The Kwolek patent disclosed that aromatic polyamides could be synthesized with a sufficiently extended chain conformation to exhibit a lyotropic nematic phase. By the time of this growing interest in polymeric mesophases, there already existed an extensive body of theory and experi-mental observations concerning the behavior of low molecular weight liquid crystals. In order to utilize this valuable background, those who study polymeric mesophases are well advised to compare the behavior observed for a polymeric liquid crystal with that of its corresponding low molecular weight analog. This chapter is written with that rationale in mind, having as its objective the presentation of certain aspects of the behavior of low molecular weight mesogens that may prove to be useful to the polymer chemist interested in polymeric mesophases.

Molecules that exhibit a liquid crystalline phase are most frequently anisodiametric in shape (e.g., rods or disks). This necessarily implies that many of the properties of these materials will be anisotropic when measured in a coordinate frame set in the molecule. For example, most mesogenic molecules are diamagnetic, so the magnetic susceptibility χ will be small in magnitude and negative. The diamagnetic anisotropy, $\chi_a = \chi_{\parallel} - \chi_{\perp}$, is usually positive, hence there will be a tendency for the molecules to be oriented with their long axes parallel to a magnetic field. Of course, if the molecules interacted individually (as in a vapor, for example), the fraction of molecules so aligned would be quite small, even for a magnetic field of 100 kG. However, in a liquid crystalline phase the interaction is cooperative, and a second-order transition to an aligned phase can occur with a field of only 10 kG. Other anisotropic properties are the dielectric constant, electrical conductivity, and viscosity. In summary, external fields have an exceptional effect upon mesophases due to their unique combination of anisotropy of the molecules and cooperativity of the molecular interactions.

In this chapter we shall consider the effect of external magnetic and electric fields and of flow fields upon polymeric nematic and cholesteric phases.

II. Magnetic-Field Effects

A. Low Molecular Weight Nematics

As indicated above, a bulk nematic phase can be aligned by imposing a magnetic field. Molecular alignment also occurs at an interface. By appropriate treatment of the supporting surface, the nematic molecules can be aligned either parallel or perpendicular to the surface, and this alignment extends for a distance into the nematic phase. If the experimental geometry is appropriately selected so that the alignment conferred by the surface differs from the preferred alignment in the magnetic field, the onset of realignment in the bulk of the nematic phase can be caused to occur suddenly at a critical field. This second-order transition, the Frederiks transition, has been used to evaluate the three elastic constants of splay, twist, and bend, which appear in the continuum theory. If the surface anchoring is firm, even after alignment in the field, a layer near the bounding surfaces will retain the original surface orientation. The thickness of this layer has been estimated by de Gennes (1971) in terms of a "magnetic coherence length," which is about 1 μm for fields of the order of 10 kG. The onset of the ordering transition becomes diffuse if the applied field is not strictly perpendicular to the initial anchoring position, and Rapini and Papoular (1969) have discussed the consequences of weak anchoring.

The first experimental study of the magnetic ordering of nematics was reported by Frederiks and Zolina (1933). They observed that the critical field is inversely proportional to the sample thickness. Shortly thereafter Zöcher (1933) showed that this observation follows as a natural consequence of continuum theory. For example, if the supporting glass plates have been rubbed so that the molecules are initially parallel to these planes and have a preferred direction, and if the magnetic field is perpendicular to the plates and to this preferred direction, the critical field H_c is given by

$$H_c = (\pi/d)(k_{11}/\chi_a)^{1/2} \tag{1}$$

where k_{11} is the elastic constant for splay, d is the sample thickness, and χ_a is the magnetic anisotropy. More extensive measurements on nematic materials were performed by Frederiks and Tsvetkov (1934).

B. Low Molecular Weight Cholesterics

A cholesteric structure is a twisted nematic in which, on traversing successive layers of the nematic structure, the director vectors specifying the preferred orientation within succesive layers trace out a helix. The helical axis is perpendicular to the layers, and hence always perpendicular to the directors representing the preferred orientation within each of the layers. If a magnetic field is applied to the cholesteric phase of a material having negative magnetic anisotropy, the helical axes align parallel to the field. Then all molecular axes are orthogonal to the field direction and there is no change of pitch. A polydomain sample is simply transformed into one having a uniform planar texture. On the other hand, if the magnetic anisotropy is positive, it is only possible to achieve the preferred alignment for part of the molecules by turning the helical axes perpendicular to the field. Those portions of the structure in which the director is along the field have lower energy and tend to expand along the helix axis. This leaves a succession of 180° walls parallel to the field, which separate the "straight" regions. With increasing field strength the separation of the walls increases, and eventually diverges logarithmically to produce an untwisted nematic structure at a critical field H_c. The relation for the critical field derived by Meyer (1968) and de Gennes (1968a) is

$$H_c = \pi^2 k_{22}/\chi_a P_0 \qquad (2)$$

where k_{22} is the twist elastic constant and P_0 is the unperturbed pitch. With the magnetic fields normally available, the cholesteric–nematic transition can only be induced in materials having a relatively large pitch. Durand et al. (1969) have studied the effect of a magnetic field on a mixture of 2% cholesteryl acetate and 98% p-azoxyanisole. Their data for the field strength dependence of the helical pitch are in good agreement with Eq. (2), as pointed out by Meyer (1969a).

A thin cholesteric layer having a Grandjean or planar texture (helical axes perpendicular to the bounding surfaces) can be prepared by placing the material between two glass slides that have been rubbed. If a magnetic field is applied along the helical axes (i.e., perpendicular to the slides), surface anchoring will prevent rotation of the helical axes. Helfrich (1971) pointed out that the directors will then undergo a periodic distortion in the two directions orthogonal to the field. This gives rise to a square grid pattern, as observed by Rondelez and Hulin (1972). A similar square grid pattern can be created in a cholesteric phase by an electric field, and Hurault (1973) has given a theoretical treatment of both cases.

A sample having planar texture can also be formed into a Cano wedge by inclining the two confining glass slides at a small angle. Anchoring at the two surfaces must then impose some local distortion of the pitch to allow an integral number of half pitches to be accommodated at every thickness. A sharp disclination line of strength $\frac{1}{2}$ separates regions differing by half a turn. Cano (1968) established that the orientation of the director differs by 180° on the two sides of a boundary that is now referred to as a Grandjean–Cano wall. The effect of a magnetic field on such a sample was treated theoretically be de Gennes (1968b), and Scheffer (1972) made some further refinements to the theory. The Orsay Liquid Crystal Group (1969) found that thicker samples also exhibit double disclinations separating regions differing by a full turn. They observed that single disclinations are unaffected by a magnetic field, while double disclinations are deformed into a zigzag shape at a threshold field that is about half that required for the cholesteric–nematic transition. An explanation for the zigzag pattern was provided by Kleman and Friedel (1969).

C. Polymeric Nematic Phases

The effect of a magnetic field on a lyotropic nematic phase of poly(p-benzamide) was first investigated by Kol'tsov et al. (1973). They inferred that the molecules have a preferred alignment parallel to the field from the NMR spectrum of the solvent. Solutions of this polymer in dimethylacet-amide containing 3% LiCl were subsequently studied more thoroughly by Papkov and co-workers (Platanov et al., 1976). In this case the molecular alignment was determined directly by infrared dichroism and x-ray diffraction performed on the solution. Considerable alignment was introduced by flow as the very thin sample cells were filled. A system of boundary walls parallel to the flow direction formed spontaneously and these only gradually diminished in number with time. Application of a 7-kG field perpendicular to the flow direction led to a realignment in the bulk of the solution. Magnetic alignment was complete in approximately 2 h for a 12% solution, which was entirely nematic, but the process was slower for 7.2 and 9.7% solutions, which were biphasic. Alignment of the surface layers remained unchanged in the field and promoted recovery of the initial texture after the magnetic field was turned off. Anchoring may also offer an explanation for the appearance of regular bands parallel to the field after the sample had been in the field 10–20 min. If observed between crossed polars with a quartz plate the areas between the bands are seen to alternate in color. After longer exposure to the magnetic field the number of bands decreased and their spacing increased. If a sample having a banded structure is rotated by 90° in the magnetic field, the bands are transformed into zigzags, as shown in Fig. 1.

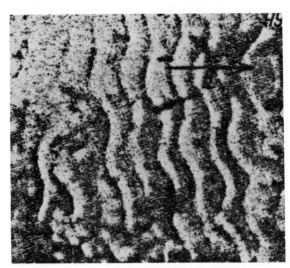

Fig. 1.   Zigzag lines 25 min after rotating a nematic sample of PBA by 90° in a magnetic field. (Reprinted with permission from *Polymer Sci. USSR* **18**, V. A. Platonov, G. D. Litovchenko, T. A. Belousova, L. P. Mil'kova, M. V. Shablygin, V. G. Kuhlichikhin, and S. P. Papkov. Copyright 1976, Pergamon Press, Ltd.)

Panar and Beste (1977) also investigated poly(*p*-benzamide) in dimethyl-acetamide with added LiCl and 1% water to reduce the rate of crystallization. They reported that a fully nematic phase becomes more transparent along the direction of a 10-kG field. If the system is biphasic with a continuous nematic phase, droplets of the isotropic phase are football-shaped. These align with their long axes along the field after standing overnight in a 10-kG field, as shown in Fig. 2.

Studies of the effects of magnetic fields on low molecular weight nematics were confined to thermotropic materials, whereas the polymeric nematic phases referred to above were all lyotropic. This was due to the fact that for a number of years few polymers forming a thermotropic nematic phase were available. However, in the past few years a number of polymers of this type have been synthesized by combining both rodlike and flexible segments in the repeating unit.

Liébert *et al.* (1981) investigated the effect of magnetic fields on the thermotropic nematic phase of a polyester having the repeating unit

$$\left[\overset{\overset{\displaystyle O}{\parallel}}{C}-(CH_2)_5-\overset{\overset{\displaystyle O}{\parallel}}{C}-O-\bigcirc-\overset{\overset{\displaystyle O}{\parallel}}{C}-O-\bigcirc-O \right]$$

The x-ray diffraction diagram of the nematic phase at 205°C indicated

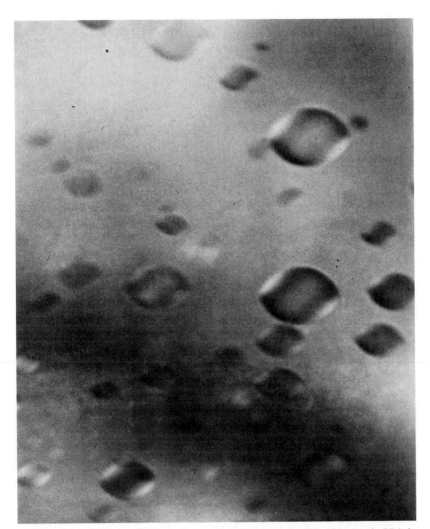

Fig. 2.   Orientation in a 10-KG field of football-shaped tactoids of isotropic PBA in a continuous nematic phase. (Reprinted with permission from Panar and Beste, *Macromolecules* **10**, 1401 (1977). Copyright 1977 American Chemical Society.)

alignment of the molecules parallel to a 3-kG field. A well-oriented fiber diagram was obtained by cooling the sample to 100°C in the field. After exposure to a field of 3 kG for 24 h the order parameter $S = \langle 3\cos^2\theta - 1\rangle/2$ was found to be 0.64 for a sample having inherent viscosity 0.29 dl/g, and 0.54 for one having a higher inherent viscosity, 0.54 dl/g. This work was extended by Hardouin *et al.* (1982) in a study of the apparent magnetic anisotropy and viscosity at low shear rate for the nematic phases of polymers of the type shown above, but having different numbers n of methylene units in the flexible dibasic acid. For a 1-min orientation time the threshold field for the onset of magnetic alignment was found to increase with increasing viscosity of the mesophase. The time course of the development of alignment in a 15-kG field was also followed. No evidence of alignment was observed after 200 min for the polymer ($n = 12$) having the nematic phase of highest viscosity. The nematic phase of next highest viscosity, for the polymer with $n = 9$, reached a plateau value of $\chi_a = 0.5 \times 10^{-7}$ emu/g, while other members of the series having less viscous nematic phases reached plateau values of about 1.1×10^{-7} emu/g. Noel et al. (1981) reported $\chi_a = 1.4 \times 10^{-7}$ emu/g for a copolyester synthesized from terephthalic acid and an equimolar mixture of the bisacetates of methylhydroquinone and pyrocatechol. Maret et al. (1981) investigated ordering effects of a magnetic field on two polyesters that they had previously characterized as thermotropic nematogens:

$$\left(\!\!-\overset{\overset{O}{\|}}{C}-(CH_2)_8-\overset{\overset{O}{\|}}{C}-O-(CH_2)_2-O-\!\!\bigcirc\!\!-\!\!\bigcirc\!\!-O-(CH_2)_2-O-\!\!\right)_x$$

$$\left(\!\!-\overset{\overset{O}{\|}}{C}-(CH_2)_8-\overset{\overset{O}{\|}}{C}-O-(CH_2)_2-O-\!\!\underset{CH_3}{\bigcirc}\!\!-CH{=}CH-\!\!\underset{CH_3}{\bigcirc}\!\!-O-(CH_2)_2-O-\!\!\right)_x$$

These polymers were heated in a capillary to the isotropic state and cooled slowly in a rather large magnetic field, 100 to 160 kG, applied along the axis of the capillary. No preferred orientation was detected in either sample after it had been cooled to room temperature in a magnetic field. The two polymers were again heated to produce the isotropic phase, and the birefringence was measured perpendicular to the applied field as the samples were cooled. The birefringence of the first polymer appears to have been much larger in the isotropic phase, but it did not increase on cooling. The birefringence of the second polymer increased to a maximum about 19° below the isotropic–nematic transition temperature, but then it decreased again before the solution became turbid. The authors conclude from these observations that in all probability neither polymer exhibits an enantiotropic nematic phase. This study was extended by Maret and Blumstein (1982) to other polymers, with some exhibiting the type of behavior described above and others giving well oriented x-ray patterns when the polymer was cooled in a magnetic field. Volino et al. (1981) investigated the effect of a magnetic field on the NMR spectrum of a polymer having the repeating unit

$$\left(O-\bigcirc-\overset{\underset{\displaystyle CH_3}{|}}{N}=\overset{\underset{\displaystyle CH_3}{|}}{N}-\bigcirc-O-\overset{O}{\underset{\|}{C}}-(CH_2)_{10}-\overset{O}{\underset{\|}{C}} \right)$$

They deduced from these data unusually large values for the order parameter S, ranging from 0.72 to 0.88. Blumstein *et al.* (1981) have also reported values of the Cotton–Mouton constant for solutions of several thermotropic polymers and low molecular weight model compounds.

At the present time no values have been reported for the elastic constants of polymeric nematic phases. As we have seen, values for the magnetic anisotropy have been reported, however, and we may confidently expect that values of k_{11} for polymeric nematic phases will appear in the near future.

D. Polymeric Cholesteric Phases

Most of the investigations of the effect of a magnetic field on polymeric cholesteric phases have involved poly(γ-benzyl-L-glutamate) (PBLG). This polymer has a positive diamagnetic anisotropy and a relatively large pitch, so the unwinding transformation should be observable. The critical magnetic field H_c has been determined by Duke and DuPré (1974a), and by Samulski and co-workers (see Sridhar *et al.*, 1974). Duke and DuPré (1974a) obtained $H_c = 5.06$ kG for a 20% solution of PBLG of molecular weight 310,000 in dioxane. Figure 3 illustrates their data plotted as P/P_0 versus H/H_c, where

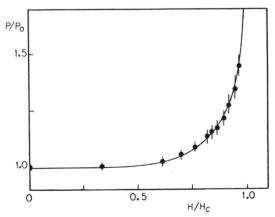

Fig. 3. Reduced pitch P/P_0 versus reduced magnetic field H/H_c for a cholesteric solution of PBLG in dioxane. (Reprinted with permission from Duke and DuPré, *Macromolecules* **7**, 374 (1974). Copyright 1974 American Chemical Society.)

P_0 is the unperturbed pitch. Sridhar *et al.* (1974) found that the kinetics of the untwisting transition could not be represented by a single exponential. They evaluated the rotational viscosity constant by following the reorientation of the nematic structure ($H > H_c$) after reducing the field below H_c and rotating the sample by 90°. Data for a 25.5% solution of PBLG of molecular weight 550,000 in CH_2Cl_2 is shown plotted as ln(tan ϕ) versus time in Fig. 4, where ϕ is the angle between the nematic director and the magnetic field. The rotational relaxation time τ was found to be proportional to H^{-2}. They also noted that τ could be reduced from hours to minutes by the addition of small amounts of trifluoroacetic acid. DuPré and Hammersmith (1974) made an optical study of the cholesteric–nematic transition, and of the alignment of the nematic phase, of chloroform solutions of PBLG in a magnetic field. They used a 12-kG field, which is considerably larger than the critical field (about 5 kG). Their solutions were held in 1-cm spectrophotometer cuvettes in an attempt to reduce the effect of surface orientation. However, after filling the cell several weeks were required for the striations to disappear. The samples were also left in the field for several weeks. Despite their attempts to reduce the surface/volume ratio, surface alignment was still evident when the aligned nematic phase was viewed perpendicular to the field direction. When viewed exactly along the field, the sample was seen to give extinction under crossed polars, but birefringence colors were evident on viewing slightly off axis. As might be anticipated, the view under crossed polars was quite sensitive to thermal gradients. Rotation of the nematic phase by 90° in the field led to reorientation, and striations appeared along the length of the cell (perpendicular to the field) early in the process.

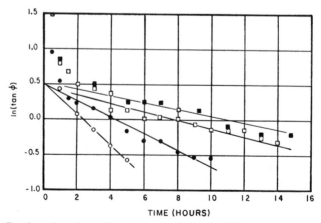

Fig. 4. Reorientation of an aligned sample of nematic PBLG in CH_2Cl_2 after rotating 90° in a magnetic field. Angle between the director and field is ϕ. Field strengths (kOe): ○ = 18, ● = 15, □ = 12, ■ = 9. (From Sridhar *et al.*, 1974.)

The reorientation was said to be complete in 10–12 min, although it required 3–4 days after rotation to regain the characteristics of a well-oriented sample.

Panar and Beste (1977) prepared a lyotropic cholesteric phase by adding several percent of $(+)$-2-methylcyclohexanone to a 20% solution of poly(p-benzamide). A system of parallel lines was seen after filling the cylindrical cell, probably as a result of flow orientation. After 20 sec in a 17-kG field the lines were transformed into a zigzag pattern, and after 40 min the transition to the nematic phase was complete.

Maret *et al.* (1981) examined the effect of a magnetic field on a series of copolymers exhibiting a thermotropic cholesteric phase. The copolyesters were synthesized using 4,4′-dihydroxyazoxybenzene with various ratios of $(+)$-3-methyladipic acid and dodecanedioic acid. The helical pitch could be increased by reducing the proportion of the chiral monomer, and only cholesteric phases having a pitch in excess of 7000 Å showed any evidence of orientation when the cooled samples were examined by x-ray diffraction.

III. Electric-Field Effects

A. Background

Chapter 8 (by Chandrasekhar and Kini, this volume) considers electrohydrodynamic instabilities in low molecular weight mesogens. Summaries of both the theoretical and experimental aspects of these effects have also been given by Chandrasekhar (1977), and by Dubois-Violette *et al.* (1978).

One might expect, by analogy with the magnetic-field transition discussed above, that a thermotropic nematic phase should show preferential alignment in an electric field because of the dielectric anisotropy. For example, the dielectric constant of p-azoxyanisole is larger perpendicular to the molecular axis ($\varepsilon_a < 0$). Hence one would expect these molecules in a nematic phase to be preferentially aligned perpendicular to an applied electric field. The electric analog of the Frederiks transition is, in fact, seen if the observations are restricted to low fields and the nematogen has low conductivity. A theoretical treatment of the electric-field transition, as well as a series of experimental results, has been given by Gruler and Meier (1972a). An electric field can also elevate the nematic–isotropic transition temperature. According to Helfrich (1970a), the elevation should be proportional to the square of the field strength.

However, in the case of electric fields other phenomena may intervene, which makes this a richer area for experimental and theoretical study. For example, if the molecules have lower symmetry than a rod, a splay or bend deformation may induce a polarization. This effect was first predicted by

Meyer (1969b), who termed this a "piezoelectric effect." The term "flexo-electric effect" now appears to be more widely used. An experimental demonstration of this effect was given by Schmidt *et al.* (1972) who observed a bowing of the director vectors when an electric field was applied to N-(p'-methoxybenzilidene)-p-n-butylaniline (MBBA).

A second class of additional electri-field effects involves hydrodynamic flow, and in this case the anisotropies of both the dielectric constant and conductivity play a role. Early observations of motion induced in a ther-motropic nematic phase by the application of an electric field were reported by Bjornstahl (1933), Frederiks and Tsvetkov (1934), Tsvetkov and Mikhailov (1938), and Naggiar (1943). If a threshold value of the voltage or field is exceeded, the flows can become organized into distinct domain patterns recognizable under the microscope. These are termed electrohydrodynamic instabilities. The first such domain pattern was reported by Williams (1963). In the case of p-azoxyanisole the instability pattern gives an "anomalous" alignment of the molecules parallel to the field because the electrical con-ductivity is larger along the molecular axis. Theoretical treatments of the flow instabilities induced by electric fields have been given by Carr (1969), Helfrich (1969b, 1970b), Dubois-Violette *et al.* (1971), and Smith *et al.* (1975).

B. Low Molecular Weight Nematics

Since electric-field-induced instabilities are treated more extensively by Chandrasekhar and Kini (Chapter 8, this volume), we shall provide only an outline of the behavior of low molecular weight nematics to allow com-parison with the observation for the analogous polymeric phases.

In order to study flow instabilities the sample is formed in a sandwich arrangement between glass slides having a transparent conductive coating and it is viewed along the field direction under a microscope. Very briefly, flow instability patterns have the following origin. Because of the dielectric and conductivity anisotropies, bend oscillations of the director vectors can lead to the buildup of a space charge. These space charges create a field perpendicular to the applied field and also cause a hydrodynamic flow. Both the dielectric and viscous torques tend to increase the bend distortion angle, and if this reinforcement is of sufficient magnitude, it can lead to a roll instability. The instabilities are subdivided into two categories by a critical frequency f_c, depending upon whether or not the space charges are able to follow the oscillations of the external field. Figure 5 gives a schematic representation of the observations for ac fields, as indicated by Dubois-Violette *et al.* (1978). If one imagines a vertical line passing through f_c, the area to the left of this line is the conduction regime. The lower curve in

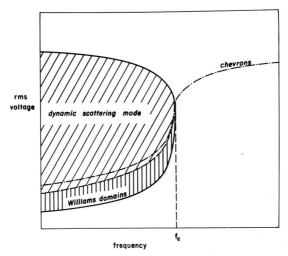

Fig. 5. Schematic voltage–frequency diagram of the different regions of electrohydro-dynamic instabilities. (After Dubois-Violette *et al.*, 1978.)

this region represents the frequency dependence of the threshold voltage for Williams domains. If the slides have been rubbed, the pattern consists of alternate bright lines and dark bands perpendicular to the rubbing direction. As explained by Penz (1971), the bright lines are due to the focusing action of counter-rotating cylindrical domains of flow. The threshold voltage is independent of sample thickness, but it increases with frequency up to the critical frequency. For a fixed frequency below f_c, on increasing the voltage one finds, in succession, Williams domains, chevrons (perhaps), and the dynamic scattering mode (DSM). Gruler and Meier (1972b) reported that for *p*-azoxyanisole at still higher voltages the DSM is replaced by a pattern of dark spots showing turbulent flow.

Above f_c there is the dielectric regime in which the space charges no longer oscillate with the field. Here the Williams domains and DSM no longer exist, but the chevron pattern does appear above a critical value of the electric field, and at still higher fields turbulence may also exist. The behavior described above is characteristic of materials having a negative dielectric anisotropy and a positive conductivity anisotropy. The types of instability patterns observable for other combinations of these signs have been discussed by Dubois-Violette *et al.* (1978).

Even for frequencies below f_c the conduction instability can be quenched by reducing the sample thickness. The pattern observed under these conditions is called the variable grating mode (VGM) because the spacing varies inversely with the applied voltage. VGM domains parallel to the rubbing direction were observed by Vistin' and Kapustin (1970) and by Vistin' (1971) for p-azoxyanisole, p-azoxyethoxybenzene, and for mixtures of these two nematogens. On the other hand, Greubel and Wolff (1971) reported VGM domains perpendicular to the rubbing direction for MBBA. Derzhanski and co-workers (1974) suggested that the VGM instability arises from flexoelectric effects, and this suggestion was subsequently confirmed by Soviet workers (Bobujlev *et al.*, 1977; Barnik *et al.*, 1978).

C. Low Molecular Weight Cholesterics

Cholesteric liquid crystalline phases also can be aligned in an electric field. The helical axes align parallel to the field if the dielectric anisotropy is negative, and this case is analogous to the effect of a magnetic field on a cholesteric material having a negative magnetic anisotropy. If the dielectric anisotropy is positive, the preferred direction of the helical axes is perpendicular to the field and at higher voltages the pitch can diverge logarithmically, as shown by Wysocki *et al.* (1968a,b, 1969). These authors investigated mixtures of cholesteryl halides and esters. Baessler and Labes (1969) concluded from ac field studies that the transition is due to interaction of permanent dipoles along the chain with the field.

As indicated earlier, a square grid pattern can be obtained by application of an electric field along the helical axes of an aligned cholesteric phase having negative dielectric anisotropy. Rondelez and Arnould-Netillard (1971), Rondelez (1973), and Arnould-Netillard and Rondelez (1974), as well as Gerritsma and van Zanten (1971) studied the Helfrich deformation for cholesterics and nematic–cholesteric mixtures. For a given frequency the threshold voltage was found to be proportional to $(d/P)^{1/2}$, where d is the sample thickness and P the helical pitch. Arnould-Netillard and Rondelez (1974), who investigated mixtures of nematic MBBA and cholesteryl nonanoate, also observed bidimensional instability patterns at frequencies above f_c. The latter are somewhat analogous to the behavior seen for nematics in the dielectric regime. For a fixed frequency below f_c, as the voltage is raised above the threshold, the helical axes rotate 90°, as shown by Kahn (1970), and at still higher voltages the DSM is observed. When the field is switched off, the system relaxes to a focal conic texture and scattering persists (storage mode). This can be cleared by an audio frequency pulse, leaving a transparent planar texture. The foregoing behavior was discovered by Heilmeier and Goldmacher (1968), and its application in conjunction with photoconductors to construct image storage panels has been described by Haas *et al.* (1973).

D. Observations for Polymers

Ringsdorf and co-workers (Finkelmann et al., 1979) investigated the electric field analog of the Frederiks transition for thermotropic methacrylate copolyesters having two types of mesogenic side chains:

$$-\overset{\overset{\text{O}}{\|}}{C}-O-(CH_2)_m-O-\bigcirc-\overset{\overset{\text{O}}{\|}}{C}-O-\bigcirc-C\equiv N$$

and

$$-\overset{\overset{\text{O}}{\|}}{C}-O-(CH_2)_n-O-\bigcirc-\overset{\overset{\text{O}}{\|}}{C}-O-\bigcirc-\bigcirc-OCH_3$$

In the examples they studied the (m, n) values were $(6, 6), (6, 2), (2, 6),$ and $(2, 2)$. The first two of these exhibited enantiotropic smectic and nematic transitions, and in the nematic phase a homeotropic texture could be achieved by applying an electric field. They did not determine the critical field strength, but they investigated the formation and relaxation time of the transition. For the first copolymer the half-time for orientation was about 10 sec at 8 V dc. The second copolymer was studied using 50 Hz ac voltage. The orientation time was less than 200 msec for 50 V, and the relaxation time was about 5 sec. This indicates that a polymeric nematic phase in which the mesogenic groups are on the side chains can have response times approaching those of low molecular weight nematics (tenths to hundredths of a second). The same authors also encountered flow instabilities, but they selected their experimental conditions so that instabilities were avoided.

The pitch of PBLG is relatively large, so a field-induced cholesteric–nematic transition should be accessible experimentally. Toth and Tobolsky (1970) studied CH_2Cl_2 and $CHCl_3$ solutions of PBLG. They observed a gentle stirring of the solution and a reduction in the amount of transmitted light at 300 V/cm, which they interpreted as a phase transition. However, they were unable to identify the mesophase formed. Duke and DuPré (1974b) applied quasi-elastic light scattering to the study of $CHCl_3$ solutions of PBLG in an electric field. As shown in Fig. 6, the line width narrowed significantly as the nematic phase formed. They determined that the cholesteric–nematic transition occurred at 350 V/cm for the system under investigation, and they obtained evidence of two different time constants in the unwinding transition.

Toyoshima et al. (1976) reported that an electric field could cause an isotropic–nematic transition in solutions of PBLG. This is an interesting

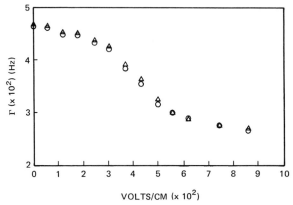

Fig. 6. Decrease of the spectral half-width Γ of light scattered from a solution of PBLG in $CHCl_3$ during the cholesteric–nematic transition in an electric field. (Reprinted with permission from Duke and DuPré, *Macromolecules* 7, 374 (1974). Copyright 1974 American Chemical Society.)

case because in the absence of a field and at a lower temperature or a higher concentration the stable phase would be cholesteric. The elevation of the transition temperature was found to be proportional to the square of the electric field. Later, Minami *et al.* (1978) reported that for PBLG in dioxane the isotropic–nematic transition at 320 V/cm was followed at 680 V/cm by the formation of a square grid pattern.

Krigbaum *et al.* (1980) investigated electro-hydrodynamic instabilities in a thermotropic nematic polymer, T2/60, provided by Tennessee Eastman Co. This is a *co*[poly(ethylene terephthalate) 1,4-benzoate] containing 60 mol % *p*-oxybenzoyl units. For each instability they compared the pattern observed for the polymer with the analogous instability exhibited by *p*-azoxyanisole. In view of the difficulty of obtaining a well aligned unique texture for polymeric nematics, it is fortunate that instability patterns can be formed in unoriented samples, although at a somewhat higher threshold voltage. Figure 7 compares the Williams domain pattern seen for T2/60 after 3 h at 275°C and 6.4 V dc with the corresponding pattern for an unoriented sample of *p*-azoxyanisole. Although Krigbaum and Lader (1980) reported this instability could be seen for T2/60 up to 100 Hz, the domain pattern formed more quickly and distinctly with dc, and the formation time was longer for lower temperatures in the nematic range. Attempts to orient 10-μm samples of this polymer by the rubbing technique were unsuccessful. No chevron or DSM pattern was observed. The variable grating mode domain pattern observed by Krigbaum *et al.* (1980) for a 4-μm sample of T2/60 after 1 h in the field is compared with the corresponding pattern in *p*-azoxyanisole in Fig. 8. The rubbing direction is horizontal, and there is some suggestion of partial

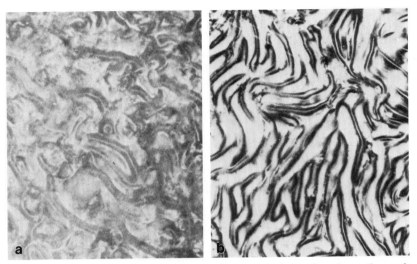

Fig. 7. Comparison of the Williams domain patterns seen in unoriented nematic samples of: (a) *p*-azoxyanisole, and (b) the copolyester T2/60. (Reprinted with permission from Krigbaum *et al.*, *Macromolecules* **13**, 554 (1980). Copyright 1980 American Chemical Society.)

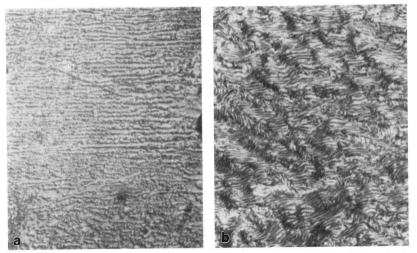

Fig. 8. Variable grating mode patterns in oriented nematic samples of low molecular weight PAA (a) and polymeric T2/60 (b). Rubbing direction is horizontal. (Reprinted with permission from Krigbaum *et al.*, *Macromolecules* **13**, 554 (1980). Copyright 1980 American Chemical Society.)

surface alignment for this thinner polymer sample. The high field turbulence patterns seen for T2/60 and p-azoxyanisole are shown in Fig. 9. This pattern appeared in the polymer after 0.5–1 min, and was not observed in the isotropic phase of T2/30, a copolyester having 30 mol % p-oxybenzoyl units. These authors suggested that the high field turbulence pattern might be suitable as a confirmatory test for the existence of a nematic phase in a thermotropic polymer.

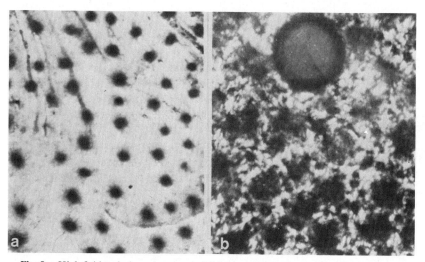

Fig. 9. High field turbulence in nematic PAA (a) at 100 V, 750 Hz, and T2/60 (b) at 150 V, 50 Hz. (Reprinted with permission from Krigbaum et al., *Macromolecules* **13**, 554 (1980). Copyright 1980 American Chemical Society.)

If the long formation time of instability patterns in T2/60 was due to its high viscosity, this problem could be overcome by studying polymer samples of low molecular weight. With this in mind Krigbaum et al. (1982) investigated the effect of electric fields on the thermotropic nematic phase of P-6, a 4,4'-dihydroxy-α-methylstilbene polymer having inherent viscosity 0.14 dl/g. A conduction regime instability pattern formed for this polymer in tenths of a second, as compared to an hour for the higher viscosity T2/60. The micrographs in Fig. 10 illustrate the variation of this pattern as the voltage is increased in 1-V steps. Figure 11 shows, for comparison, the fluctuating Williams domain (FWD) pattern in MBBA, which has been extensively studied by Balomey and Dimitropoulos (1976) and Hirakawa and Kai (1977). Slightly above the Williams domain threshold voltage the domains begin to undulate, as illustrated in Fig. 11a. With increasing voltage these undulations increase in amplitude and at voltage V^* a switching motion characteristic

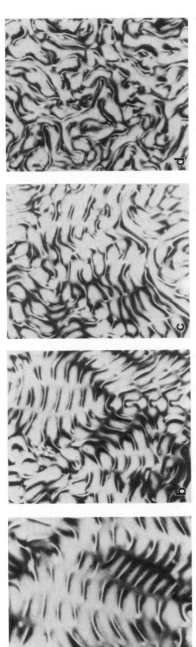

Fig. 10. Fluctuating Williams domain (FWD) pattern in a 20-μm thick sample of P-6 at 195°C. The frequency is 30 Hz and the voltages for micrographs are: (a) 7 V, (b) 8 V, (c) 9 V, and (d) 10 V.

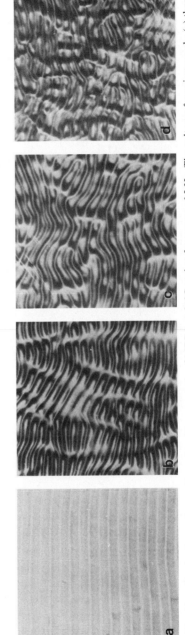

Fig. 11. Conduction regime patterns in a 20-μm sample of MBBA at 20°C and at frequency of 5 Hz. The voltages for micrographs (a) through (d) are: 5.5 V, 7 V, 8 V, and 10 V.

of the FWD sets in (Fig. 11b). With further increase in voltage the remaining portions of the original Williams domains (shown horizontal in Fig. 11a) tilt at larger angles, and the distance between crossovers decreases (Fig. 11c and d). Eventually the pattern in MBBA becomes disordered and the DSM sets in. The similarity between Figs. 10b and 11c is rather striking if one takes into account that the Chatelain rubbing technique did not result in a well-oriented sample of P-6. It was found for P-6 that V^* at a fixed temperature increases with frequency up to a critical frequency, decreases with temperature for a fixed frequency, and for fixed temperature frequency increases with sample thickness. It will be recalled that the threshold voltage for the Williams domain pattern is independent of sample thickness. In contrast with our previous experience with T2/60, P-6 did exhibit a DSM pattern, as shown in Fig. 12. Moreover, the formation time of this pattern in P-6 is a few seconds, which permitted us to obtain the micrograph of the DSM precursor shown in Fig. 13. It is interesting that this is more regular than the FWD pattern that existed at somewhat lower voltages. So far as we are aware, it has not been possible to photograph a DSM precursor pattern for

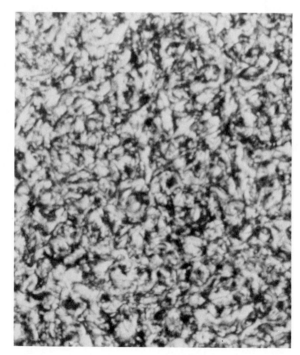

Fig. 12. DSM pattern in a 10-μm sample of P-6 at 195°C with an applied voltage of 20 V at 30 Hz.

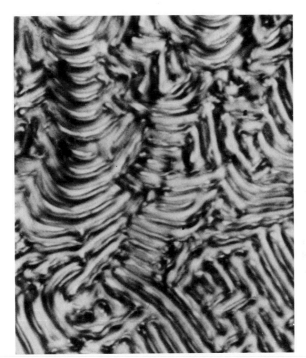

Fig. 13. Micrograph of the pre-DSM pattern taken under the conditions of Fig. 13, but a few seconds earlier.

any low molecular weight nematogen. This suggests that extension of studies of electrohydrodynamic instabilities to polymeric systems may provide some novel observations, since some parameters may exhibit values beyond the range available for low molecular weight nematogens.

IV. Viscous Flow

A. Background

Chapter 5 by de Gennes and Chapter 8 by Chandrasekhar and Kini present theoretical considerations concerning the viscosity of mesophases, while experimental techniques for the study of lyotropic cholesteric mesophases are described by Asada in Chapter 9.

The flow behavior of nematic and cholesteric mesophases is treated by the continuum theory formulated by Oseen (1933), and Zöcher (1933). Later Ericksen (1960, 1961) and Leslie (1966, 1968) developed the conservation laws and constitutive equations. The differential equations of Leslie were solved by Finlayson (1974) for a nematic phase under shear flow with various boundary conditions. No detailed solutions of the Ericksen–Leslie equations have been obtained for cholesteric phases. Helfrich (1969a) proposed that at low shear rates flow along the cholesteric helical axis occurs by permeation (i.e., with no motion of the helical structure) and that the velocity profile is flat in this case rather than parabolic. Chandrasekhar and co-workers (see Kini *et al.*, 1975) showed that the essential features of Helfrich's model could be derived from the Ericksen–Leslie theory. Flow normal to the helical axes of a cholesteric was treated by Leslie (1969). Doi (1980) has given a molecular theory of the dynamic behavior of concentrated solutions of rodlike macromolecules. This was recently extended by Marrucci (1982) to include the effects of the molecular field arising from nonuniformity of the directors or from an external magnetic field.

Consideration of the hydrodynamic components of the stress tensor leads to six coefficients of viscosity, five of which must be independent if Onsager's reciprocity relation is assumed to be applicable. Hence a complete description of the hydrodynamic behavior of a nematic or cholesteric material requires specification of at least five of the Leslie coefficients μ_i. The treatment of Marrucci mentioned above predicts the magnitudes of these coefficients for solutions of rodlike polymers as a function of the order parameter S. The experimental evaluation of these coefficients requires careful control of the boundary conditions, which include the molecular ordering and strength of anchoring at the surfaces, and the absence of disclination lines in the sample. Fisher and Frederickson (1969) showed that the viscosity of p-azoxyanisole, as measured in a capillary viscometer, is profoundly affected by the orientation at the capillary walls.

B. Low Molecular Weight Mesophases

Tsvetkov (1939) evaluated $\mu_2 - \mu_3$ by measuring the torque on a stationary nematic sample in a rotating magnetic field. Miesowicz (1946) obtained three viscosity constants using an oscillating plate viscometer with a magnetic field applied along three directions: parallel to flow, parallel to the velocity gradient, and perpendicular to both the flow and velocity gradient. Gähweller (1973) investigated the orientation profile of the directors during flow through a tube having a rectangular profile. He performed measurements in the presence and absence of a magnetic field, and he also studied the high field case, but with the field oriented at an arbitrary angle with respect to the flow coordinates. Combination of these results allowed Gähwiller to evaluate all five viscosity coefficients of MBBA, p-n-hexyloxybenzylidene-p'-aminobenzonitrile, and a nematic ternary mixture. Other experimental methods have been applied to evaluate some of the coefficients. For example, Léger-Quercy (1970) determined four of the viscosity coefficients for p-azoxyanisole by quasi-elastic light scattering measurements, while Martinoty and Candau (1971) used the reflection of shear waves at a solid–nematic interface to measure viscosity.

Porter *et al.* (1966) made a detailed comparison of the literature data concerning the flow behavior of low molecular weight nematics, cholesterics, and smectics. There is a slight decrease in viscosity on going from the isotropic to the nematic phase, whereas there is a large increase for the cholesteric phase if flow occurs along the helical axes of an aligned sample. Further, this latter increase is quite shear dependent, approaching a factor of 10^6 at low shear rates. Candau *et al.* (1973) studied the Poiseuille flow of cholesteric–nematic mixtures over a range of pitch values. In their geometry the helical axes are radial, so flow occurs normal to the helical axes. No unusual shear dependence was noted and the observed viscosity was about the same magnitude as that of a nematic.

Some additional observations of interest may be noted. Porter *et al.* (1966) reported that the monotropic transition of cholesteryl acetate becomes enantiotropic under shear. Gähwiller (1973) observed that nematic materials having a positive dielectric anisotropy exhibit anomalous flow properties as the temperature is reduced below that of the nematic–isotropic transition. He interpreted this anomaly as arising from a reversal of the sign of μ_3. Pieranski and Guyon (1973) reported a shear rate dependent flow instability that they attributed to a coupling of orientation and flow.

C. Polymeric Mesophases

One might expect, from the broad relaxation time spectrum characteristic of polymers and their long relaxation times, that the rheological behavior of polymeric mesophases would differ significantly from that of low molecular weight liquid crystals. Polymeric mesophases exhibit surface orientation and, moreover, are easily oriented by flow, even at quite low shear rates. For example, it may require a week to dissipate the orientation produced by flow during the filling of a cell.

Papkov and co-workers (see Kuhlichikhin *et al.*, 1979) have performed the only determination of the viscosity coefficients for a polymeric mesophase. They investigated poly(p-benzamide) (PBA) in dimethylacetamide with 3% added LiCl. They performed couette and falling ball measurements simultaneously, using capillary flow to provide orientation for the falling ball measurement, rather than using a magnetic field. While the flowing molecules appeared to be well aligned in the shear field, the average angle ϕ between the molecular axis and the flow direction could not be reduced below 30° at the highest shear rates studied. A third viscosity coefficient was determined by measuring the yield point of a homeotropic texture (molecules perpendicular to the surface) in a channel between two chambers. The values obtained for $\eta_\perp$, $\eta_{||}$, and η_c for an 11.8% solution of PBA were 12, 7, and 25 P. These may be compared with the values of η_a, η_b, and η_c obtained for p-azoxyanisole at 122°C by Miesowicz: 0.034 (extrapolated to 122°C), 0.024, and 0.092 P. Upon dividing the three values in each set by the smallest, one obtains the ratios 1.7, 1.0, and 3.6 for PBA and 1.4, 1.0, and 3.8 for p-azoxyanisole. As Papkov and co-workers (see Kuhlichikhin *et al.*, 1979) noted, $\eta_\perp$ and $\eta_{||}$ are not strictly equal to η_a and η_b since the disorientation angle could not be reduced below 30° by shear flow orientation. Despite this reservation, it is interesting that the viscosity ratios are so similar for two materials having a 300-fold difference in the magnitudes of their viscosity coefficients.

The same workers found, as shown in Fig. 14, that lower values of ϕ could be obtained during the recovery from flow at low shear rates. If the shear rate was below 10 sec^{-1}, ϕ fell below 15°C after the flow was stopped, and then increased to a plateau of 20–25°C as a structure of parallel domains formed. This domain pattern was visible under the microscope, or it could be studied by light scattering, and the domains were stable for some time. For shear rates in the range 10 to 30 sec^{-1}, ϕ was reduced momentarily during the relaxation process, but no domains were formed and ϕ increased continuously during the remainder of the disorientation process. Finally, only disorientation was observed during recovery if the flow rate exceeded 30 sec^{-1}. This type of domain formation during relaxation from flow appears to be unique to polymeric mesophases. Papkov and collaborators (see

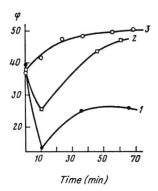

Time (min)

Fig. 14. Variation of the angle ϕ between the nematic director of PBA and the direction of flow during recovery from shear at (1) 1 sec^{-1}, (2) 16 sec^{-1}, and (3) 90 sec^{-1}. (Reprinted with permission from *Polymer Sci. USSR* **18**, V. G. Kuhlichikhin, V. A. Platonov, L. P. Braverman, T. A. Belousova, and S. P. Papkov. Copyright 1976, Pergamon Press, Ltd.)

Kuhlichikhin *et al.*, 1976) have suggested that piezoeoectric effects may be responsible for the formation of domains in this case.

Hermans (1962) performed the earliest study of the concentration dependence of the viscosity of rodlike polymers at various shear rates. He found that, for low shear rates, the viscosity of poly(γ-benzyl-L-glutamate) in *m*-cresol exhibited a sharp maximum with increasing concentration. This maximum has been interpreted as indicating the first appearance of the anisotropic phase. Subsequent work revealed a broad minimum following the maximum, and this was suggested to signal the completion of the transformation to the anisotropic phase. Both Matheson (1980) and Aharoni (1980) have considered the concentration dependence of the viscosity in terms of existing theory (Simha, 1940, 1949; Doi, 1975). Both point out that addition of a second phase leads to a viscosity increase, so the first appearance of the anisotropic phase should occur at a lower concentration than the corresponding to the viscosity maximum. There may (or may not) be a predicted change of slope at the concentration corresponding to the appearance of the anisotropic phase. Several earlier experimental studies (Iizuka, 1974; Kiss and Porter, 1978; Aharoni, 1979) had reported "anomalous" shoulders prior to the maximum. The viscosity maximum is predicted to occur when phase inversion leads to a continuous anisotropic phase. Aharoni and Walsh (1979) had demonstrated that in the case of the polyisocyanates the anisotropic phase was present at concentrations below that of the viscosity maximum. Balbi *et al.* (1980) investigated solutions of PBA in dimethyl-acetamide with added LiCl. They studied two samples that had been polymerized in different ways. One of these showed, prior to the maximum, a change of slope and the appearance of the anisotropic phase, while the other polymer sample showed no change of slope or appearance of the anisotropic phase prior to the maximum. The two viscosity treatments differ in their interpretation of the viscosity minimum. Both treatments involve a number of assumptions and approximations that may be more nearly valid in some cases than others. While the status of the theory is still somewhat unsatisfactory, it is clear from both theory and experiment that extrema in the viscosity–concentration plot should not be relied upon for the construction of a phase diagram.

As the shear rate is increased the viscosity maximum becomes less pronounced and finally disappears. This is illustrated by the data of Kiss and Porter (1980a) shown in Fig. 15 for PBLG in *m*-cresol. No viscosity maximum is evident when the shear rate exceeds 1000 sec^{-1}. However, the same authors showed that a maximum is also found in the concentration dependence of the first normal stress difference. This maximum persists at high shear rates, as seen in Fig. 16 (open circles). Further, the sign of the first

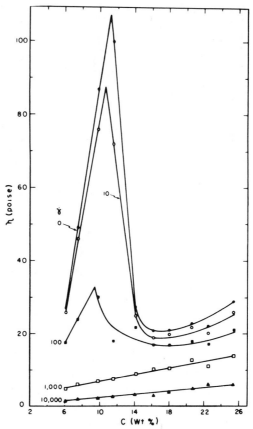

Fig. 15.   Concentration dependence of the viscosity of cholesteric PBLG in *m*-cresol measured for a range of shear rates. (From Kiss and Porter, 1980a.)

normal stress difference measured at increasing shear rates was found (Kiss and Porter, 1978, 1980a) to change from positive to negative, and back to positive again, as illustrated in Fig. 17. The original work (Kiss and Porter, 1980a) should be consulted for details concerning their efforts to eliminate any possible artifacts in these measurements. The magnitudes of the extrema (positive and negative) were found to increase linearly with concentration, and the values extrapolated to zero at the concentration at which the anisotropic phase disappears. These phenomena are observed at concentrations above the minimum in the plot of viscosity versus concentration, which indicates that the presence of a biphasic system is not a necessary criterion for the appearance of a negative first normal stress difference. Kiss and Porter (1980b) performed rheo-optical studies on *m*-cresol solutions of two samples

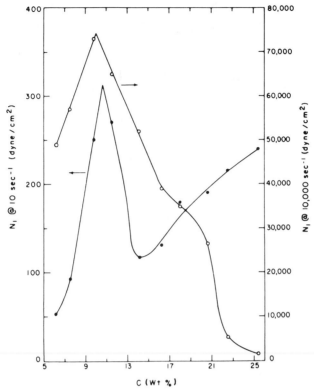

Fig. 16. First normal stress difference versus concentration of cholesteric PBLG in *m*-cresol at shear rates of 10 sec^{-1} (●) and 10^4 sec^{-1} (○). (From Kiss and Porter, 1980a.)

of PBLG having molecular weights 150,000 and 350,000. Only the sample of higher molecular weight exhibited a range of shear rates in which the first normal stress difference is negative. Photomicrographs of this polymer under steady shear show irregular regions elongated transverse to the flow direction at these same shear rates. They propose, as a schematic model, that the molecules are aligned parallel to the flow at low shear, but turn 45° to the flow direction at medium to high shear rates. Alternate wedge-shaped planes are visualized as being oriented at −45° and +45° to the flow direction, and the wedges are proposed to form a physical connection between the cone and plate. The irregular transverse striations are attributed to these planes being viewed edge-on. It should be noted that surface orientation prior to shearing plays no role in this model. At still higher shear rates the striations break into tumbling domains, and photographs taken between crossed polars become featureless.

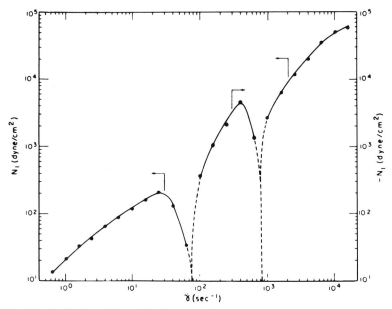

Fig. 17. Oscillation of sign of the first normal stress difference with shear rate shown by cholesteric PBLG in *m*-cresol. (From Kiss and Porter, 1980a.)

All of the foregoing refers to a sample under steady shear conditions. Both the high and low molecular weight polymers recovered after shearing by forming striations perpendicular to the flow direction. For the polymer of molecular weight 150,000 the striation pattern had better definition, but degenerated more rapidly as the shear rate was increased over the range 3.3 to 100 sec^{-1}. These observations for cholesteric PBLG differ from those mentioned above for nematic PBA. In the latter case domains formed most completely and were most stable to dissipation on recovery from shear at very low rates.

The observation of a parallel banded structure in PBLG under various types of external fields are quite numerous. Go *et al.* (1969) reported a structure of lines perpendicular to a magnetic field of 9.6 kG in CH_2Cl_2 solutions of PBLG. Sridhar *et al.* (1974) noted that under high magnetic fields the pattern of lines perpendicular to the applied field was distorted into a "herringbone" pattern, the amplitude of which increased with field strength. Toth and Tobolsky (1970) reported that a very slight mechanical shearing produced a pattern of parallel bands oriented perpendicular to the shear direction. These bands had different colors when viewed under crossed polars. Finally, very extensive rheo-optical studies have been performed by Asada and co-workers; these will be reported in Chapter 9 of this volume.

D. Flow-Induced Phase Transitions

The effect of an external flow field upon the isotropic–nematic transition of rodlike macromolecules was investigated theoretically by Marrucci and Ciferri (1977). They predicted that for solutions of rods having axial ratio x, there should exist a critical extensional velocity gradient $G_c = 2/x^3$ above which the isotropic phase is no longer stable. For extensional flow fields available in the laboratory it is predicted that a flow-induced transition should be observable if $x = 500$ but not if $x = 100$. Marrucci and Sarti (1979) extended this treatment, with certain approximations, to chains consisting of N flexibly linked rods having axial ratio x. They found the critical value of the extensional flow gradient to be $G_c = 8/(3Nx^3)$. This indicates that the likelihood of a flow-induced transition is much more dependent upon the axial ratio of the rodlike segments than upon the number of such segments in the chain. If $x = 100$ a transition is predicted to occur even for dilute solutions of chains consisting of relatively few segments. These authors also predicted that if N is large, the flow-induced transition to a nematic phase should be accompanied by a conformational transition of the molecules to fully extended form.

Valenti and Ciferri (1978) and Baird et al. (1979) sought evidence for a flow-induced transition in isotropic solutions of the polyterephthalamide of p-aminobenzhydrazide (Monsanto X-500). All their measurements involved shear flow, and one might anticipate that chain entanglements would be necessary for this type of flow to cause a phase transition. In this connection it should be noted that Onogi et al. (1980) concluded from birefringence measurements of cholesteric aqueous hydroxypropylcellulose solutions that elongational flow is about three times as effective as shear flow in orienting the molecules. Valenti and Ciferri (1978) found for low shear rates a change of slope when the concentration dependence of the viscosity was plotted in a log–log manner. This change of slope occurred at progressively lower concentrations for samples of higher molecular weight. At higher concentrations and higher shear rates the flow became unstable. For a fixed shear rate the concentration dependence of the viscosity exhibited a maximum and minimum. This behavior, as well as the change in slope at low shear rate, is illustrated for three samples of X-500 in Fig. 18. Baird et al. (1979) observed for the same polymer a change in slope of the concentration dependence of the flow birefringence when the data were plotted in a similar manner. The concentrations corresponding to the change in slope were in

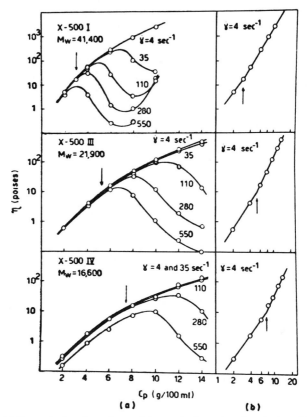

Fig. 18. (a) Concentration dependence of the steady shear viscosity of polyamide–hydrazide X-500 measured at various shear rates, and (b) a log–log plot of the viscosity of X-500 at low shear rates versus concentration. (From Valenti and Ciferri, 1978.)

approximate agreement with those observed for the viscosity, and there was some evidence for a second change of slope at still higher concentration, as shown in Fig. 19. These workers also confirmed the previous observation by Valenti and Ciferri (1978) of flow instabilities at higher shear rates by cone and plate and capillary viscometer data. Both Chapoy and la Cour (1980) and Valenti et al. (1981) found a second change of slope at lower concentration. These authors interpret the slope change at lower concentration to the onset of entanglements, and suggest that the slope change at higher concentration may possibly indicate the onset of the phase transition. Although flow instabilities might be expected to accompany a transition to a nematic phase under high shear conditions, in no case has direct evidence been obtained for a flow-induced transition to a nematic phase.

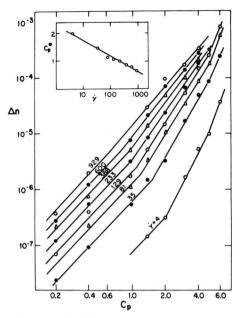

Fig. 19.   Flow birefringence of X-500 versus concentration (log–log plot) for various shear rates. (From Baird *et al.*, 1979.)

References

Aharoni, S. M. (1979). *Macromolecules* **12**, 94.

Aharoni, S. M. (1980). *Polymer* **21**, 1413.

Aharoni, S. M., and Walsh, E. K. (1979). *Macromolecules* **12**, 271.

Arnould-Netillard, H., and Rondelez, F. (1974). *Mol. Cryst. Liq. Cryst.* **26**, 11.

Baessler, H., and Labes, M. M. (1969). *J. Chem. Phys.* **51**, 5397.

Baird, D. G., Ciferri, A., Krigbaum, W. R., and Salaris, F. (1979). *J. Polym. Sci., Polym. Phys. Ed.* **17**, 1649.

Balbi, C., Bianchi, E., Ciferri, A., Tealdi, A., and Krigbaum, W. R. (1980). *J. Polym. Sci., Polym. Phys. Ed.* **18**, 2037.

Balomey P. H., and Dimitropoulos, C. (1976). *Mol. Cryst. Lig. Cryst.* **36**, 75.

Barnik, M. I., Blinov, L. M., Trufanov, A. N., and Umanski, B. A. (1978). *J. Phys. (Orsay, Fr.)* **39**, 417.

Bawden, F. C., and Pirie, N. W. (1937). *Proc. R. Soc. London, Ser. B* **123**, 274.

Bjornstahl, Y. (1933). *Z. Phys. Chem. Abt. A* **175**, 17.

Blumstein, A., Maret, G., and Vilasagar, S. (1981). *Macromolecules* **14**, 1543.

Bobujlev, J. P., and Pikin, S. A. (1977). *Zh. Eksp. Teor. Fis.* **72**, 361.

Candau, S., Martinoty, P., and Debeauvais, F. (1973). *C. R. Hebd. Seances Acad. Sci., Ser. A* **277**, 769.

Cano, R. (1968). *Bull. Soc. Fr. Mineral. Cristallogr.* **91**, 20.

Carr, E. F. (1969). *Mol. Cryst. Liq. Cryst.* **7**, 253.

Chandrasekhar, S. (1977). "Liquid Crystals," p. 165. Cambridge Univ. Press, London and New York.
Chapoy, L. L., and la Cour, N. F. (1980). *Rheol. Acta* **19**, 731.
de Gennes, P. G. (1968a). *Solid State Commun.* **6**, 163.
de Gennes, P. G. (1968b). *C. R. Hebd. Seances Acad. Sci., Ser. B* **266**, 571.
de Gennes, P. G. (1971). *Mol. Cryst. Liq. Cryst.* **12**, 193.
Derzhanski, A. I., Petrov, A. G., Khinov, P., and Markovski, B. (1974). *Bulg. J. Phys.* **1**(2), 165.
Doi, M. (1975). *J. Phys. (Orsay, Fr.)* **36**, 607.
Doi, M. (1980). *Ferroelectrics* **30**, 247.
Dubois-Violette, E., de Gennes, P. G., and Parodi, O. (1971). *J. Phys. (Orsay, Fr.)* **32**, 305.
Dubois-Violette, E., Durand, G., Guyon, E., Mannerville, P., and Pieranski, P. (1978). *Solid State Phys. Suppl.* **14**, 147.
Duke, R. W., and DuPré, D. B. (1974a). *J. Chem. Phys.* **60**, 2759.
Duke, R. W., and DuPré, D. B. (1974b). *Macromolecules* **7**, 374.
DuPré, D. B., and Hammersmith, J. R. (1974). *Mol. Cryst. Liq. Cryst.* **28**, 365.
Durand, G., Leger, L., Rondelez, F., and Veyssie, M. (1969). *Phys. Rev. Lett.* **22**, 227.
Elliott, A., and Ambrose, E. J. (1950). *Discuss. Faraday Soc.* **9**, 246.
Ericksen, J. L. (1960). *Arch. Ration. Mech. Anal.* **4**, 231.
Ericksen, J. L. (1961). *Trans. Soc. Rheol.* **5**, 23.
Finkelmann, H., Naegle, D., and Ringsdorff, H. (1979). *Macromol. Chem.* **180**, 803.
Finlayson, B. A. (1974). "Liquid Crystals and Ordered Fluids" (J. F. Johnson and R. S. Porter, eds.), Vol. 2, p. 211. Plenum, New York.
Fisher, J., and Frederickson, A. G. (1969). *Mol. Cryst. Liq. Cryst.* **8**, 267.
Flory, P. J. (1956). *Proc. R. Soc. London, Ser. A* **234**, 60, 73.
Frederiks, V., and Tsvetkov, V. (1934). *Physik. Z. Sowjetunion* **6**, 490.
Frederiks, V., and Zolina, V. (1933). *Trans. Faraday Soc.* **29**, 919.
Friedel, G. (1922). *Ann. Phys. (Leipzig)* **18**, 273.
Gähwiller, C. (1973). *Mol. Cryst. Liq. Cryst.* **20**, 301.
Gerritsma, C., and van Zanten, P. (1971). *Phys. Lett.* **28**, 593.
Go, Y., Ejira, S., and Fukada, E. (1969). *Biochim. Biophys. Acta* **175**, 454.
Greubel, W., and Wolff, V. (1971). *Appl. Phys. Lett.* **19**, 213.
Gruler, H., and Meier, G. (1972a). *Mol. Cryst. Liq. Cryst.* **16**, 299.
Gruler, H., and Meier, G. (1972b). *In* "Liquid Crystals," Part II (G. H. Brown and M. M. Labes, eds.), p. 279. Gordon & Breach, New York.
Haas, W. E., Adams, J. E., Dir, G. A., and Mitchell, C. W. (1973). *Proc. Soc. Inf. Disp. 14/4, Fourth Quarter* p. 121.
Hardouin, F., Achard, M. F., Gasparoux, H., Liébert, L., and Strzelecki, L. (1982). *J. Polym. Sci., Polym. Phys. Ed.* **20**, 975.
Heilmeier, G. H., and Goldmacher, J. E. (1968). *Appl. Phys. Lett.* **13**, 132.
Helfrich, W. (1969a). *Phys. Rev. Lett.* **23**, 372.
Helfrich, W. (1969b). *J. Chem. Phys.* **51**, 4092.
Helfrich, W. (1970a). *Phys. Rev. Lett.* **24**, 201.
Helfrich, W. (1970b). *J. Chem. Phys.* **52**, 4318.
Helfrich, W. (1971). *J. Chem. Phys.* **55**, 839.
Hermans, J., Jr. (1962). *J. Colloid Sci.* **17**, 638.
Hirakawa, K., and Kai, S. (1977). *Mol. Cryst. Liq. Cryst.* **40**, 261.
Hurault, J. P. (1973). *J. Chem. Phys.* **59**, 2068.
Iizuka, E. (1974). *Mol. Cryst. Liq. Cryst.* **25**, 287.
Kahn, F. J. (1970). *Phys. Rev. Lett.* **24**, 209.
Kini, U.D., Ranganath, G. S., and Chandrasekhar, S. (1975). *Pramana* **5**, 101.
Kiss, G., and Porter, R. S. (1978). *J. Polym. Sci., Polym. Symp.* **65**, 193.

Kiss, G., and Porter, R. S. (1980a). *J. Polym. Sci, Polym. Phys. Ed.* **18**, 361.

Kiss, G., and Porter, R. S. (1980b). *Mol. Cryst. Liq. Cryst.* **60**, 267.

Kleman, M., and Friedel, J. (1969). *J. Phys. (Orsay, Fr.)* **30**, Suppl. C4, 43.

Kol'tsov, L. G., Bel'nikevich, N. G., Gribanov, A. V., Papkov, S. P., and Frenkel, S. Y. (1973). *Vysokomol. Soedin., Ser. B* **15**, 645.

Krigbaum, W. R., and Lader, H. J. (1980). *Mol. Cryst. Liq. Cryst.* **62**, 87.

Krigbaum, W. R., Lader, H. J., and Ciferri, A. (1980). *Macromolecules* **13**, 554.

Krigbaum, W. R., Grantham, C. E., and Toriumi, H. (1982). *Macromolecules* 15, 192.

Kuhlichikhin, V. G., Platanov, V. A., Braverman, L. P., Belousova, T. A., and Papkov, S. P. (1976). *Polym. Sci. USSR (Engl. Transl.)* **18**, 3031.

Kuhlichikhin, V. G., Vasil'yeva, N. V., Platonov, V. A., Malkin, A. Y., Belousova, T. A., Khanchich, O. A., and Papkov, S. P. (1979). *Polym. Sci. USSR (Engl. Transl.)* **21**, 1545.

Kwolek, S. L. (1972). U.S. Patent 3,671,542 (assigned to E. I. duPont).

Léger-Quercy, L. (1970). Thesis, Univ. Paris.

Lehmann, O. (1889). *Z. Phys. Chem., Stoechiom. Verwandschaftsl.* **4**, 462.

Leslie, F. M. (1966). *Q. J. Mech. Appl. Math.* **19**, 357.

Leslie, F. M. (1968). *Arch. Ration. Mech. Anal.* **28**, 265.

Leslie, F. M. (1969). *Mol. Cryst. Liq. Cryst.* **7**, 407.

Liébert, L., Strzelecki, L., van Luyen, D., and Levulut, A. M. (1981). *Eur. Polym. J.* **17**, 71.

Maret, G., and Blumstein, A. (1982). *Mol. Cryst. Liq. Cryst.* (in press).

Maret, G., Blumstein, A., and Vilasagar, S. (1981). *Polym. Prepr., Am. Chem. Soc., Div. Polym. Chem.* **22**, 246.

Marrucci, G. (1982). *Mol. Cryst. Liq. Cryst. (Lett.)* **72**, 153.

Marrucci, G., and Ciferri, A. (1977). *Polym. Lett.* **15**, 643.

Marrucci, G., and Sarti, G. C. (1979). *In* "Ultrahigh Modulus Polymers" (A. Ciferri and I. M. Ward, eds.), p. 137. Appl. Sci. London.

Martinoty, P., and Candau, S. (1971). *Mol. Cryst. Liq. Cryst.* **14**, 243.

Matheson, R. R., Jr. (1980). *Macromolecules* **13**, 643.

Meyer, R. B. (1968). *Appl. Phys. Lett.* **12**, 281.

Meyer, R. B. (1969a). *Appl. Phys. Lett.* **14**, 208.

Meyer, R. B. (1969b). *Phys. Rev. Lett.* **22**, 918.

Miesowicz, M. (1946). *Nature (London)* **158**, 27.

Minami, N., Aikawa, Y., and Sukigara, M. (1978). *Mol. Cryst. Liq. Cryst. (Lett.)* **41**, 189.

Naggiar, V. (1943). *Ann. Phys. (Leipzig)* **18**, 5.

Noel, C., Monnerie, L., Achard, M. F., Hardouin, F., Sigaud, G., and Gasparoux, H. (1981). *Polymer* **22**, 578.

Onogi, Y., White, J. L., and Fellers, J. F. (1980). *J. Non-Newtonian Fluid Mech.* **7**, 121.

Orsay Liquid Crystal Group (1969). *J. Phys. (Orsay, Fr.)* **30**, Suppl. C4, 43.

Oseen, C. W. (1933). *Trans. Faraday Soc.* **29**, 883.

Panar, M., and Beste, L. F. (1977). *Macromolecules* **10**, 1401.

Penz, P. A. (1971). *Mol. Cryst. Liq. Cryst.* **15**, 141.

Pieranski, P., and Guyon, E. (1973). *Solid State Commun.* **13**, 435.

Platonov, V. A., Litovchenko, G. D., Belousova, T. A., Mil'kova, L. P., Shablygin, M. V., Kuhlichikhin, V. G., and Papkov, S. P. (1976). *Polym. Sci. USSR (Engl. Transl.)* **18**, 256.

Porter, R. S., Barrall, E. M., and Johnson, J. F. (1966). *J. Chem. Phys.* **45**, 1452.

Rapini, A., and Papoular, M. (1969). *J. Phys. (Orsay, Fr.)* **30** Suppl. C4, 54.

Reinitzer, F. (1888). *Monatsch. Chem.* **9**, 421.

Rondelez, F. (1973). Thesis, Univ. Paris.

Rondelez, F., and Arnould-Netillard, H. (1971). *C. R. Hebd. Seances Acad. Sci., Ser. B* **273**, 549.

Rondelez, F., and Hulin, J. P. (1972). *Solid State Commun.* **10**, 1009.

Scheffer, T. J. (1972). *Phys. Rev. A***5**, 1327.

Schmidt, D., Schadt, M., and Helfrich, W. (1972). *Z. Naturforsch., A* **27A**, 277.

Simha, R. (1940). *J. Phys. Chem.* **44**, 25.

Simha, R. (1949). *J. Res. Natl. Bur. Stand. (U.S.)* **42**, 409.

Smith, I. W., Galerne, Y., Lagerwall, S. J., Dubois-Violette, E., and Durand, G. (1975). *J. Phys. (Orsay, Fr.)* **36** (C1), 237.

Sridhar, C. G., Hines, W. A., and Samulski, E. T. (1974). *J. Chem. Phys.* **61**, 947.

Toth, W. J., and Tobolsky, A. V. (1970). *Polym. Lett.* **8**, 53.

Toyoshima, Y., Minami, N., and Sukigara, M. (1976). *Mol. Cryst. Liq. Cryst.* **35**, 325.

Tsvetkov, V. (1939). *Acta Physicochim. USSR* **10**, 557.

Tsvetkov, V., and Mikhailov, G. M. (1938). *Acta Physicochim. USSR* **8**, 77.

Valenti, B., and Ciferri, A. (1978). *Polym. Lett.* **16**, 657.

Valenti, B., Alfonso, G. C., Ciferri, A., Giordani, P., and Marrucci, G., (1981). *J. Appl. Polym. Sci.* **26**, 3643.

Vistin', L. K. (1971). *Sov. Phys.—Dokl. (Engl. Transl.)* **15**, 908.

Vistin', L. K., and Kapustin, A. P. (1970). *Sov. Phys.—Crystallogr. (Engl. Transl.)* **14**, 638.

Volino, F., Martins, A. F., Blumstein, R. B., and Blumstein, A. (1981). *C. R. Hebd. Seances Acad. Sci., Ser. II* **292**, 829.

Williams, R. (1963). *J. Chem. Phys.* **39**, 384.

Wysocki, J., Adams, J., and Haas, W. (1968a). *Phys. Rev. Lett.* **20**, 1024.

Wysocki, J., Adams, J., and Haas, W. (1968b). *Phys. Rev. Lett.* **21**, 1791.

Wysocki, J., Adams, J., and Haas, W. (1969). *Mol. Cryst. Liq. Cryst.* **8**, 471.

Zöcher, H. (1933). *Trans. Faraday Soc.* **29**, 945.

11

Liquid Crystal Display Devices

G. Baur

Fraunhofer-Institut für Angewandte Festkörperphysik
Freiburg, Federal Republic of Germany

I. Introduction

Liquid crystal displays are achieving great popularity as low voltage, low power devices that are visible over a wide range of ambient light levels. At present such devices are produced in quantities of many millions per month for application in wristwatches, calculators, and clocks. In the near future displays of larger size, e.g., 30×15 cm^2 for applications in automotive dashboards or displays for higher information content up to television sizes, will be available.

II. Design of Displays

The most useful electro-optic effects that can be applied in liquid crystal displays are based on light scattering or light absorption by polarizers or dissolved dyes. A liquid crystal display consists of three parts, the display cell, the liquid crystal layer, and a reflector or a back light. The display cell is usually made from glass plates coated with a transparent electrode. Front and rear plates are spaced 6–15 μm apart. The most important constituent of the display is the liquid crystal, which fills the space between the plates. For all the present applications low molecular weight liquid crystals are

309

used. A comprehensive overview of components and materials is given by Gray in Chapter 1 of this volume.

If scattering effects are used in order to display information, an electro-hydrodynamic turbulence or defects are induced by applying an electric field. If the display is operated in a reflective mode, a mirror is put behind the cell. Then the information appears as a scattering pattern on a metallic reflector. If the display is operated in a transmissive mode, the mirror is replaced by a blind sheet combined with a light source. In this case the non-addressed segments are dark and the addressed segments appear as a bright scattering pattern. A schematic representation is given in Fig. 1a,b.

Electro-optic absorption effects are based on a field-induced deformation of the liquid crystal layer, starting either from homogeneous, homeotropic, or hybrid alignment, or from twisted structures. If the cell is placed between two polarizer sheets, there is a field-dependent light absorption in the second polarizer due to the phase difference between the ordinary and the extraordinary ray, as indicated in Fig. 1c,d.

If pleochroic dyes are dissolved in the liquid crystal layer, an acceptable contrast can be achieved by combining the cell with only one polarizer. In the special case of twisted structures with a small number of turns of the cholesteric helix in the layer, no polarizers are needed at all. Such electro-optic effects employing pleochroic dyes are called "guest–host effects." The order parameter of the dyes dissolved in the liquid crystal matrix or in the polarizer sheets should be at least 0.8. If pleochroic dyes can be oriented in polymer liquid crystals with an order parameter considerably higher than 0.8, such polymers can be suitable for the fabrication of polarizer sheets.

The display cells are combined with a diffuse or cholesteric reflector or with a back light. The optical properties of reflectors of the cholesteric type are discussed in detail by Scheffer (1975). Cholesteric reflector sheets based on polysilane side-chain polymers have been proposed and developed by Finkelmann et al. (1978) and realized on a commercial scale by Kreuzer (1981). Either a combination of a light source with a waveguide or a fluorescent light collector is used as a back light. In order to get high contrast over a wide range of ambient light levels, including darkness, back lights coated with a semitransparent reflector are frequently used. This kind of display is called transflective.

An overview of the different display types and different modes of operation is given in Table I. There is no significant mass production of displays based on light scattering effects but some engineering models have been developed using dynamic scattering (DS) (Heilmeier et al., 1968), the cholesteric–nematic phase transition (Wysocki et al., 1968) or the phase transition in smectics (Maydan et al., 1972; Taylor and Kahn, 1974). Electro-optic effects that belong to the second group are tunable birefringence

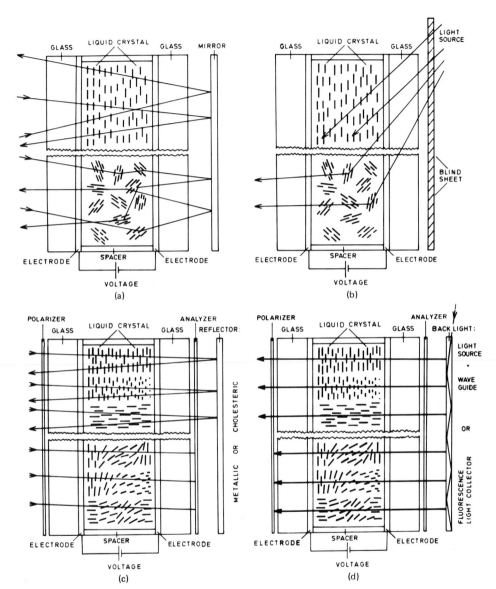

Fig. 1. Schematic diagram of different types of liquid crystal displays. Light scattering: (a) reflective mode, (b) transmissive mode. Light absorption: (c) reflective mode, (d) transmissive mode.

TABLE I

Summary of Different Types of Liquid Crystal Displays
and Different Modes of Operation

1. LIGHT SCATTERING EFFECTS	2. LIGHT ABSORPTION BY POLARIZERS
	OR DISSOLVED DYES
DYNAMIC SCATTERING	TUNABLE BIREFRINGENCE (DAP)
CHOLESTERIC-NEMATIC PHASE TRANSITION (BISTABILITY EFFECT)	TWISTED NEMATIC
PHASE TRANSITION IN SMECTICS	GUEST – HOST EFFECT

	MODE OF OPERATION	
REFLECTIVE	TRANSFLECTIVE	TRANSMISSIVE

(Schiekel and Fahrenschon, 1971), the twisted nematic effect (TN) (Schadt and Helfrich, 1971) and the guest–host effect (Heilmeier and Zanoni, 1968; White and Taylor, 1974). In the past, the most useful candidate for practical applications was the TN effect and this also appears to hold true in the near future, although the first commercial guest–host displays have now become available.

III. Optimization of Displays

The optimization of a liquid crystal display is demonstrated by considering the example of the well-known twisted nematic (TN) cell.

The configuration of a TN cell is illustrated in Fig. 2. The molecules at the surfaces are aligned homogeneously parallel but the directors at the surfaces are oriented perpendicular to each other. This fixed surface orientation induces a twist of 90°. If the optical path difference (the product of cell thickness and refractive index anisotropy, $d \times \Delta n$) is large compared with the wavelength of the light, the plane of polarization is rotated by 90° on

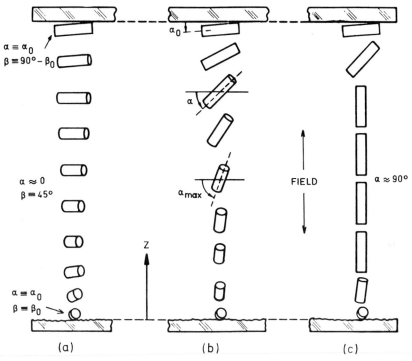

Fig. 2. Diagram of the orientation of the liquid crystal axis in a cell (a) with no applied field, (b) with about twice the critical field, and (c) with several times the critical field. Note slight permanent tilt (α_0) and turn (β_0) at the surfaces.

traversing the helical structure. Therefore the cell is transparent between crossed polarizers and dark between parallel polarizers.

If an electric field is applied to a TN cell with a twist of 90°, deformation starts at the well known threshold voltage

$$V_0 = \pi \left[\frac{k_{11} + (k_{33} - 2k_{22})/4}{\varepsilon_0 \, \Delta\varepsilon} \right]^{1/2}$$

Somewhat above this voltage the cell becomes transparent in the case of parallel polarizers and dark in the case of crossed polarizers.

The information rate that can be displayed by liquid crystals is limited by the multiplexing capability of the display. Therefore electro-optic effects having a strong nonlinearity of the electro-optic characteristics for a wide range of viewing angles are required. The optical characteristics themselves are a complicated function of material and cell parameters. Optimization of the optical response can be performed in two steps. First, the optimum combination of material parameters should be evaluated by computer model calculations and second, liquid crystal mixtures that possess these properties to a good approximation must be developed, if possible.

A. Model Calculations

In the case of model calculations it seems reasonable first, to calculate tilt (α) and twist (β) angles in a sequence of sublayers as a function of the applied voltage and second, the optical transmission for different angles of observation, as demonstrated by Berreman (1975), van Doorn (1975), and Gharadjedaghi and Robert (1976).

A definition of the angles α and β is illustrated in Fig. 2. In part (a) the orientation of the liquid crystal axis is given with no applied field, in (b) with about twice the critical field, and in (c) with several times the critical field.

The deformation profile is dependent on the dielectric constants $\varepsilon_{\|}$ and $\varepsilon_{\perp}$, the elastic constants for splay, twist, and bend (k_{11}, k_{22}, k_{33}), the total twist $\beta = (90° - 2\beta_0)$, the tilt angle α_0 at the surface of the substrates, and the applied voltage. In addition the optical response depends on the refractive

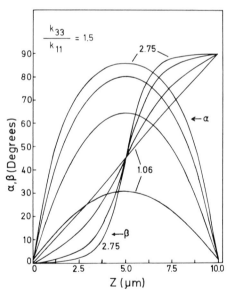

Fig. 3. Tilt (α) and twist (β) angle as a function of the position z within the layer for a 10-μm cell. The curves are calculated with $\varepsilon_{\|} = 10.5$, $\varepsilon_{\perp} = 10.0$, $k_{11} = 10.9 \times 10^{-12}$ N, $k_{22} = 9.5 \times 10^{-12}$ N, $k_{33} = 15.4 \times 10^{-12}$ N. (From Baur, 1980.)

indices n_e and n_0 and the ratio λ/d of wavelength to cell thickness. For display applications, a finite tilt at the surfaces is required to avoid areas of opposite tilt.

The deformation profiles are calculated for various combinations of k_{33}/k_{11}, k_{22}/k_{11}, and $\Delta\varepsilon/\varepsilon_\perp$ using a program of D. W. Berreman. All calculations are performed for 10-μm cells and a pretilt $\alpha_0 = 1°$. Results are shown in Fig. 3. As an example, tilt and twist angles α and β are plotted as functions of position within the cell thickness. Numbers shown with each curve are the reduced voltage V/V_0. It is sometimes helpful to examine the director distribution in order to understand the complicated transmission curves.

To characterize the performance of a display, it is necessary to know the transmission for a large range of viewing angles, as demonstrated by Baur *et al.* (1975), Baur (1980, 1981), van Doorn *et al.* (1980), and Kahn and Birecki (1980a,b). Therefore, the transmission curves for normal incidence, and for eight different viewing angles in four different planes of observation are evaluated. The viewing angles are characterized by the off-axis angle θ and an angle φ defined by the plane of observation and the plane containing the director in the center of the cell and the z axis, as shown in Fig. 4.

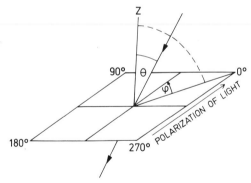

Fig. 4.   Schematic diagram to explain the viewing angles chosen for the calculation of the transmission characteristics. The off-axis angle is θ; φ, the angle defined by the plane of observation and the plane containing the director in the center of the cell and the Z axis. (From Baur, 1981.)

The results are shown in Fig. 5b. The ratio k_{33}/k_{11} is 1.5 in part (a) of the figure, and 0.5 in part (b). It is obvious that the slopes become steeper for smaller values of k_{33}/k_{11} and, in addition, the distance between curves for normal incidence ($\theta = 0°$, $\varphi = 0°$) and off-axis incidence ($\theta = 45°$, $\varphi = 180°$) becomes smaller.

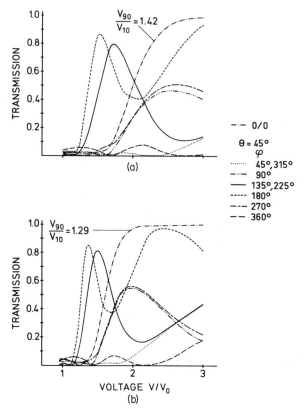

Fig. 5. Transmission characteristics of a TN cell, calculated with $\varepsilon_{\parallel} = 25.4$, $\varepsilon_{\perp} = 5.4$; $n_e = 1.728$, $n_0 = 1.528$; $\lambda = 632.8$ nm; $d = 10$ μm. (a): $k_{33}/k_{11} = 1.5$; (b): $k_{33}/k_{11} = 0.5$. (From Baur, 1981.)

Results calculated with k_{33}/k_{11} and $\Delta\varepsilon/\varepsilon_{\perp}$ fixed, but with a Δn ranging from 0.2 to 0.1, are given in Fig. 6. For smaller Δn the transmission curves are steeper and the well-known bounce is less pronounced. This result indicates that the optical characteristics can be improved considerably by minimizing the product $d \times \Delta n$.

The previous section considered only a classic twisted nematic cell in the limits of the Mauguin approximation. In this approximation the inequality

$$d \times \Delta n \gg \lambda/2$$

must be satisfied in order that light waves traveling through a TN cell be linearly polarized.

If this inequality is not satisfied, for smaller $d \times \Delta n$ plane polarized incident light emerges as elliptically polarized when transmitted through the

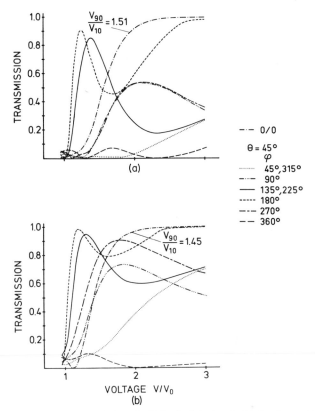

Fig. 6. Transmission characteristics of a TN cell. Calculated with $k_{33}/k_{11} = 1.5$: $\varepsilon_{\parallel} = 10.5$, $\varepsilon_{\perp} = 10.0$; $\lambda = 632.8$ nm; $d = 10$ μm. (a): $n_e = 1.728$, $n_0 = 1.528$; (b): $n_e = 1.62$, $n_0 = 1.52$.

cell. The optical properties of such structures have been considered by a number of authors. A closed solution for the light intensity transmitted through a structure having a pitch small compared with the cell thickness has been given by Gooch and Tarry (1975). If both the polarizer and analyzer are parallel to the director at the first surface, the transmitted intensity is

$$T = \frac{\sin^2\left[\frac{1}{2}\pi(1 + u^2)\right]^{1/2}}{(1 + u^2)}$$

with

$$u = 2 \times d \times \Delta n/\lambda$$

There are minima in transmission for

$$u = 2 \times d \times \Delta n/\lambda = \sqrt{3}, \sqrt{15}, \sqrt{35}, \sqrt{63}, \ldots,$$

as shown in Fig. 7.

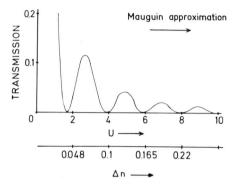

Fig. 7. Transmission of a twisted nematic structure as a function of $u = 2d \times \Delta n/\lambda$. Second scale ($\Delta n$) is calculated with $d = 10~\mu$m and $\lambda = 550$ nm. (From Baur, 1981.)

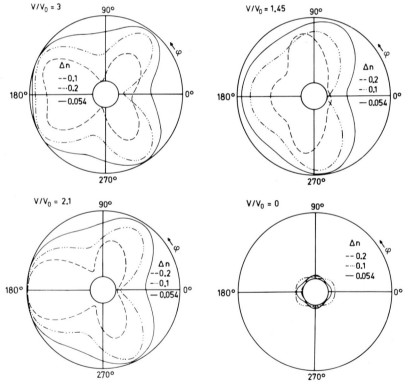

Fig. 8. Transmission characteristics ($\theta = 45°$) of a TN cell calculated with $k_{33}/k_{11} = 0.5$, $k_{22}/k_{11} = 0.4$: $\varepsilon_{\|} = 10.5$, $\varepsilon_{\perp} = 10.0$; $d = 10~\mu$m; $\lambda = 632.8$ nm. Parameter indicated in curves is Δn.

In contrast to the Mauguin approximation, the transmitted intensity in the off-state is only a minimum for discrete combinations of d, Δn, and λ.

Considering the transmission curves calculated for different values of Δn, it is expected to obtain optimized optical characteristics if the equality $d \times \Delta n = \frac{1}{2}\sqrt{3}$ is satisfied, as demonstrated by Pohl *et al.* (1981). Therefore transmission curves are calculated as a function of the applied voltage for a 10-μm cell with $\Delta n = 0.054$. This combination of cell thickness and Δn corresponds to a first transmission minimum in the off-state centered at $\lambda = 632.8$ nm. Results are plotted in Fig. 8. To illustrate the transmitted intensity for a large range of viewing angles, curves have been evaluated for an off-axis angle $\theta = 45°$ and the azimuthal angles φ in steps of $10°$, with $k_{33}/k_{11} = 1.5$, $\Delta\varepsilon = 20$, and Δn ranging from 0.2 to 0.054. The parameter is the reduced voltage. The dashed lines are calculated with $\Delta n = 0.2$ and 0.1, respectively. The solid line is calculated with $\Delta n = 0.054$. It is obvious that the range of observation angles is considerably improved if the optical path difference, $d \times \Delta n$, is adapted to the first transmission minimum in the off-state.

The results are summarized in Table II. $V_x(\theta, \varphi)$ is the percent transmission at a viewing angle characterized by θ and φ. The figures given there are a

TABLE II

Optical Performance Calculated with
Different Combinations of k_{33}/k_{11}, k_{22}/k_{11},
$\Delta\varepsilon/\varepsilon_\perp$, and Δn^a

$\dfrac{k_{33}}{k_{11}}$	$\dfrac{k_{22}}{k_{11}}$	$\dfrac{\Delta\varepsilon}{\varepsilon_\perp}$	$\Delta n = 0.2$ $\dfrac{V_{90}(0,0)}{V_{10}(0,0)}$	$\dfrac{V_{90}(0,0)}{V_{10}(45,180)}$	$\dfrac{V_{50}(0,0)}{V_{10}(45,180)}$	$\Delta n = 0.1$ $\dfrac{V_{90}(0,0)}{V_{10}(0,0)}$	$\dfrac{V_{90}(0,0)}{V_{10}(45,180)}$	$\dfrac{V_{50}(0,0)}{V_{10}(45,180)}$	$\Delta n = 0.054$ $\dfrac{V_{90}(0,0)}{V_{10}(0,0)}$	$\dfrac{V_{90}(0,0)}{V_{10}(45,180)}$	$\dfrac{V_{50}(0,0)}{V_{10}(45,180)}$
1.5	0.9	$\dfrac{3.7}{0.05}$	1.42 / 1.51	1.95 / 1.85	1.60 / 1.48	1.44 / 1.45	1.98 / 1.77	1.66 / 1.43	1.60 / 1.52	1.95 / 1.56	1.51 / 1.25
	0.4	$\dfrac{3.7}{0.05}$	1.30 / 1.43	1.75 / 1.77	1.53 / 1.44	1.36 / 1.39	1.87 / 1.68	1.57 / 1.38	1.48 / 1.44	1.88 / 1.54	1.43 / 1.24
1.0	0.9	$\dfrac{3.7}{0.05}$	1.40 / 1.48	1.86 / 1.72	1.54 / 1.38	1.40 / 1.37	1.94 / 1.64	1.59 / 1.33	1.57 / 1.42	1.92 / 1.50	1.52 / 1.21
	0.4	$\dfrac{3.7}{0.05}$	1.34 / 1.42	1.75 / 1.68	1.47 / 1.37	1.33 / 1.36	1.8 / 1.55	1.53 / 1.30	1.45 / 1.38	1.82 / 1.45	1.47 / 1.19
0.5	0.9	$\dfrac{3.7}{0.05}$	1.39 / 1.40	1.72 / 1.53	1.43 / 1.25	1.38 / 1.32	1.76 / 1.49	1.48 / 1.23	1.52 / 1.28	1.76 / 1.32	1.42 / 1.11
	0.4	$\dfrac{3.7}{0.05}$	1.29 / 1.32	1.59 / 1.42	1.38 / 1.21	1.28 / 1.25	1.53 / 1.32	1.37 / 1.20	1.40 / 1.22	1.60 / 1.22	1.38 / 1.11

^{a} The cell thickness $d = 10$ μm, the pretilt $\alpha_0 = 1°$.

measure of optical performance and multiplexing capability. The smaller the numbers corresponding to a fixed combination of material parameters, the better is the optical performance of the display. From a theoretical point of view the optical characteristics of a TN display can be optimized by using a material with small k_{33}/k_{11} and k_{22}/k_{11}; also $\Delta\varepsilon/\varepsilon_\perp$ should not be larger than required by the given addressing scheme. In addition, the combination of cell thickness and refractive index anisotropy should be matched to the first minimum of transmission corresponding to the wavelength of the maximum sensitivity of the human eye. In order to achieve switching times less than 100 msec, viscosities of about 0.2 P at room temperature are required.

B. Material Parameters

In the previous section the influence of material parameters on the optical characteristics was discussed and the optimized features of a useful material were evaluated. Components and mixtures are available that at least partly

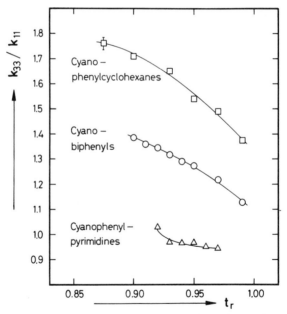

Fig. 9. The ratio k_{33}/k_{11} versus reduced temperature t_r for binary mixtures of terminally polar compounds. The pentyl and heptyl components of the cyanophenylalkylcyclohexanes, the cyanoalkylbiphenyls, and the cyanophenylalkylpyrimidines are in molar proportions of 40% : 60%. (From Scheuble and Baur, 1981a,b.)

fulfill these requirements. However, it seems very difficult to optimize all the material parameters in the same mixture. The most serious problem is to influence the elastic constants of a liquid crystal, but promising attempts to overcome these difficulties have been made by Schadt and Müller (1979), Bradshaw et al. (1980), Scheuble et al. (1981) and Scheuble and Baur (1981a,b).

In order to get a small ratio of bend to splay elastic constants, k_{33}/k_{11}, several terminally polar nematics and a large number of esters that are terminally nonpolar have been investigated by Scheuble in our laboratory. There is a great difference in the k_{33}/k_{11} ratios of the terminally polar compounds, even though their molecular dimensions are almost the same (pentyl and heptyl components are always in molar proportions of 40%:60%). Replacing the pyrimidine ring with a phenyl ring yields an increase in k_{33}/k_{11} from 1.0 to 1.3 and there is a further increase from 1.3 to about 1.7 between the cyanobiphenyl mixture and the cyanophenylcyclohexane mixture as shown in Fig. 9. In contrast to the terminally polar compounds, the terminally nonpolar esters listed in Table III exhibit only slight variation in

TABLE III

Chemical Composition of the
Esters Investigated

k_{33}/k_{11}, as shown in Fig. 10. The variation of k_{33}/k_{11} for the different esters that are terminally substituted with alkyl or alkoxy groups is very small. The ratio changes from about 1.25 to about 1.1. The lowest ratios are observed for the esters that are terminally substituted with an alkanoyloxy group. The k_{33}/k_{11} values of NP 1008, H_{33}, and H_{53} are close to 1.0 or less. With decreasing temperature the nematic phase of H_{33} and H_{53} is followed by the smectic phase. As expected, k_{33}/k_{11} diverges at the nematic–smectic phase transition. For all the other mixtures k_{33}/k_{11} is close to unity and is only slightly temperature dependent. In addition, the elastic constants can be improved by blending the terminally nonpolar esters with terminally polar compounds. As an example, the different esters are doped with the same proportion of a heptylcyanophenylcyclohexane (PCH 7). Results are given in Fig. 11. The lowest value of k_{33}/k_{11} obtained with doped alkanoyloxy substituted esters is about 0.6. This is not too far away from the optimum values in the model calculations.

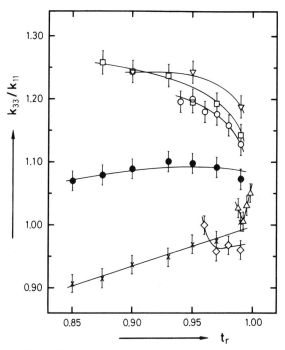

Fig. 10. The ratio k_{33}/k_{11} versus the reduced temperature t_r for the esters given in Table III: NP 1052 ($\triangledown$), ZLI 1745 ($\square$), ME 135/5 ($\bigcirc$), BCO 35/5 ($\bullet$), H_{33} ($\triangle$), H_{53} ($\diamond$), and NP 1008 ($\times$). (From Scheuble and Baur, 1981b.)

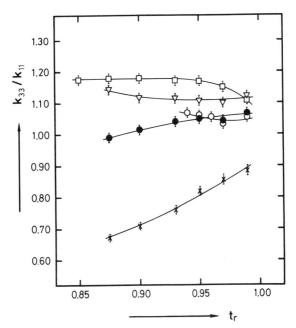

Fig. 11.   The ratio k_{33}/k_{11} versus the reduced temperature t_r for the esters given in Table III doped with 7.5% PCH 7: ZLI 1745 ($\square$), NP 1052 ($\triangledown$), ME 135/5 ($\bigcirc$) BCO 35/5 ($\bullet$), and NP 1008 ($\times$). (From Scheuble and Baur, 1981b.)

IV. Back Light Layout

The combination of a light source with a waveguide is an appropriate method of illuminating a liquid crystal display at low level ambient light conditions. If the waveguide is replaced by a fluorescent light collecting plate, high brightness of the displayed information can be achieved over a wide range of ambient light levels.

The combination of a liquid crystal display with such a light collecting plate, the fluorescence activated-display (FLAD), has been developed in our laboratory by Baur and Greubel (1977). The fluorescent plate is a plastic sheet appropriately doped with fluorescent dyes, and having V-shaped notches or a printed pigment pattern on the back. The edges of the plate are coated with a reflective layer. The high brightness at the segments of the light collecting plate is achieved by spectral conversion and also by concentration of the diffuse ambient light in the fluorescent plate.

Diffuse ambient light entering the plate from the front, or possibly also from the back, is absorbed and converted to fluorescent light. Most of the

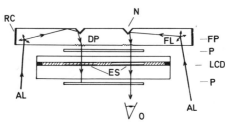

Fig. 12. Schematic diagram of a fluorescence activated LCD. The abbreviations are: reflector, RC; notch, N; fluorescent plate, FP; dispersing pattern, DP; fluorescent light, FL; electrode segment, ES; polarizer, P; and ambient light, AL.

fluorescent light, which is emitted isotropically, is trapped in the plate because of total reflection and reflection at the edge mirrors. The trapped light is deflected by the printed pigment, as shown schematically in Fig. 12. From the mode of operation described, it is obvious that the brightness of the display is selfadapting to the ambient light conditions. Depending on the color temperature of the ambient light, dyes with peak emission wavelengths ranging from green to red are efficient.

For a refractive index of 1.5, 75% of the fluorescent light is trapped in the plate and the brightness enhancement should be proportional to the ratio of collecting area and segment area, but in practice there are limitations. The plate acts as a waveguide for the fluorescent light, which must travel distances of 1 to 50 cm without significant loss before the light is deflected at notches or at a printed pigment pattern. To achieve this without significant loss in intensity, plates with perfect optical surfaces fabricated from highly transparent poly(methyl methacrylate) (PMMA) are required.

The problem of light traveling in the plate is much more serious if fluorescent dyes are dissolved in the matrix, because of the overlap of absorption and emission bands. Unfortunately, the overlap is larger for a dye dissolved in a rigid matrix than for a liquid matrix, because of an increase in the relaxation time of the dye in going from a liquid to a solid matrix. This problem can be overcome partly by doping the matrix with highly mobile polar molecules such as dimethyl sulfoxide (DMSO) or camphor. If the PMMA matrix is doped with about 7% DMSO as indicated by the dashed line in Fig. 13, the fluorescence band is shifted back almost to that observed with a monomer matrix, as shown by Sah *et al.* (1980). A similar effect should be obtained if liquid crystalline polymers such as cross-linked side-chain polymers are used as a matrix for the fluorescent dyes.

For a long time the stability of fluorescent dyes in a PMMA matrix has been a serious problem. Results of a systematic investigation of a large number of dyes in PMMA with different molecular weights and glass temperatures are given in Fig. 14. The decrease of absorbance and the

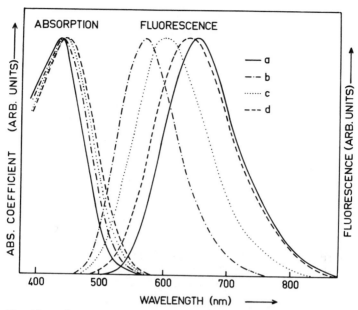

Fig. 13. Absorption and fluorescence spectra of 4-dimethylamino-4-nitrostilbene at room temperature: (a) in monomer methyl methacrylate (MMA), (b) in poly(methyl methacrylate) (PMMA), (c) in PMMA doped with 4.9wt% of dimethyl sulfoxide, (d) in PMMA doped with 11.2wt% of dimethyl sulfoxide. (From Sah *et al.*, 1980.)

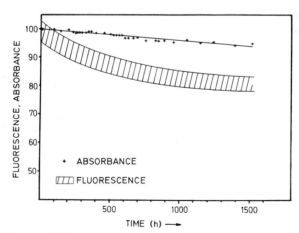

Fig. 14. Decrease of the absorbance and the fluorescence intensity of a collector plate irradiated with a light intensity of 1 kW/m².

fluorescence intensity with irradiation time indicates that the stability of fluorescent dyes is now comparable to the stability of pleochroic dyes, liquid crystal materials, and polarizer sheets.

Figure 15 illustrates a laboratory model of a FLAD clock as a kind of summary and as proof that liquid crystal displays can be improved considerably by using suitable liquid crystal materials, stable pleochroic and fluorescent dyes, and highly purified polymers. In addition, it is to be expected that the application of polymer liquid crystals will stimulate new ideas and offer further possibilities in the display field.

Fig. 15. Laboratory model of a FLAD clock.

Acknowledgments

The author wishes to thank W. Fehrenbach for performing computer calculations, V. Wittwer for stability measurements, and G. Meier and B. S. Scheuble for helpful discussions and critical reading of the manuscript.

References

Baur, G. (1980). *In* "The Physics and Chemistry of Liquid Crystal Devices" (G. J. Sprokel, ed.), pp. 61–78. Plenum, New York.
Baur, G. (1981). *Mol. Cryst. Liq. Cryst.* **63**, 45–58.
Baur, G., and Greubel, W. (1977). *Appl. Phys. Lett.* **31**, 4–6.
Baur, G., Windscheid, F., and Berreman, D. W. (1975). *Appl. Phys.* **8**, 101–106.
Berreman, D. W. (1975). *J. Appl. Phys.* **46**, 3746–3751.
Bradshaw, M. J., McDonnell, D. G., and Raynes, E. P. (1980). *Liq. Cryst., Int. Liq. Cryst. Conf., 8th, Kyoto, 1980*, p.326.

Finkelmann, H., Koldehoff, J., and Ringsdorf, H. (1978). *Angew. Chem.* **90**, 992–993.

Gharadjedaghi, F., and Robert, J. (1976). *Rev. Phys. Appl.* **11**, 467–473.

Gooch, C. H., and Tarry, H. A. (1975). *J. Phys. D* **8**, 1575–1584.

Heilmeier, G. H., and Zanoni, L. A. (1968). *Appl. Phys. Lett.* **13**, 91–92.

Heilmeier, G. H., Zanoni, L. A., and Barton, L. A. (1968). *Proc. IEEE* **56**, 1162–1171.

Kahn, F. J., and Birecki, H. (1980a). *In* "The Physics and Chemistry of Liquid Crystal Devices" (G. J. Sprokel, ed.) pp. 79–124. Plenum, New York.

Kahn, F. J., and Birecki, H. (1980b). *In* "The Physics and Chemistry of Liquid Crystal Devices" (G. J. Sprokel, ed.), pp. 125–142. Plenum, New York.

Kreuzer, F.-H. (1981). *Freib. Arbeitstag. Fluessigkrist.* **11**, 5.

Maydan, D., Melchior, H., and Kahn, F. J. (1972). *IEEE Conf. Rec. 1972, Conf. Disp. Devices*, pp. 166–169.

Pohl, L., Weber, G., Eidenschink, R., Baur, G., and Fehrenbach, W. (1981). *Appl. Phys. Lett.* **38**, 497–499.

Sah, R. E., Baur, G., and Kelker, H. (1980). *Appl. Phys.* **23**, 369–372.

Schadt, M., and Helfrich, W. (1971). *Appl. Phys. Lett.* **18**, 127–128.

Schadt, M., and Müller, F. (1979). *Rev. Phys. Appl.* **14**, 265–274.

Scheffer, T. J. (1975). *J. Phys. D* **8**, 1441–1448.

Scheuble, B. S., and Baur, G. (1981a). *Freib. Arbeitstag. Fluessigkrist. Freiburg* **11**, 8.

Scheuble, B. S., and Baur, G. (1981b). *Proc. Eurodisplay '81, Eur. Disp. Res. Conf., 1st, München*, pp. 21–24.

Scheuble, B. S., Baur, G., and Meier, G. (1981). *Mol. Cryst. Liq. Cryst.* **68**, 57–67.

Schiekel, M., and Fahrenschon, K. (1971). *Appl. Phys. Lett.* **19**, 391–393.

Taylor, G. N., and Kahn, F. J. (1974). *J. Appl. Phys.* **45**, 4330–4338.

van Doorn, C. Z. (1975). *J. Appl. Phys.* **46**, 3738–3745.

van Doorn, C. Z., Gerritsma, C. J., and de Klerk, J. J. M. J. (1980). *In* "The Physics and Chemistry of Liquid Crystal Devices" (G. J. Sprokel, ed.), pp. 95–104. Plenum, New York.

White, D., and Taylor, G. N. (1974). *J. Appl. Phys.* **45**, 4718–4723.

Wysocki, J., Adams, J., and Haas, W. (1968). *Phys. Rev. Lett.* **20**, 1024–1025.

12

Recent Advances in High-Strength Fibers and Molecular Composites

D. C. Prevorsek

Allied Corporation
Morristown, New Jersey

I. Introduction

A closely packed ensemble of parallel extended polymer chains should exhibit the highest achievable specific strength. These unique characteristics of uniaxially oriented polymers are the basis for the technology of

329

Copyright © 1982 by Academic Press, Inc.
All rights of reproduction in any form reserved
ISBN 0-12-174680-1

ultrahigh-strength organic polymeric fibers. In reality, fibrous structures of significant dimensions that approach their theoretical maximum strength have not yet been prepared. However, the gap between the theoretical limit and experimental results has been greatly reduced during recent years.

Although it was recognized very early that the strength of man-made polymeric fibers is considerably below the theoretical limit, attempts to produce very strong fibers remained for many years rather unsuccessful. A strength of 1 GPa appeared to be a limit that was only rarely exceeded in laboratory experiments, but could not be achieved on a commercial scale.

It was the discovery of very strong polyaramid fibers produced from anisotropic solution, and the realization of the importance of the liquid crystalline state in the preparation of highly ordered extended chain fibers that provided an unprecedented impetus to the research in ultrahigh-strength fibers. Especially during the last ten years, many schemes have been proposed that lead to fibers approaching or exceeding 4 GPa. In this chapter we review the most important recent developments in ultrahigh-strength fibers with emphasis on the role of the liquid crystalline state in the technology of fiber formation.

II. Role of the Liquid Crystalline State

The liquid crystalline state is of great importance in the preparation of uniaxially oriented polymeric structures exhibiting ultrahigh modulus and strength. Its role is immediately recognizable by inspecting the structural model of Fig. 1, representing an ensemble of molecules exhibiting maximum theoretical modulus and strength. For a given finite molecular weight this model resembles a nematic structure, except that in the case of high-strength fibers the distance between the chain ends is very large. Also, some polymeric systems may exhibit a high degree of three-dimensional local order in spite of random spacing of chain ends.

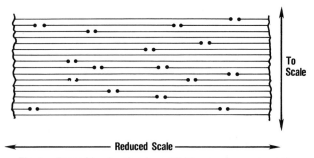

Fig. 1.   Ensemble of molecules exhibiting maximum strength.

Years of systematic scientific and technological research demonstrated that, at least in principle, it is possible to convert almost every polymer or polymeric precursor that can be post-polymerized to fibrous structures exhibiting nearly perfect molecular orientation with chain ends and other imperfections distributed sufficiently at random to avoid a large concentration of stress. In fact, high-modulus and high-strength organic polymeric fibers exhibiting structures resembling that in Fig. 1 have been prepared from a variety of polymers. On one extreme we have very rigid, insoluble, and unfusible graphite fibers, and, on the other, flexible, low melting, soluble polyethylene. Semirigid and rodlike polymers exhibiting thermotropic and/or lyotropic behavior fall in the range between these two extremes.

A linear polymer such as polyethylene (PE), because of a small molecular cross section and very strong backbone, is an ideal molecule for the achievement of high theoretical strength. However, the tendency of PE to crystallize in folded-chain morphology renders the task of designing a process leading to structures approaching that of Fig. 1 very challenging. A considerable level of chain folding persists on crystallite surfaces, even after the polymer is nearly completely uniaxially oriented and fully drawn, and this greatly reduces the fiber strength. Nevertheless, even fibers from flexible molecules produced by present commercial processes, such as PE, nylon, and PET, exhibit to some extent structural elements approaching those of liquid crystalline nematic order.

After many years of laborious research on fiber morphology it is now generally accepted that fibers that are melt-spun and drawn from flexible low-melting linear polymers can best be represented by the two-phase composite model shown in Fig. 2. According to this model the fibers consist

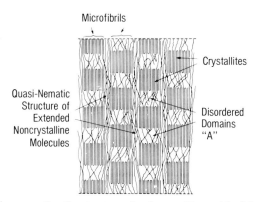

Fig. 2. Model representing the structure of melt-spun fibers of flexible polymers such as PET, aliphatic polyamides, polyethylene. (Reprinted from Prevorsek *et al.*, 1973, p. 152, by courtesy of Marcil Dekker, Inc.)

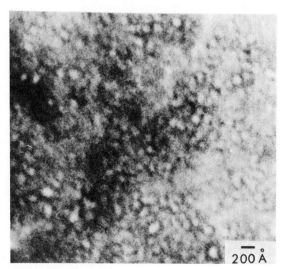

Fig. 3. Transmission micrograph of a thin cross section in an unstained sample of highly drawn PET fiber. (Reprinted from Prevorsek *et al.*, 1973, p. 153, by courtesy of Marcel Dekker, Inc.)

of microfibrils surrounded by an extended chain matrix in which the molecules are highly oriented but under considerable strain (Peterlin, 1979; Prevorsek *et al.*, 1973). This matrix of extended interfibrillar molecules, which is clearly shown in the electron micrograph in Fig. 3, has such a high orientation as to approach the nematic phase model in Fig. 1. The structure of the microfibrils, on the other hand, consists of a series of crystalline and amorphous domains with many chain folds accumulated in the boundary between the two elements of the microfibril (Peterlin, 1979).

During the process of drawing, the volume fraction of the microfibrils decreases, whereas that of the extended drawn matrix increases. The increase in fiber strength on drawing is to a large degree a result of transformation of the microfibrils into extended chain interfibrillar molecules (Peterlin, 1979; Prevorsek *et al.*, 1977). The extent of structural change in the microfibrils during drawing appears to vary from polymer to polymer. These changes appear to be almost negligible in PET and nylon and quite significant for PE.

The problem of preparing fiber structures from flexible polymers that are close to the ideal model in Fig. 1 is that stretching to the maximum draw ratio is not sufficient to eliminate the microfibrillar two-phase structure and stacked chain folding, leading to a high concentration of stress (see Fig. 4). In spite of these difficulties, which originate in the normal crystallization behavior of flexible chain polymers, it is possible to produce polymer struc-

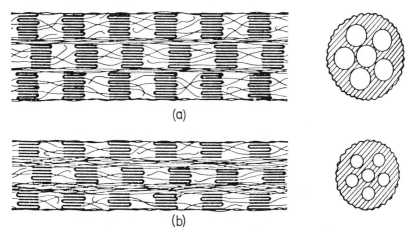

Fig. 4.   Model representing the longitudinal and cross-sectional views of (a) a low and (b) a high draw ratio fiber. Fiber properties such as strength, diffusion, and shrinkage force are determined by the matrix. The dispersed phase and the microfilbrils primarily provide dimensional stability. The model applies to nylon 6, nylon 66, and PET fibers. (Reprinted from Prevorsek *et al.*, 1973, p. 154, by courtesy of Marcel Dekker, Inc.)

tures similar to that in Fig. 1 from PE in many ways, such as solid state extrusion (Zachariades *et al.*, 1979), drawing of fibers from dilute solutions (Pennings and Meihingen, 1979), and gel spinning (Smith and Lemstra, 1979).

The chain folds that are so persistent in flexible molecules can be reduced also by increasing the rigidity of the polymer molecule. If a large section of the molecule is straight and rigid, the polymers begin to exhibit liquid crystalline behavior. The increase in the rigidity of polymer molecules is invariably accompanied by two phenomena: a reduction in solubility and an increase in the melting point. Therefore, the rigid-rod polymer that is an ideal material for producing ultrahigh-strength, ultrahigh-modulus fibers may not be processed either from the melt or solution. A great deal of chemical and physicochemical research was required to overcome these problems and to design and synthesize polymers that are semirigid to a degree sufficient to give liquid crystalline behavior in the melt (Preston, 1979) or in solution (Schaefgen *et al.*, 1979) at temperatures below their de-composition temperatures.

An example of a very rigid ultrahigh-modulus fiber is graphite fiber. Although this material in final form is neither fusible nor soluble, it can be prepared with a high degree of orientation and molecular order.

TABLE I

Processes for the Production of High-Modulus, High-Strength Fibers

Examples of Chemical Composition	Process Description
Graphite	Spinning and stretching of flexible polymeric precursor, such as PAN, followed by subsequent conversion to graphite structure (Reynolds 1973; Bacon, 1973)
	Melt spinning of isotropic or mesophase pitches, followed by subsequent conversion to graphite structure (Singer, 1979)
Aromatic polyamides, aromatic polyhydrazides, etc.	Spinning from anisotropic solutions, followed by heat treatment
	Spinning from isotropic solutions, followed by heat treatment
Aromatic polyesters, etc.	Spinning from anisotropic melt followed by heat treatment
	Spinning from anisotropic melt of low molecular weight polymer followed by solid state polymerization in fiber form
Polyethylene, polypropylene	Solid state extrusion
	Hydrostatic extrusion
	Surface growth from dilute solution
	Gel spinning

An overview of processes used to produce high-modulus fibers is presented in Table I. One cannot stop wondering about the variety of processes leading to highly uniaxially oriented structures with minimal contents of clustered defects. Although man-made high-strength fibers have been in use since the 1930s, it seems that the 1960s and 1970s were particularly fruitful in the development of fiber technology and the exploitation of the easily orientable, relatively fold-free liquid crystalline state needed to produce high-strength fibers.

III. High-Modulus, High-Strength Fibers from Thermotropic Polymers

A. Introduction

Polymers that exhibit liquid crystalline order in solution or in the melt can potentially produce materials with a high degree of molecular orientation and excellent mechanical properties. Several successful technologies based on these characteristics of rodlike polymers have been developed, and the technology of aramid fibers spun from anisotropic solutions attracted a great deal of attention. Comprehensive reviews of this subject and of the chemistry and physics of lyotropic systems have appeared in the literature (Schaefgen *et al.*, 1979; Ciferri and Valenti 1979).

The technology of thermotropic polymers, on the other hand, is still in the early phase of development. At present, most of the developments have been in the area of plastics. Research in the area of fibers and injection molded plastics, and extensive patent coverage, suggest that major development may also soon follow in these fields.

In this section we shall review the process for formation of fibers from a liquid crystalline melt, as well as some relevant rheological, physicochemical, and mechanical characteristics of these systems.

B. *Process for Making High-Tenacity Fibers*

High molecular weight rodlike polymers cannot be subjected to large extensional or shear flows without chain breakage, and only relatively low molecular weight thermotropic polymers are processable from the melts. Therefore, most thermotropic polymers currently known cannot be prepared and extruded into fibers at a sufficiently high molecular weight to achieve a high level of mechanical strength.

In a typical fiber-forming process a low molecular weight thermotropic polymer is extruded into a fiber that is, after solidification, almost fully oriented, but because of its low molecular weight has relatively low strength (0.25–0.70 GPa). Between the die exit and the solidification point in the spin-way the fiber undergoes considerable thinning involving elongational flow. The extent of thinning is controlled by extrudate and take-up velocity, while the maximum achievable thinning is controlled by the melt temperature and the molecular weight of the polymer.

The spinning process is followed by a heat treatment of essentially relaxed fibers near the flow temperature, but substantially below the temperature at which the fibers begin to melt (Luise, 1980; Schaefgen, 1978). During this step the molecular weight increases, but is seldom recorded because the post-polymerized fibers are often difficult to dissolve.

Post-polymerization is usually carried out in an inert atmosphere by flushing the fiber with preheated nitrogen or in a vacuum. It is also desirable to increase the temperature gradually as the polymerization progresses. As a result of post-polymerization the fiber strength sometimes increases by a factor of six to eight. The modulus, on the other hand, remains much less affected by this treatment but may still increase by a factor of two (Luise, 1980). This increase in modulus suggests that the heat treatment leads not only to an increase in molecular weight, but must also result in an improvement in the order and lateral packing of the molecules in the fiber.

The requirement that the extruded fibers be post-polymerized to achieve high mechanical properties restricts the choice of thermotropic polymers to those having end groups (or functional derivatives of such end groups) that allow further polymerization upon heat treatment. Examples of such end groups are —OH, —COOH, —NH$_2$, and —CHO.

In spite of considerable patent literature regarding thermotropic polymers and their conversion into fibers, there are only a few cases where the post-polymerization step is well documented. An example of thermotropic polyester processing is represented by poly(phenylhydroquinone terephthalate). This polymer was disclosed in a Du Pont patent (Payet, 1979) and reinvestigated by Jackson (1980). This author reports that the melt-spun fiber having inherent viscosity $\eta_i = 3.0$ before spinning has, after solidification, a strength of 0.45 GPa, which increases to 3.2 GPa after a short (30 min) heat treatment at 340°C. The complete data of Jackson's experiments are summarized in Table II.

In addition to these data, Jackson (1980) reports the η_i and molecular weight of this fiber. It is estimated that the molecular weight of about 5000, which corresponds to about 16 repeating units, is sufficient to produce

TABLE II

Properties of Poly(phenylhydroquinone
terephthalate) Fibers^{a,b}

	Heat-Treatment Time (min)		
	As-Spun	30	60
Denier/filament	1.6	1.4	1.4
Strength (GPa)	0.45	3.2	3.7
Elongation (%)	0.9	4.2	4.3
Elastic modulus (GPa)	52	98	108

^{a} From Jackson (1980).

^{b}

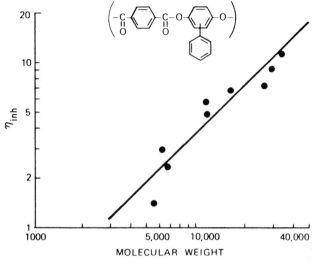

Fig. 5.   Correlation of the inherent viscosity with molecular weight of poly(phenylhydro-quinone terephthalate). (From Jackson, 1980.)

fibers with tenacities of up to 0.71 GPa. This is very low in comparison with flexible polyesters such as PET, which require a molecular weight of about 30,000 to achieve similar fiber properties. After solid state polymerization the molecular weight of the poly(phenylhydroquinone terephthalate) reached approximately 40,000 ($\eta_i \sim 10$) (Jackson, 1980). The inherent viscosity–molecular weight relationship for this polymer is presented in Fig. 5.

C. Selection and Optimization of Composition

Since an ideal fiber structure is nematic, rodlike polymers represent, at least in principle, an ideal system for the preparation of high-modulus and high-strength fibers. The problem is in the processability of the high molecular weight rigid-rod polymers, because very often the melting point exceeds the decomposition temperature.

A logical candidate to give an anisotropic nematic melt is poly(p-phenylene); however, the melting temperature T_m (467°C) and the glass transition T_g (267°C) of this polymer are too high with respect to the decomposition temperature to allow controlled experiments in the melt.

The same dilemma exists with other rigid-rod homopolymers. In order to develop a fiber-forming thermotropic polymer it is therefore necessary to lower the glass transition and melting temperatures without destroying the tendency to form a liquid crystalline melt. This can be achieved by modification of the chemical structure using unsymmetrically substituted nuclei, flexible links, or nonlinear comonomers as described in more detail by Ciferri (Chapter 3 of this volume). In fact, a complete replacement of a linear rigid moiety by a modified nucleus usually leads to a complete disruption of the liquid crystalline order in the melt. Considering also the changes in T_m and T_g of the polymeric system that accompany these modifications, it is desirable to carry out these studies with polymers allowing a large choice of monomers and compositions.

Condensation polymers such as polyesters are particularly suitable because they frequently allow a continuous alteration of composition. This permits great flexibility in the design of the polymer in obtaining optimal processing and property characteristics. A polymer optimization study is best represented by constructing a phase diagram as a function of the chemical composition of the polymers under consideration.

An example reported by Griffin and Cox (1980) is illustrated below for a series of polymers derived from chlorohydroquinone and a mixture of three dicarboxylic acids: 4,4'-oxydibenzoic acid, terephthalic acid (TA), and isophthalic acid (IA).

The content of 4,4'-oxydibenzoic acid is maintained constant at 0.25 mol while the ratio of terephthalic and isophthalic acids is varied over the entire range of 0 to 100%.

Using DSC measurements in combination with hot stage microscopy, Griffin and Cox constructed the phase diagram in Fig. 6. Shown there are the transition temperatures versus composition for the replacement of para units (terephthalate) with the meta (isophthalate) links.

Starting with 100% IA, the melt is isotropic and the onset of optical anisotropy is observed at a TA:IA ratio of 15:85. Further increase in the TA content increases the stability of the mesophase, as reflected by an increase in the temperatures of nematic to isotropic transition, T_{NI}. In addition, the authors found a narrow biphasic region, similar to observations with lyotropic polymers. A clear resolution of the relaxation associated with T_g by DSC is possible only in the range of TA content below 50%. When TA content is increased above 50%, the T_g relaxation becomes suppressed and values of T_g can no longer be assigned with certainty. The temperatures at which these polymers can be plastically deformed depends strongly on the TA:IA ratio. For the isotropic polymer containing no TA, the deformation temperature is $\sim 100°$ below the melting point of the polymer. However, with increasing TA content the development of the mesophase and the difference between the melting point and deformation temperature decreases. For the thermotropic polymer containing 30% TA, the deformation temper-

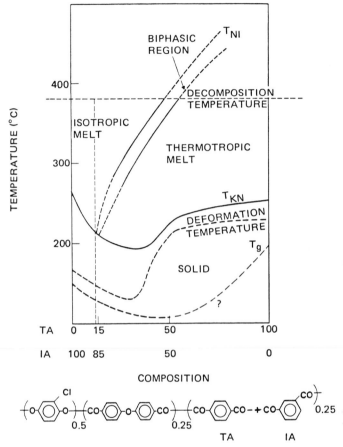

Fig. 6. Transition temperature versus composition for the TA:IA series. (From Griffin and Cox, 1980.)

ature is only 50° below the crystalline–nematic transition temperature. For liquid crystalline polymers containing more than 50% TA, this difference is further reduced to about 25°. This resistance to deformation is attributed to molecular order, which supresses the large scale relaxation at temperatures significantly below the apparent melting point of the polymer (Griffin and Cox, 1980).

The DSC data and deformation temperatures as presented in Fig. 6 are complemented by the results of dynamic visco-elastic measurements shown in Fig. 7 (Griffin and Cox, 1980). The isotropic (TA:IA = 0:100) polymers show a strong mechanical loss and a sharp decrease in modulus characteristic

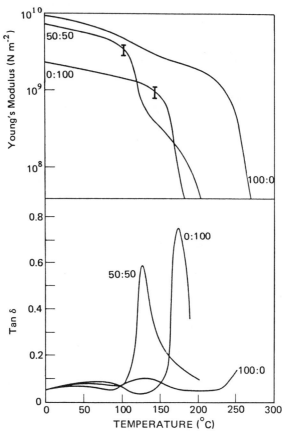

Fig. 7. Dynamic mechanical analysis of TA:IA series. Composition as in Fig. 6. (From Griffin and Cox, 1980.)

of amorphous polymers. The position of the $(\tan \delta)_{max}$ for the 50:50 composition is shifted $\sim 100°$ below that of the 100% IA polymer, but the peak is lower. The modulus–temperature curve for this polymer indicates two relaxation mechanisms, one at T_g, and another broader one at $\sim 175°C$. This shows that for this polymer partial order suppresses the large-scale molecular motions until the T_{CN} melting temperature is reached.

The polymer containing 100% TA and exhibiting $\sim 15\%$ crystallinity shows only a small mechanical loss at 125°C, and the modulus begins to decrease significantly only near the T_{CN} transition at 250°C. In this case, conventional crystallinity may be a significant factor restricting the relative motion of the rodlike molecules (Griffin and Cox, 1980).

D. Composition–Melt–Viscosity Relationship

In processes for fiber forming from the melt, the melt viscosity and its shear-rate dependence are the parameters of greatest importance. If the melt viscosity is plotted as a function of composition over a range in which the polymer exhibits a transition from the isotropic to an anisotropic melt, the onset of liquid crystallinity is reflected by a sharp decrease in the melt viscosity. For the case of PET modified with p-hydroxybenzoic acid (PHB), this phenomenon is illustrated in Fig. 8 (Jackson and Kufhuss, 1976).

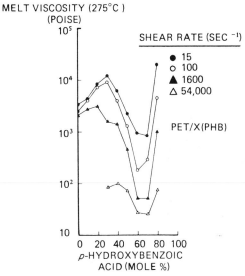

Fig. 8. Melt viscosity of PET modified with p-hydroxybenzoic acid. (From Jackson and Kufhuss, 1976.)

At very low PHB content the melt viscosity first increases with increasing content of PHB. This trend is reversed at a PHB content of about 30%, which marks the onset of liquid crystallinity. Depending on the shear rate, the decrease in viscosity amounts to one to two orders of magnitude. The viscosity reaches a minimum at a PHB content of about 60 mol %, when a sudden increase is observed. It is possible that at this point the melt is already fully oriented under shear and further addition of PHB increases the effective axial ratio of the rodlike segment.

For the poly(chlorohydroquinone-4,4'-oxydibenzoate-*co*-terephthalate-*co*-isophthalate) series discussed above the behavior is very similar. At a

TA:IA ratio of about 15:85, where the onset of liquid crystallinity is observed, the melt viscosity starts to decrease abruptly. At a shear rate of 1000 sec^{-1} it drops by three orders of magnitude before leveling off at a TA:IA ratio of about 50:50. For the lower shear rates, the plateau of melt viscosity is shifted to a higher content of TA, or more ordered melts, as shown in Fig. 9.

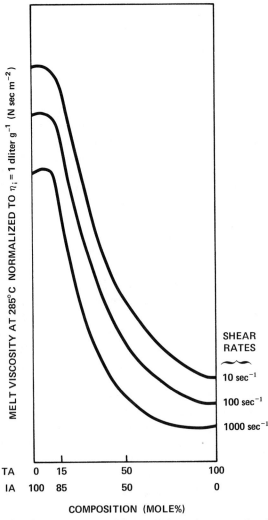

Fig. 9. Melt viscosity versus composition at various shear rates for the TA:IA series. Composition as in Fig. 6. (From Griffin and Cox, 1980.)

E. Factors Affecting Filament Uniformity and
Orientation after Extrusion

Fiber processes involving high swelling of the extrudate at the exit from the die are difficult to control. Filament nonuniformity, expressed as fluctuations in diameter and orientation function along the extruded filament, increases with increasing die swell. In addition, die swell adversely affects the time of continuous spinning without filament breakage at the die. In this respect thermotropic polymers show a distinct advantage over the isotropic melts because they can be extruded practically without a detectable die swell. It is possible that this lack of die swell is a result of the laminar flow of rodlike molecules involving no elastic energy storage.

This advantageous characteristic of the thermotropic melts is to some extent offset by the high sensitivity of viscosity to shear rate. If the melt viscosity is highly dependent upon shear rate, it is usually difficult to control the filament diameter across the bundle in multifilament spinning.

This problem can be anticipated in the spinning of thermotropic melts, which frequently exhibit a decrease in viscosity at shear rates as low as 0.1 sec^{-1}. The isotropic homologs of some series may, on the other hand, not show any shear thinning until shear rates of 100 to 1000 sec^{-1} are reached.

In the execution and analysis of spinning experiments using anisotropic melts it is also necessary to be aware of large thermal and stress history effects. These phenomena could play a significant role in process optimizations and scaling up.

For illustration we present in Fig. 10 the effect of thermal history on the viscosity of the 60:40 copolymer of p-hydroxybenzoic acid and PET. In these experiments the polymer was charged into the capillary rheometer at 210, 240, and 300°C and then cooled to 210°C before measuring the viscosity at different shear rates. Note the large decrease in viscosity with increasing preheating temperature. Initially, the heating at 300°C decreased the viscosity by a factor of ten, but most of this decrease was regained after 20 min at 210°C (Wissbrun, 1980).

These changes in the rheological behavior are attributed to structural changes in the melt, melting of crystallites, destruction of domains of intermediate order, etc. This hypothesis was verified by a study conducted at Celanese Laboratories that involved x-ray analysis of fibers spun from Eastman Kodak Polyester X76 (PET modified with p-hydroxybenzoic acid). Fibers spun at 230°C showed, at room temperature, weak but sharp crystalline x-ray reflections, while no such reflections were visible for fibers extruded at 280°C.

With regard to shear history, Wissbrun (1980) also reports the effects of shear thickening and a form of thixotropy that is occasionally observed in

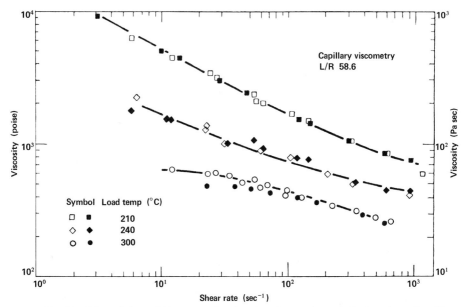

Fig. 10. Effect of thermal history on the viscosity of a copolymer of hydroxybenzoic acid (HBA) and poly(ethylene terephthalate) (PET); composition 60/40 HBA/PET. (From Wissbrun, 1980.)

dynamic testing. This latter effect was observed in variable frequency experiments. Viscosities are higher when measurements are taken at increasing frequency and lower if the frequency is decreasing. The opposite trend is observed with storage modulus.

Shear history effects were attributed by Wissbrun (1980) to the existence of a structure that requires a considerable deformation for its destruction. It is possible that the "domains" of different local orientation, discernible by microscopic investigation are responsible for these phenomena. It has been proposed that the rest structure illustrated in Fig. 11 would require considerable shearing to reach the steady state shear orientation.

After extrusion the solidified filaments exhibit high orientation and are converted to high-strength fibers by a simple solid state polymerization process involving no subsequent drawing of the fiber. In order to achieve a high degree of orientation it is essential that, in addition to the liquid crystalline (nematic) behavior, the polymer also exhibit relaxation times long in comparison with the time needed to cool the molten filament below its solidification temperature.

Measurements of the x-ray orientation angle, which is often as low as $10°$, demonstrate that high orientation can be achieved by spinning molten thermotropic polymers into the ambient air. These findings imply that the

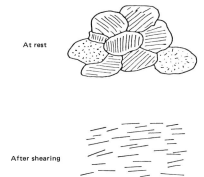

At rest

Fig. 11. Schematic diagram of hypothetical domain structure in liquid crystal polymers, showing regions of locally parallel orientation, but no correlation of orientation between adjacent domains. (From Wissbrun, 1980.)

After shearing

relaxation times for thermotropic polymers must be very long, despite their relatively low molecular weight. A set of data illustrating the increase in the relaxation time as a function of the composition of a homologous series involving an isotropic polymer and its anisotropic derivatives are shown in Fig. 12. The lowest curve represents PET ($\eta_i = 0.60$) while the other curves are PET/PHB copolyesters ($\eta_i = 0.60$), and the content of p-hydroxybenzoic acid is expressed as mole fraction Y. Note that only the PET/PHB copolyesters containing 40 to 60 mol % of PHB are liquid crystalline, and that their relaxation times are much more affected by frequency (shear rate) than those of more flexible isotropic homologs.

The long relaxation times characteristic of all thermotropic polymers are, along with low viscosity and low die swell, a major factor contributing to (1) the easy processability of these polymers into fibers and (2) the very high strength that can be achieved without subsequent stretching of these polymers.

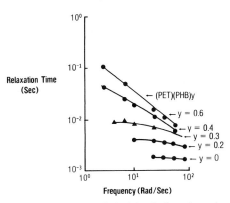

Fig. 12. Relaxation times of PET modified with p-hydroxybenzoic acid (PHB). $Y = $ PHB content. (From Jackson, 1980.)

IV. Ultrahigh-Modulus, Ultrahigh-Strength Fibers by Solution Spinning and Drawing

A. Introduction

Attempts to prepare rigid, rodlike polymers lead to systems that decompose below the temperatures required for melt processing. Hence these polymers can only be processed from solution. Although rigid, rodlike polymers are in general difficult to dissolve, systematic solvent research has led to the discovery of several polymer–solvent systems that are suitable for preparation of fibers. In favorable cases the solutions of rodlike polymers exhibit a spontaneous isotropic–nematic transition predicted by Flory's statistical theory (Flory, 1956). This transition is observed if the critical concentration is below that at which the polymer precipitates from the solution. The ultrahigh-modulus high-strength aramid fibers of the Kevlar type are an example of fiber technology in which the characteristics and properties of anisotropic solutions of rigid, rodlike polymers are a key factor.

Since chain rigidity suppresses polymer solubility, it is desirable for technological reasons to increase chain flexibility and thus expand the choice of solvents for solution spinning. However, an increase in chain flexibility increases the critical polymer concentration at which a spontaneous isotropic–nematic transition takes place. Consequently, for the more flexible polymers, we should frequently encounter conditions where the polymers crystallizes from solutions before the critical concentration for formation of an anisotropic phase is reached. Marrucci and Sarti (1979) developed a theory predicting that for semirigid polymers that do not form an anisotropic solution under quiescent conditions, an isotropic–nematic transition may be induced by shear or elongational flow. Extensive studies with semirigid poly(terephthalamide of p-aminobenzhydrazide) (X-500) support the view that these fibers probably fall into this category (Ciferri and Valenti, 1979). Both rigid Kevlar and semirigid X-500 can be spun to yield highly oriented high-modulus fibers without subsequent stretching except that higher elongational rates are required to achieve high levels of orientation for X-500.

It must not be overlooked that in general two criteria must be met to achieve a high degree of orientation in as-spun fibers. First, the polymer must be easily orientable by shear or elongational flow and second, this orientation must be preserved until the chain mobility in the filament is forzen by the coagulation process. The first criterion is met by the existence of an anisotropic solution, either formed spontaneously or flow induced. The second requires a relatively long relaxation time, which must be of same order or longer than the time needed to convert the spinning dope into a "solidified" filament in which the molecules are unable to undergo long-range molecular motion. Only under these conditions is it possible to "freeze in" the orientation induced by shear and elongation flow for semirigid and flexible molecules.

While reduction in chain rigidity offers a larger choice of solvents, it also introduces restrictions on the spinning process, such as high elongational rates in the spin-way and short coalescence times. It will be shown below that both these criteria could be successfully met with semirigid X-500 in DMSO–LiCl solutions.

It is reasonable to assume that the production of ultrahigh-modulus fibers is not restricted to rigid and semirigid polymers. Structures similar to that shown in Fig. 1 might also be achieved with more flexible molecules but this will undoubtedly require processes that are different from those used to produce Kevlar and X-500 fibers.

Experiments with more flexible polymers such as PAN and nylon 66 failed, so far, to produce ultrahigh-modulus fibers by solution spinning. This can be attributed to (1) inability to achieve in the spin-way elongational flow sufficient to produce the required level of orientation and/or (2) inability to freeze in the flow-induced orientation due to relaxation times in solution that are too short in comparison with the time needed to solidify the filaments. A very promising solution for overcoming these problems is offered by spinning high molecular weight flexible polymers from a dilute solution, and then drawing these gel fibers to extremely high draw ratios. The morphology and the properties achieved with polyethylene (120 GPa modulus and 3.8 GPa strength) suggest that the fibers so produced approximate the idealized structure in Fig. 1. Although gel spinning has been so far successfully demonstrated only with polyethylene, it is expected that with some modification this method will also be applicable to other higher melting flexible polymers.

We have presented in this section an overview of the technologies used to produce ultrahigh-modulus, ultrahigh-strength fibers from rigid, semirigid, and flexible polymers.

B. Spinning from Nematic Solutions

Aromatic polyamides such as poly(1,4-benzamide) (PBA) and poly(1,4-phenyleneterephthalamide) (PPD-T) exist in solution preferentially in highly extended chain conformations. This is reflected in very high values (1.7) of the exponent a in the Mark–Houwink relationship ($[\eta] = KM^a$, where K and a are constants and M is the molecular weight), as well as in high values of chain persistence length (300–400 Å) (Arpin and Strazielle, 1977). Above a critical concentration of polymer in solution, these polymers exhibit lyotropic (nematic) mesophases that are very important in the formation of ultrahigh-modulus, ultrahigh-strength fibers. The spinning of anisotropic solutions into fibers can lead to products of nearly perfect orientation without subsequent stretching of the extruded and solidified filaments. This result suggests that the lyotropic mesophases consist of highly oriented polymer chains. It should also be noted that the degree of orientation achievable by spinning of the lyotropic (nematic) dopes can exceed that of fully drawn melt-spun fibers from flexible polymers such as poly(ethylene terephthalate) or aliphatic polyamides such as nylon 6 or nylon 66. Strength and modulus of fibers produced from anisotropic solutions can be very high (3.0–3.5 GPa and 600–1000 GPa, respectively). This implies a large degree of structural perfection in addition to a nearly perfect orientation. It appears, therefore, that fibers produced from anisotropic solutions can under specific conditions approach the idealized structure of Fig. 1, and Kevlar-type aramid fibers represent, to date, the most important technological development based on the chemistry and physics of liquid crystalline high molecular weight polymers.

Spinning of fibers from nematic dopes and the properties of such fibers have been extensively reviewed. Since the appearance of the reviews by Schaefgen et al. (1979) and Ciferri and Valenti (1979) some new relevant findings appeared in the scientific or patent literature. Totally aromatic rigid-rod polymers have been reported by Wolfe and Arnold (1981) and by Wolfe et al. (1981). These polymers, which will also be considered in Section V.B, have the structural formulas given in Table VI. The presence of only aromatic and aromatic heterocyclic units assures excellent thermal and thermo-oxidative stability, which is important in aerospace applications. The elastic modulus of poly(p-phenylenebenzobisthiazole) (PBT) may reach 250 GPa, which is twice the modulus of Kevlar-49 (Allen et al., 1981). However, strength (2.4 GPa) and elongation to break (1.5%) appears inferior to those of Kevlar. Spinning is performed using solutions containing from 5 to 10% polymer in methanesulfonic, chlorosulfonic, or polyphosphoric acids.

Considerable interest is also developing for copolymers containing a rigid block (e.g., PBA) and a block that may be either flexible (e.g., Nylon 6, Takayanagi et al., 1980) or semirigid (e.g., X-500, Asrar et al., 1982).

Another important development is the spinning of cholesteric dopes of cellulose in aqueous solutions of a cyclic amine oxide, such as N-methylmorpholine N-oxide, to yield fibers with strengths of the order of 1 GPa (Franks and Varga, 1979). This technology appears to hold the greatest promise and a plant section is already in operation.

Cholesteric rodlike polymers, such as PBA containing a few mol % of L-valine, have been reported (Krigbaum *et al.*, 1979). These polymers could allow the fabrication of films having high modulus along two perpendicular directions in the plane of the film ("molecular" composites, see Ciferri, Chapter 3, this volume).

C. *Flow-Induced Transitions to a Nematic Phase*

In the section above we reviewed the importance of the nematic lyotropic phase in the orientation and mechanical properties of rodlike aromatic fibers. However, fibers of very similar mechanical properties can be obtained from polymers such as polyterephthalamide of *p*-aminobenzhydrazide (X-500), which does not readily form an anisotropic solution and whose persistence length, 50 Å in dimethyl sulfoxide (DMSO), falls between those of the aromatic polyamides PBA and PPD-T and those of flexible randomly coiled polymers (Ciferri and Valenti, 1979). Since conventional spinning techniques fail to produce high-strength, high-modulus fibers from flexible random-coil polymers, it was proposed that in the case of X-500 the high degree of orientation and high fiber modulus are a result of its intermediate chain rigidity. This may be sufficient to cause, under the influence of shear and/or elongational flow in the spin line, a transition from an anisotropic into a nematic state.

These conclusions are supported by recent results of comprehensive studies of solution and fiber properties of X-500 in DMSO. The most important findings (Valenti *et al.*, 1981) are reviewed below.

Figure 13 shows a plot of fiber modulus versus filament extension ratio in the post-coagulation–washing step. In the graph, $V_0 =$ linear flow rate at the spinnerette, while V_1 and V_F are the linear take-up rates at the end of the coagulation bath and the washing bath, respectively. Note that:

(1) The modulus–extension ratio data fall in two distinct domains. At a V_1/V_F ratio below a critical value of 1.5 the modulus increases almost linearly with V_1/V_F, while above this critical value the modulus remains essentially unaffected by further increases of V_1/V_1.

(2) The modulus achievable by increasing V_1/V_F decreases with increasing V_0.

(3) The maximum modulus attained, 15 GPa, is very high compared with those for flexible coil polymers, but considerably smaller than that of fibers spun from anisotropic solution of PBA (50 GPa).

The orientation that exists in as-spun fibers is, according to Valenti *et al.* (1981), a prerequisite for development of ultrahigh modulus fibers by solid state post-spinning treatments (heat stretching).

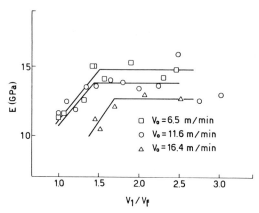

Fig. 13. Effect of extension ratio during coagulation V_1/V_f on elastic modulus of X-500 fibers. Only V_1 is varied while V_0 is kept constant at indicated values. (From Valenti *et al.*, 1981.)

In H_2SO_4, X-500 gives a nematic phase in the quiescent state, and therefore in this solvent the solvent-induced rigidity of X-500 is sufficient to lead to a spontaneous isotropic–nematic transition. A nematic phase is, however, not observed with X-500 in DMSO–LiCl solutions, because the polymer crystallizes from the solution before the critical concentration C_p^* for isotropic–nematic transition is reached. Although in DMSO–LiCl the rigidity of X-500 is insufficient to form a nematic phase, it can be inferred that the system is probably close to such a transition. These conclusions are supported by a systematic study of the effect of polymer concentration and shear rate on the viscosity. Ciferri and Valenti (1979) reported that, at sufficiently high shear rates, the DMSO–LiCl solutions of X-500 show a behavior resembling that of anisotropic solutions of PBA, i.e., the plots of viscosity, η, on C_p show distinct maxima and minima, while at low shear rates the viscosity of X-500 shows a monotonic increase with increasing C_p (see Fig. 14). Very recently Valenti *et al.* (1981) showed that the dependence of $\log \eta$ versus C_p at low shear rates can be divided into three linear segments exhibiting different slopes, which change at well defined values of C_p. The change in slope at C_{p1}^* has been attributed to the formation of entanglements, as observed with flexible random-coil polymers. With X-500 in DMSO–LiCl systems, C_{p1}^* occurs at $\sim 1 \, \text{g}/100 \, \text{ml}$. The interpretation of C_{p2}^* (at $5 \, \text{g}/100 \, \text{ml}$) is more speculative, but it may reflect a flow-induced transition from an isotropic to a nematic state at low shear rates.

It is important to note that the inflexion point on $(\log \eta)$–C_p plots observed at higher shear rates occurs just after the critical "entanglement" concentration C_{p1}^* is reached. Ciferri and Valenti (1979) proposed that after the entanglement concentration is reached, a sufficiently strong shear gradient

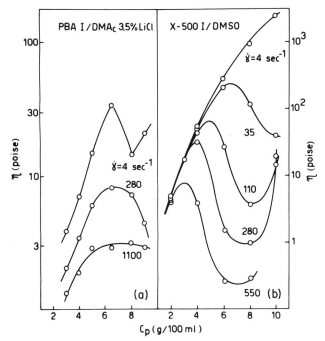

Fig. 14. Steady state shear viscosity–polymer concentration plots for (a) PBA and (b) X-500 at 20°C. Shear rate is indicated. (From Ciferri and Valenti, 1979.)

can lead to large orienting effects. The lowering of the viscosity could reflect the formation of anisotropic clusters in which individual molecules are highly oriented, leading to flow instability.

In summary, these studies show that semirigid polymers such as X-500 in DMSO can be spun into fibers having a high modulus of 15 GPa, which may be increased to 67 GPa by subsequent thermal treatments. Although the data suggest but do not confirm the existence of a flow-induced isotropic–nematic transition, the studies with semirigid X-500 produced several findings that are relevant for the technology of ultrahigh-modulus, ultrahigh-strength fibers:

(1)   Anisotropic solutions and use of very rigid polymers may be desirable but are not essential for the preparation of ultrahigh-modulus fibers.

(2)   For more flexible molecules a lowering of viscosity is observed at relatively high shear rates if the polymer concentration is above C_{p1}^*.

(3)   From viscosity–C_p plots it can be inferred that high-modulus fibers can be obtained from isotropic solutions of semirigid polymers such as X-500 if the processing conditions (concentration, shear rate) involve an "entangled" polymer solution.

The obvious question that arises is, where is the limit in molecular flexibility in producing idealized fiber structures as shown in Fig. 1, and what modifications in the fiber process are required to achieve similar molecular perfection with a random-coil polymer? Some of these questions will be answered in Section IV.D.

D. Gel Spinning and Drawing of Polyethylene

It has been shown that a variety of processing methods lead to fibers exhibiting moduli and/or strengths that are considerably higher than those of commercially available fibers.

Many successful attempts have been reported during recent years leading to polyethylene fibers exhibiting exceptionally high modulus and strength. For example, Capaccio and Ward (1974) prepared high-modulus poly-ethylene filaments with Young's modulus of the order of 70 GPa by melt spinning and drawing. High-modulus fibers can also be prepared by solid state extrusion as shown by Southern and Porter (1970). However, the strength of fibers produced by this technique is usually quite low.

Fibers of exceptionally high modulus (100–120 GPa) and tensile strength (3–4 GPa) have also been prepared in a Couette-type apparatus by Pennings and Meihingen (1979). These authors developed a new method of producing fibers from a dilute solution by first absorbing the polymer on the surface of the rotor of the Couette apparatus and then withdrawing and crystallizing this polymer in a fiber form by surface growth.

Very recently the researchers at the central laboratory of DCM in Geleen (Smith and Lemstra, 1979) and at the Department of Polymer Chemistry, State University of Groningen (Kalb and Pennings, 1980) reported an alternate route to the production of high-strength and high-modulus poly-ethylene filaments that approach the modulus and strength of fibers prepared by solution growth techniques (Smith *et al.*, 1979). The approach is based on the discovery that high molecular weight fibers spun from dilute solution can be drawn to a much higher draw ratio than fibers that are melt-spun or spun from concentrated solution. Systematic research (reported in subsequent articles) revealed the heretofore neglected role of chain entanglement in the drawability of high molecular weight polymers. The results, which are of great theoretical and technological importance, are reviewed below.

1. Process and Raw Materials

Polyethylene samples used in all solution spinning reported to date were high molecular weight Hostalen GUR having $\bar{M}_n = 2 \times 10^5$, and $\bar{M}_w = 1.5 \times 10^6$, or Hi-flex 1900 having $\bar{M}_w = 4 \times 10^6$. Polymer solutions

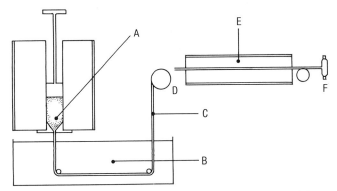

Fig. 15. Schematic diagram of the gel spinning apparatus (Smith and Lemstra, 1980b). A, solution of PE at elevated temperature; B, quench bath; C, solidified gel filament; D, take-up device; E, elevated temperature drawing chamber; F, take-up device.

were obtained by dissolving the polymers in paraffin oil, decalin, or dodecane. In order to minimize degradation of polyethylene, the dissolution was carried out under nitrogen and in the presence of an antioxidant. Solutions of polyethylene were spun using the experimental setup shown schematically in Fig. 15 or by means of a Reifenhouser extruder SOL3-25. The extrusion temperature varied from 130 to 175°C for dilute solutions and was fixed at 120°C for concentrated solutions. The spun filaments were quenched in cold water to form gel fibers and collected on a winder.

Drawing was carried out by: (1) stretching a gel fiber in a hot air oven and simultaneously removing the solvent to yield an essentially solvent-free highly oriented fiber, or (2) drying the gel filaments at room temperature and then extracting the last traces of solvent with ethanol.

The drawing experiments were carried out over a range of temperatures (70–143°C) using a constant temperature or a temperature gradient.

The process variables include: solvent (paraffin oil, decalin, and dodecane); solvent concentration; porosity and orientation of dried gels; and draw temperature, draw temperature profile, and draw ratio.

Presumably the numerous process variables were not well optimized. Nevertheless, the authors were able to identify the key variables that control the mechanical properties of the final fiber and to deduce interesting conclusions about the structure of the extruded gel and the mechanism of drawing.

The concentrations of the solutions spun and quenched into gel fibers appear to be the most important process variable. Smith and Lemstra (1979) varied the concentration of Hostalen GUR in decalin. Their results are summarized in Table III. Fibers extruded from the melt or from concentrated

TABLE III

Effect of Solvent Concentration on Maximum Draw Ratio[a]

Spinning	Drawing		
Solvent (%)	Solvent (%)	Temp (°C)	Maximum Ratio
0	0	120	5
10–50	10–15	120	5
98	98	120	32
98	0	120	22

[a] From Smith *et al.* (1979).

solution exhibited only small differences in maximum draw ratio, while fibers extruded from dilute (2%) solution could be drawn to a ratio six times higher ($\sim 32 \times$). In order to resolve whether the high draw ratio of fibers extruded from dilute solutions is due to a particular structure of these fibers or to the presence of solvent, another experiment was carried out in which the solvent in fibers spun from dilute solutions was completely removed before drawing. This was accomplished by extracting decalin with methanol and drying. This fiber could be drawn to a draw ratio of 22, which is somewhat less than that of the wet fibers spun from the same concentration, but still four times higher than the draw ratio of fibers spun from the melt or concentrated solution (Smith and Lemstra, 1979). Based on these results, the authors concluded that (a) the key factor affecting the draw ratio of fibers is not the presence of solvent but a particular feature of the structure of undrawn fibers; (b) the presence of solvent is helpful but not essential because it lowers the drawing stress.

The authors suggested that the observed effect of solvent is related to reduced chain entanglement in fibers extruded from dilute solution, which affects the maximum draw ratio of fibers. This hypothesis finds further support in the studies of Smith *et al.* (1981). For a polymer concentration exceeding the value for onset of coil overlap, the molecular weight between entanglements $(M_e)_{soln}$ is given approximately by

$$(M_e)_{soln} \cong (\rho/c)M_e = M_e/v_2$$

Here ρ is the bulk density of the polymer, C is the polymer concentration, v_2 is the polymer volume fraction, and M_e is the molecular weight between entanglements in the undiluted polymer melt.

Taking precautions that the evaporation of solvent does not appreciably alter the interpenetration of molecules, they prepared a series of polyethylene

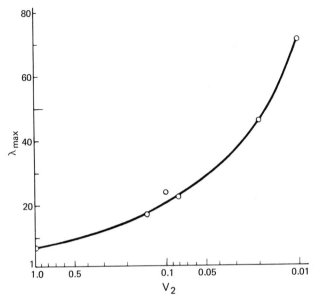

Fig. 16. Maximum draw ratio versus initial polyethylene volume fraction. (From Smith *et al.*, 1981.)

gels from decalin using Hostalen GUR ($M_w = 1.5 \times 10^6$). Varying the initial polyethylene volume fraction from 0.001 to unity, they determined the maximum draw ratio of this gel film at 120°C. The results, summarized in Fig. 16, clearly support the hypothesis that the drawability is closely related to the volume fraction of the solvent in the solution from which the gel is formed.

In the analysis of polymer drawability one must not neglect the role of temperature, which usually increases the maximum draw ratio. This effect is illustrated in Table IV for polyethylene films cast from solutions of various concentrations and given the maximum draw at 90, 120, and 130°C.

Smith *et al.* (1981) explained these data by invoking rubber elasticity theory. They used the dependence of the maximum draw ratio of a permanent network (λ_{max}) on the number N_c of statistical chain segments between cross-linkings, which is given by

$$\lambda_{max} = (N_c)^{1/2}$$

Assuming further that the majority of the entanglements trapped in the polymeric solid N_e act as permanent cross-links,

$$N_c \approx N_e$$

TABLE IV

Maximum Draw Ratio of Solution Cast
Polyethylene Fibers: Effect of Temperature
and Concentration[a]

Initial Polyethylene Volume Fraction	Temperature (°C)		
	90	120	130
1.00	5.5	6.9	9.4
0.14	12	17	28
0.10	14	24	32
0.08	13	23	36
0.02	27	46	62
0.01	36	72	[b]

[a] From Smith et al. (1981).
[b] Unstable draw, films of polyethylene with $\bar{M} = 3.5 \times 10^6$ cast from a 1% v/v solution could be drawn to $\lambda_{max} = 110$ at 130°C.

Here N_e is the number of statistical chain segments between entanglements, which equals M_e/m_s, m_s being the molecular weight of a statistical chain segment. For polyethylene m_s is about 140. Introducing the concentration dependence of M_e in solution (M_e/v_2) we obtain

$$\lambda_{max} = (N_e/v_2)^{1/2} \qquad \text{or} \qquad \lambda_{max} = \lambda_{max}^1 v_2^{1/2}$$

where λ_{max}^1 is λ_{max} at $v_2 = 1$, equal to $(N_e)^{1/2}$. The molecular weight between entanglements in polyethylene melt is about 1900 (Porter and Johnson, 1966), which yields for N_e, the number of statistical chain segments between entanglements, 13.6 and for the draw ratio, $\lambda_{max}^1 = 3.7$. When the maximum draw ratios obtained at 90, 120, and 130°C are plotted against $v_2^{1/2}$ one observes a linear relationship between the two quantities, in agreement with the above relation.

Kalb and Pennings (1980) also considered the effect of molecular weight on the maximum draw ratio. This can be qualitatively illustrated as follows. Complete extension (λ_{max}) of a macromolecule can be approximated by

$$\lambda_{max} \approx L/2R_G$$

where L is the length of the fully extended molecule and R_G is the radius of gyration. Note that $L = n \cdot l$, where n is the degree of polymerization and l is the length of the repeating unit. Since

$$R_G = Cl(n/6)^{1/2}$$

where C is the characteristic ratio for coil expansion, it follows that

$$\lambda_{max} = (6n)^{1/2}/2C$$

Considering that an average value of C is 6.8 and that the weight of the repeating unit of polyethylene is 14, Kalb and Pennings (1980) obtained for a completely disentangled polymer

$$(\lambda_{max})_0 \approx M^{1/2}/20.8$$

2.   Effect of Microporosity and Mechanical Properties

Draw ratios approaching the theoretical limit can be achieved only with fibers that are microporous, or that contain considerable solvent during drawing. This is not surprising because the drawability of polymers is influenced by the mobility of polymer molecules as expressed by the relation for nonrecoverable creep,

$$\dot{\varepsilon} = A \exp[-(U_a - \sigma v)kT]$$

where $\dot{\varepsilon}$ is the rate of creep, U_a is the activation energy for segmental jumps, σ is the local stress, v is the activation volume, and T, k, and A are the absolute temperature, Boltzmann constant, and a pre-exponential factor that can be further analyzed. Both the applied stress and activation volume decrease the barrier for segmental jumps. However, the activation volume decreases with increasing molecular weight as shown for polyethylene by Wilding and Ward (1978). In addition, for high molecular weight polyethylene, the nonrecoverable creep does not occur below a critical value of $\sigma v/kT$. Since U_a seems to increase, while v tends to decrease with increasing orientation, the drawing stress increases rapidly with increasing draw ratio. Therefore, the maximum draw ratio is much more difficult to achieve with high molecular weight polyethylene than with low molecular weight polymer. This difficulty can be overcome by increasing the free volume through microporosity or by lowering the activation energy for segmental jump by using a plasticizer (a solvent in the case of gel fibers). Furthermore, the drawing zone can be extended by using a temperature gradient in the drawing step. The draw ratio and drawing stress increase because of increasing orientation in the fiber (Kalb and Pennings, 1980).

Since nonporous fibers cannot be drawn to high draw ratios, we limit the discussion to the experiments with fibers that were spun from a 2% by weight solution in decalin at 130°C. The fibers were either drawn immediately or extracted with ethanol to remove the last traces of solvent and then dried before drawing (Smith and Lemstra, 1980a). Both fibers were drawn to the

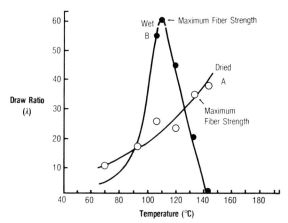

Fig. 17. Maximum draw ratio versus drawing temperature for (A) dried and (B) wet gel fibers of high molecular weight polyethylene ($\bar{M}_\omega = 1.1 \times 10^6$). (Reproduced from Smith and Lemstra, 1980a, *Polymer* **21**, 1351, by permission of the publishers, IPC Business Press Ltd. ©.)

maximum draw ratio at various temperatures between 70 and 143°C, at a strain rate of $\sim 15 \text{ sec}^{-1}$. The results presented in Fig. 17 show that for the dried fibers the maximum draw ratio increases with the temperature, almost to the melting point of the polymer. Although the maximum achievable draw ratio was ~ 40 at 143°C, at this high temperature drawing is no longer very effective because of the excessive relaxation of the molecules during drawing. Thus the draw ratio–property data fall on a different curve (lower line) than those of fibers drawn between 70 and 133°C. With wet fibers, on the other hand, the maximum draw ratio reaches a very distinct maximum at about 105°C. At the optimal drawing temperature the maximum draw ratio of wet fibers approaches 60, which is 50% above the value obtained with the dried fiber prepared from the same undrawn precursor (Smith and Lemstra, 1980a). Note also that for wet fibers the maximum strength correlates with maximum draw ratio, while this is not the case for the dry fibers.

The importance of drawability of this system is best illustrated by a plot of mechanical properties as a function of draw ratio. The pertinent data obtained by Smith and Lemstra (1980b) are presented in Table V. Note that up to a draw ratio of 30 the modulus and strength (under given drawing conditions) increase almost linearly with draw ratio, while the elongation at break (and creep) decrease. For draw ratios above 30 the modulus and strength continue to increase but the increments are smaller.

The structure of undrawn gel fibers changes upon drawing continuously with the draw ratio. Using standard x-ray and electron microscopy techniques, these changes can be followed up to a draw ratio of 30. Above this

TABLE V

Mechanical Properties of Solution–Spun/Drawn
Polyethylene Filaments[a]

Draw Ratio	Modulus (GPa)	Tensile Strength (GPa)	Strain at Break
1.0	2.4	0.06	0.722
2.8	5.4	0.27	0.108
7.3	17.0	0.73	0.076
8.4	17.6	0.81	0.083
8.6	25.1	1.17	0.090
9.8	27.5	1.39	0.070
10.3	28.3	1.52	0.074
10.4	28.1	1.27	0.065
11.3	23.9	1.32	0.071
12.1	37.5	1.65	0.074
13.1	40.9	1.72	0.063
13.9	32.6	1.45	0.065
15.7	41.2	1.80	0.068
25.5	68.3	2.87	0.059
31.7	90.2	3.04	0.060

[a] From Smith and Lemstra (1980b).

stretch ratio the structural changes cannot be resolved by these techniques, although the changes in mechanical properties are still very significant. X-ray diffraction indicates that the undrawn fibers exhibit considerable crystallinity with no preferred orientation. Optical microscopy reveals a distinct lamellar morphology. Analysis of WAXS patterns and birefringence for fibers with draw ratios 1.5, 15.7, and 31.7 show that the orientation function and birefringence increases monotonically with the draw ratio (Smith and Lemstra, 1980b). Structural changes are also indicated by DSC thermograms. The undrawn dried monofilaments melt at 133°C, while filaments with draw ratios of 15.7 and 31.7 exhibit melting peaks at 141 and 145.5°C, respectively, according to Smith and Lemstra (1980b). The change in melting temperature with draw ratio is monotonic, as shown by the data in Fig. 18. As indicated above for the x-ray diffraction data, DSC also fails to resolve structural changes occurring at draw ratios above 30. This suggests that the drawing of fibers between a draw ratio of about 30 and a maximum that may be as high as 60 or more involves negligible changes in molecular and crystallite orientation and short range order. The improvements in the mechanical properties and strength achieved by these extreme drawing conditions are most likely due to the removal of loops and chain folds, and improvement in long-range order that has not yet been identified.

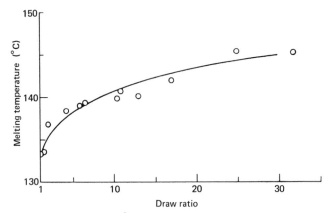

Fig. 18. DSC melting temperatures for solution spun drawn polyethylene filaments versus draw ratio. (From Smith and Lemstra, 1980b.)

E. Conclusions

Ultrahigh-modulus, ultrahigh-strength fibers can be prepared from a variety of polymers by solution spinning. The final fiber properties depend upon the characteristics of the polymer chain. For optimal results the chain must: (a) assume in its extended chain form a nearly linear conformation, (b) allow a close packing of molecules, and (c) exhibit high modulus and strength under tension. Chain rigidity, on the other hand, is a key factor determining process characteristics.

With rigid, rodlike polymers that form thermodynamically stable nematic solutions, anisotropic dopes are readily spun in fibers of nearly perfect orientation. By gradually introducing some chain flexibility a situation is reached where the critical concentration exceeds the concentration at which the polymer precipitates from solution. Under favorable conditions, semi-rigid polymers can be spun from isotropic solution into high-modulus fibers, provided the elongational flow rates in the spinning line are sufficient to achieve high orientation, and the coagulation times are short in comparison with the relaxation time to preserve this orientation. In addition, post-spinning treatments are required to enhance the mechanical properties.

Both theoretical and experimental data suggest that these systems may undergo a flow-induced phase transition into an anisotropic state. Whether this phenomenon actually occurs is an intriguing scientific problem, which deserves further investigation. For the purpose of this review it is sufficient to stress the technological consequences of these findings, namely: (1) ultrahigh-modulus fibers are not restricted to rodlike polymers forming anisotropic solutions, and (2) further research may show that more flexible polymers than X-500 may, with proper process modifications, lead to satisfactory results.

With decreasing chain rigidity, the process requirements for achieving ultrahigh-modulus, ultrahigh-strength fibers become more severe, and probably culminate with the gel spinning of PE. In many respects PE represents the opposite extreme of chain flexibility as compared with rigid-rod PBA or PPD-T. For the rodlike aramid Kevlar molecular design, properties of the solution, and choice of the solvent are the critical factors in forming an anisotropic dope, and the spinning process and post-treatments are relatively simple. By contrast, the gel spinning process is rather complicated. The key factors in preparing ultrahigh-modulus, ultrahigh-strength PE fibers are: (1) very high molecular weight (1,000,000 or more); (2) spinning dilute solutions with minimal chain entanglement into essentially unoriented gel fibers with minimal chain entanglement; and (3) stretching plasticized and/or porous fibers to almost theoretical limits by a drawing process involving a temperature gradient as well as simultaneous removal of solvent and decrease in free volume (porosity). Although PE is currently the only example of successful gel spinning, it is expected that developments in ultrahigh-modulus fibers will not end here.

V. Molecular Reinforcement

A. Introduction

The strength of an isolated extended chain of molecules exceeds by an order of magnitude or more the strength of fibers produced from the same polymer. This situation exists because we are unable to prepare fibers in which the polymer molecules are perfectly aligned and without flaws, as shown schematically in Fig. 1. Fiber-reinforced composites exhibit additional losses in strength due to adhesion problems at the fiber–matrix interface and concentration of stress at discontinuities such as fiber ends. By increasing fiber orientation we gain strength while decreasing the mechanical properties in the transverse direction, and the ultrahigh-strength fibers consistently show a great tendency to fibrilate. There seems to be very little hope that this problem can be overcome in the near future.

In the search for a radical solution of these problems no approach appears to be more attractive and promising than the development of composites consisting of a dispersion of rodlike molecules in the matrix of an amorphous homolog that contains a sufficient number of flexible bonds to adopt a random-coil configuration. Chemical similarity would provide conditions for good wetting and adhesion between the two species, and the concentration of stress at the chain end would be minimized.

The principal problem with the design of such systems is, of course, thermodynamic equilibrium. Additional problems arise from the high melting point and poor solubility of rodlike polymers. Considering the high melting point of rodlike molecules, it appears that the chances for arriving at a homogeneous dispersion of rigid rods in a random coil homolog are much greater using a ternary solution than proceeding via the melt.

Statistical thermodynamics of ternary systems comprising a solvent, a rodlike solute, and a random-coil chain were developed by Flory (1978). These theoretical developments show that (1) the anisotropic phase rejects the coiled molecules with a selectivity approaching that of a pure crystal and (2), for a sufficiently high axial ratio required to achieve reinforcement, a high degree of selectivity occurs even at relatively low concentration. Since this theory, when closely examined, also provides guidelines for overcoming this problem, it represents a very effective tool in present research on molecular composites.

Two systems of this type have attracted attention, one based on rodlike poly(p-phenylenebenzobisthiazole) (PBT) and more flexible amorphous poly[2-5(6)-benzimidizole] (ABPBI) along with several other pairs of rodlike and flexible polymers (Husman *et al.*, 1979; Hwang *et al.*, 1981, and the other involving rigid wholly aromatic polyamides, such as poly(p-phenylene-terephthalamide) dispersed in a matrix of nylon 6 or nylon 66 (Takayanagi *et al.*, 1980). Although not completely successful, these initial results are very encouraging, suggesting that a breakthrough in this technology may not be far way. A review of research in this field is, however, very desirable because it suggests directions that may lead to systems and structures with properties considerably better than those presently achieved.

B. Composites Based on PBT and ABPBI, etc.

The composite films investigated by Husman *et al.* (1979) were obtained from 2% solution (by volume) of various polymer pairs in methanesulfonic acid. The processing technique involved: (a) vacuum casting, (b) precipitation, and (c) shear quenching. The chemical composition of the rodlike and coiling polymers used in these studies is shown in Table VI.

Scanning electron microscopy revealed that for rodlike polymer content of up to 50 wt %, the films consist of two distinct phases. The rodlike polymers tend to form discrete saucer-shaped particles that become elongated as the fibers are stretched. Since the volume fraction of the dispersed phase is

TABLE VI

Chemical Structure of Rodlike and Coiling Polymers[a]

	Chemical Structure	Acronym
Rodlike polymers		PDIAB
		PBO
		PBT
Coil-like polymers		M-PBI
		AB-PBI
		PEPBO
		PPBT

[a] From Husman *et al.* (1979).

larger than that of the rodlike polymer, it was inferred that the dispersed particles consist of a composite of rod and coil molecules. The calculated volume fraction of rodlike polymers in the dispersed phase appears to be constant and independent of the rod content in the fiber.

Of the blends studied by Husman *et al.* (1979), the AB-PBI/PDIAB type was most thoroughly characterized. The mechanical properties of fibers containing 0, 10, 20, and 30 wt % of rods in the unoriented or stretched state

TABLE VII

Mechanical Properties of AB-PBI/PDIAB Blends[a]

% Rod Polymer	Stretch % Area Reduction	Modulus (GPa)	Strength (MPa)
0	None	1.03	79.92
0	Mechanical/20	2.00	134.36
0	Solvent/60	3.37	105.42
10	None	2.00	70.28
10	Mechanical/37	4.60	161.92
10	Solvent/57	6.86	315.56
20	None	1.58	44.10
20	Mechanical/37	2.38	82.68
20	Solvent/70	7.17	253.55
30	None	1.25	36.52
30	Mechanical/20	2.16	71.66
30	Solvent/60	8.96	189.48

[a] From Husman et al. (1979).

are shown in Table VII. Stretching was achieved by retesting the samples broken in a tensile test. This procedure was repeated until the samples were too short for satisfactory clamping and retesting. Some specimens were also plasticized with methanol to give increased ductility during stretching. However, the mechanical properties listed in the table were obtained using samples in which methanol had been evaporated after stretching (Husman et al., 1979).

The results in Table VII show significant increases in modulus and strength with the addition of rodlike polymer to the random-coil matrix. A modulus of almost 7 GPa in a solvent-stretched sample achieved by adding only 10% rodlike polymer indicates significant reinforcement, especially because it is accompanied by a threefold increase in the fiber strengths.

The authors estimated the effective modulus of the dispersed phase from the morphological data characterizing the average size of the dispersed domains in these fibers. Pertinent data used in these calculations are presented in Table VIII.

From the data in Tables VII and VIII the effective modulus of the dispersed particles was calculated using the Halpin–Tsai equations:

$$E/E_M = (1 + \zeta \eta v_f)/(1 - \eta v_f)$$

$$\eta = \left(\frac{E_f}{E_M} - 1 \right) \Big/ \left(\frac{E_f}{E_M} + \zeta \right)$$

TABLE VIII

Morphology of AB-PBI/PDIAB Blends[a]

	Composition		
Conglomerate	90/10	80/20	70/30
Length (cm $\times 10^{-4}$)			
Initial	1.78	3.81	5.84
Mechanically stretched	2.54	5.08	7.62
Solvent stretched	3.81	5.08	6.35
Thickness (cm $\times 10^{-4}$)			
Initial	.635	1.27	1.91
Mechanically stretched	.508	.889	1.27
Solvent stretched	.254	.508	.762
Width (cm $\times 10^{-4}$)			
Initial	1.78	3.81	5.84
Mechanically stretched	1.52	3.30	5.08
Solvent stretched	1.02	1.14	1.27
Aspect Ratio (L/D)			
Initial	2.8	3.0	3.1
Mechanically stretched	5.0	5.7	6.0
Solvent stretched	1.5	10	8
Volume % in film	18%	35%	53%
Volume % rod in conglomerate	56%	57%	57%

[a] From Husman *et al.* (1979).

where E is the composite or film modulus, E_M the corresponding matrix modulus, E_f the corresponding effective reinforcement modulus, ζ the measure of reinforcement dependent on boundary conditions (for these calculations, taken as twice the aspect ratio of dispersed particles), and v_f the volume fraction of reinforcement. The results are tabulated in Table IX. Note that

TABLE IX

Calculated Conglomerate Modulus of
AB-PBI/PDIAB Blends[a]

Composition	Treatment	$\bar{E}_F$ (GPa)
90/10	Initial	16.15
90/10	Mechanically stretched	33.35
90/10	Solvent stretched	26.23
80/20	Solvent stretched	15.43
70/30	Solvent stretched	14.94

[a] From Husman *et al.* (1979).

the modulus of the dispersed phase increases with increasing aspect ratio, indicating increasing orientation with stretching (Husman *et al.*, 1979). This analysis showed that in the cast films the dispersed phase consists of composite aggregates exhibiting a high degree of composite-type reinforcement, but very limiting potential for processing.

In order to gain insight into the factors controlling the morphology of the cast fibers, Hwang *et al.* (1981) undertook a study of the thermodynamics of the ternary system. The ABPBI solutions were isotropic at all concentrations, whereas the PBI solutions were anisotropic above a critical concentration C^*. The authors proceeded to establish the existence of a critical concentration for the ABPBI/PBT polymer blend beyond which the solution consists of an anisotropic nematic domain dispersed in the isotropic solution. The calculated phase diagram for aspect ratio $X_2 = X_3 = 300$, along with experimental points, is presented in Fig. 19. The triangular points indicate

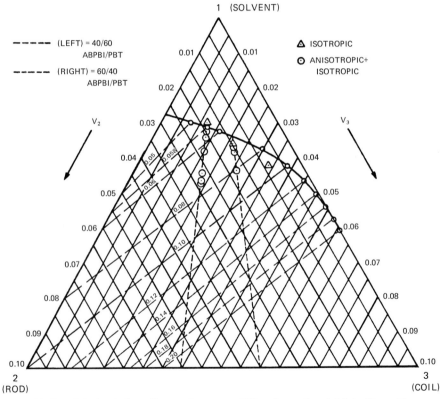

Fig. 19. Calculated phase diagram ($x_2 = x_3 = 300$) and experimental data. (From Hwang *et al.*, 1981.)

the experimentally determined C^*, which agree within experimental error with the values predicted by theory.

As expected, the ABPBI/PBT solutions in methanesulfonic acid are thermotropic. A biphasic solution containing a dispersed anisotropic phase becomes isotropic when heated to the nematic–isotropic transition temperature. The concentration dependence of this transition temperature is shown in Fig. 20 for 40/60 and 60/40 ABPBI/PBT blends (Hwang *et al.*, 1981).

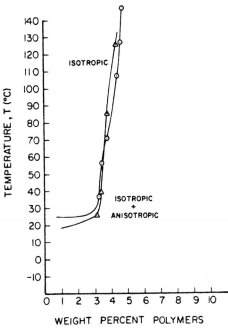

Fig. 20. Temperature–concentration phase diagram for 40/60 ($\odot$) and 60/40 ($\triangle$) AB-PBI/PBT blends. Solvent is 97, MSAI 3 CSA.. (From Hwang *et al.*, 1981.)

Several important conclusions were derived from these results:

(a) A quantitative verification of Flory's theory, which can be used as a guideline for future experiments.

(b) Criteria for achieving a homogeneous composite fiber (film) from solution are (i) absence of rodlike polymer aggregates, and (ii) a high degree of molecular orientation of rodlike molecules in the random-coil matrix.

(c) Criteria in (b) may be achievable if composite fibers or film are processed from polymer solutions at or near the critical concentration. At this stage the solution is still isotropic, and the rodlike polymer should be easily oriented by external shear stress (mixing).

C. Composites Based on Aromatic and Aliphatic Polyamides

1. Chemical Background

Takayanagi *et al.* (1980) studied the morphology and properties of several blends of rigid-rod polyaramides or their block copolymers and nylon 6 or nylon 66 as matrix polymers.

Poly(*p*-phenyleneterephthalamide) (PPTA) and poly(*p*-benzamide) (PBA) were prepared according to the methods described in U.S. Patent 3,817,941 (Du Pont, 1974). PPTA–nylon 66 block copolymer was prepared by first treating *p*-phenylenediamine with terephthalyl chloride in a mixed solvent of hexamethylphosporamide (HMPA), *N*-methylpyrrolidone (NMP), and LiCl [40 ml HMPA, 20 ml NMP, 2 g LiCl]. Then adipyl chloride, hexamethylenediamine, and triethylamine were added to the PPTA solution. The polymerization reaction was carried out at 0°C for 5 h. The product was poured into water and repeatedly washed with water and methanol. The molecular weight of the PPTA was 4200.

$$H_2N-\langle\bigcirc\rangle-NH_2 + ClOC-\langle\bigcirc\rangle-COCl \text{ (excess)} \xrightarrow{\text{HMPA–NMP–LiCl}}$$

$$ClOC-\langle\bigcirc\rangle-CO + NH-\langle\bigcirc\rangle-NH-CO-\langle\bigcirc\rangle-CO)_n Cl$$

I

$$\textbf{I} + H_2N + CH_2)_6 NH_2 + ClOC + CH_2)_4 COCl \xrightarrow{Et_3N}$$

PPTA–Nylon 66 block copolymer

PPTA–Nylon 6 block copolymer was prepared by dissolving nylon 6 in a mixed solvent of HMPA (70 ml), NMP (30 ml), and LiCl (3 g) and terminated with excess terephthalyl chloride using triethylamine.

$$H_2N + \text{nylon 6} + COOH + ClOC-\langle\bigcirc\rangle-COCl \text{ (excess)} \xrightarrow{Et_3N}$$

$$ClOC-\langle\bigcirc\rangle-CO-NH + \text{nylon 6} + COOH + ClOC-\langle\bigcirc\rangle-COCl$$

Then *p*-phenylenediamine was added to form PPTA blocks. The reaction product was precipitated in water and repeatedly washed with water and methanol.

PBA–Nylon 6 block copolymer was prepared by dissolving low molecular weight PBA and nylon 6 in a mixed solvent of HMPA (60 ml), NMP (30 ml), and LiCl (3 g) and treating with triphenyl phosphite in pyridine. The reaction product was precipitated in water and washed repeatedly with water and methanol.

Polymer composites of aromatic and aliphatic polyamides were prepared by extruding the sulfuric acid solutions of polymers into a large amount of water and methanol. Specimens for mechanical testing were prepared by compression molding nylon 6 composites at 240°C and nylon 66 composites at 290°C.

2. Effect of Molecular Weight on Morphology

The systems investigated were prepared under conditions that may, because of mixing and turbulence, be far from equilibrium. The morphology of the samples was investigated by examining the fractured surfaces of blends by electron microscopy. The specimens were broken in liquid nitrogen. For the samples prepared from 7% PPTA having $M_v = 34,000$ and 93% nylon 6, the elongated PPTA microfibrils were uniformly dispersed. Their diameter was about 30 nm. When PPTA of $M_v = 4500$ was used, the morphology was similar except that the microfibril diameter was reduced to 15 nm and the number of microfibrils was increased. Extrusion of the sulfuric acid solution of both components into water resulted in local orientation of microfibrils; the nylon matrix, on the other hand, was unoriented. Examination of fracture surfaces showed the absence of long sections of microfibrils of PPTA protruding out of the nylon 6 matrix, suggesting good adhesion between the microfibrils and the matrix (Takayanagi et al., 1980).

Based on these observations, the authors concluded that rigid poly-aramides blend with aliphatic polyamides to form a segregated microfibrillar composite. The microfibrils have diameters ranging from 10 to 30 nm, depending on the molecular weight of the polyaramides. Although these diameters are much smaller than the diameters of fibers used in fiber-reinforced composites, the present systems are still far from a homogeneous composite. However, in comparison with the samples discussed in the previous section, the microfibrils in these polyamide composites have a considerably higher L/D ratio. A morphological model consistent with the electron microscopic observations and thermal properties of these systems is presented in Fig. 21.

Polyaramides dispersed in a nylon matrix act as nucleating agents (Takayanagi et al., 1980). The effect increases with increasing molecular weight of the polyaramide. Shown in Fig. 22 are the wide angle x-ray diffraction intensity curves for nylon 6 and its blends with PPTA of different

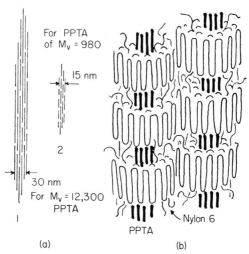

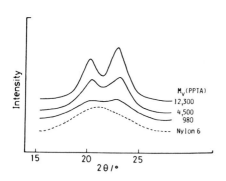

Fig. 21. Schematic diagrams of (a1) a microfibril of high molecular weight PPTA and (a2) one of low molecular weight PPTA; (b) superstructure of a polymer composite reinforced by microfibrils of PPTA (heavy lines). Nylon 6 molecules induced by microfibrils to crystallize in folded chain crystals are indicated by thin lines. (Reprinted from Takayanagi *et al.*, 1980, p. 608, by courtesy of Marcel Dekker, Inc.)

Fig. 22. Wide-angle x-ray diffraction curves for nylon 6 and its blends with PPTA of various molecular weights. All samples were quenched from 250°C to room temperature under identical conditions. (Reprinted from Takayanagi *et al.*, 1980, p. 600, by courtesy of Marcel Dekker, Inc.)

molecular weights. The samples were quenched from the melt under the same conditions.

Other evidence for nucleation of crystallization of nylon 6 by PPTA can be deduced from the DSC curves obtained during cooling of the melt. The sample was kept molten at 260°C for 20 min, and the cooling rate was 10°C/min. An exotherm corresponding to the crystallization of nylon 6 was observed. (Fig. 23). Nylon 6 crystallized at 180–185°C, whereas blends with PPTA of high molecular weight ($M_v = 4{,}500$ and 12,300) showed onset of

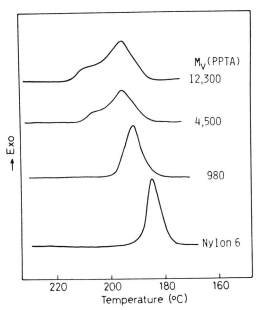

Fig. 23. DSC cooling curves for nylon 6 and PPTA/nylon 6 (5/95) blends with PPTA with various molecular weights. Samples were held in the melt at 260°C for 20 min and the cooling rate was 10°C/min. (Reprinted from Takayanagi *et al.*, 1980, p. 602, by courtesy of Marcel Dekker, Inc.)

crystallization at a temperature around 210°C, which is very close to the melting temperature of nylon 6. This was followed by a main crystallization exotherm located around 195°C. The two-stage crystallization in the latter case may be explained by the fact that the active sites located on the surface of PPTA microfibrils initiate crystallization of nylon 6 near the melting temperature of nylon 6, and lamellar crystallization of nylon 6 takes place in the remaining space at the second stage. The curve for the blend of nylon 6 with PPTA of $M_v = 980$ is located between the two cases of nylon 6 and its blend with high molecular weight PPTA.

Crystallization isotherms obtained by dilatometry also showed accelerated crystallization of nylon 6 with rigid polyaramides. The full curves in Fig. 24 show the crystallization isotherms at 210 and 212°C for a blend of PBA and nylon 6 (5/95) and for nylon 6 (dashed lines). The patterns of crystallization are common for the isotherms at both temperatures: the isotherms of nylon 6 agree with those calculated by the Avrami equation with Avrami's constant $n = 4$, whereas those of the blends with PBA are represented by $n = 2$. Avrami's equation is

$$a = \exp(-kt^n)$$

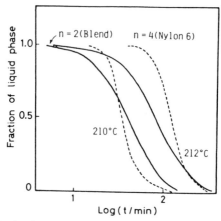

Fig. 24. Crystallization isotherms at 210 and 212°C for nylon 6 (−−) and its blend with PBA (nylon 6/PBA = 95/5) (——). (Reprinted from Takayanagi *et al.*, 1980, p. 602, by courtesy Marcel Dekker, Inc.)

where a is the volume fraction of the liquid phase and k is a crystallization constant. The value of n is dependent on the crystallization kinetics; $n = 4$ corresponds to homogeneous nucleation and growth of three-dimensional spherulites, whereas $n = 2$ corresponds to heterogeneous nucleation followed by two-dimensional disk-shaped growth or homogeneous nucleation followed by fibrous growth (Takayanagi *et al.*, 1980).

3. Mechanical Properties

The molecular weight of the rigid-rod component has a marked effect on the mechanical properties of the blend. For a composition ratio of PPTA to nylon 6 of 5/95, this effect is shown in Fig. 25 and Table X. On increasing the molecular weight of PPTA from 980 to 12,300, we see a consistent increase in modulus, yield stress, and breaking stress. The elongation of break, on the other hand, decreases. In spite of the small content of PPTA, the effect of the rodlike molecules on the stress–strain behavior of the composite approaches that of the glass-reinforced plastics containing as much as 30% of chopped glass fibers.

These results can be explained by assuming that the strengths of the microfibrils are strongly dependent on the molecular weight, and that the strong high-modulus PPTA microfibrils restrict the ductility of the matrix more than their weak low molecular weight counterparts.

Extensibility of the composites can be improved without loss in strength if the rigid segments of PBA or PPTA are chemically bonded to the flexible chain blocks of nylon 6 or nylon 66. Morphologically this leads to finer microfibrils and a stronger interface between the microfibrils and the matrix.

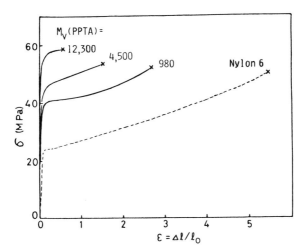

Fig. 25. Plots of nominal stress σ versus strain ε for nylon 6 (--) and for 5/95 blends of PPTA/nylon 6 with various molecular weights of PPTA (—) (Reprinted from Takayanagi *et al.*, 1980, p. 609, by courtesy of Marcel Dekker, Inc.)

TABLE X

Mechanical Properties of Nylon 6 and the Polymer
Composites of PPTA/Nylon 6[a]

Polymer Composite	$\bar{M}_v$ of PPTA	E (GPa)	σ_y (MPa)	σ_B (MPa)	ε_B (%)
Nylon 6	—	0.91	24	51	5.3
PPTA/nylon 6 (5/95)	980	1.45	40	52	2.7
PPTA/nylon 6 (5/95)	4,500	1.59	46	54	1.6
PPTA/nylon 6 (5/95)	12,300	1.67	58	59	0.6

[a] From Takayanagi *et al.* (1980).

This effect is shown in Fig. 26 for blends of pure PPTA and nylon 66–PPTA block copolymers with nylon 66. In all cases the total content of the PPTA in the blend is 5% by weight. The molecular weight of PPTA homopolymers in the blend is 4500 and 980 for curves A and B, respectively. When block copolymers having rigid segments of the same length are blended with nylon 66, the ductility of the samples increases markedly while the strength and modulus remain essentially unaffected.

These results show that (1) the present systems are morphologically far from a true homogeneous reinforcement, but (2) even in the case of microfribilar morphology, the levels of reinforcement achieved are very high in comparison with fiber reinforced materials.

In order to estimate the effective strength of the microfibrils in these systems, some of the samples were stretched to achieve uniaxial orientation

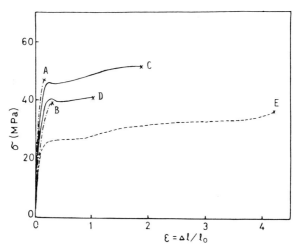

Fig. 26. Stress–strain curves (A, B) of physical blends of PPTA and nylon 66; (C, D) blends of the block copolymers of PPTA and nylon 66; and (E) nylon 66 only. The total blend ratio of PPTA/nylon 66 is 5/95. The composition of the block copolymer of curve C is PPTA/nylon 66 = 28/72, and that of curve D is 37/63. The molecular weight of the PPTA block in the block copolymer is 4200. Molecular weights of PPTA for the samples of curves A and B are 4500 and 980, respectively. (Reprinted from Takayanagi *et al.*, 1980, p. 611, by courtesy of Marcel Dekker, Inc.)

and allow a simple property analysis. Stress–strain diagrams of an oriented blend of PPTA/nylon 6 having a 3/97 weight ratio and a nylon 6 sample stretched to the same extension ratio of 3.1 are shown in Fig. 27 (Takayanagi *et al.*, 1980). Improvements in strength and modulus resulting from a 3% addition of PPTA ($M_v = 34,000$) are outstanding. The modulus of 1.18 GPa for oriented nylon 6 is increased to 3.35 GPa for the oriented blend with a PPTA/nylon 6 ratio of 3/97. The ultimate strength of 220 MPa for oriented nylon 6 is increased to 340 MPa for the oriented polymer composite. The ultimate elongation of the oriented blend is 20%, while that for oriented nylon is 28%. Orientation is seen to be clearly effective in reinforcement, when comparison is made with the unoriented sample.

If the simple assumption of the volume additivity of the ultimate strengths of PPTA and nylon 6 is adopted, the strength of a PPTA microfibril is estimated to be 4 GPa. This is comparable to or higher than the highest tenacity, 3.46 GPa, of commercial PPTA fiber, (Du Pont Kevlar 49). Thus even when discontinuity of the PPTA microfibrils in the specimen is neglected, the full strength of PPTA may be realized in a polymer composite.

The same observation applies to the effective microfibril modulus, 75 GPa, which falls between the modulus of Kevlar 29 with $E = 61$ GPa and Kevlar 49 with $E = 120$ GPa.

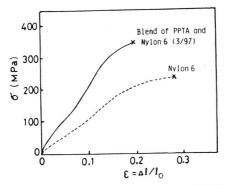

Fig. 27.   Stress–strain curves of an oriented physical blend of PPTA/nylon 6 = 3/97 (——) and of nylon 6 (– –). The film was hot-drawn at 170°C to three times the original length. Molecular weight of PPTA is 34,000. (Reprinted from Takayanagi *et al.*, 1980, p. 614, by courtesy of Marcel Dekker, Inc.)

4. Conclusions

The results of the above studies demonstrate the potential and difficulties of homogeneous reinforcement. The data presented, along with the theoretical background and molecular design (such as block copolymers of rigid rods with flexible homologs), indicate the directions that may ultimately lead to products of great technological importance.

References

Allen, S. R., Filippov, A. G., Farris, R. J., Thomas, E. L., Wong, C.-P., Berry, G. C., and Chenevey, E. D. (1981). *Macromolecules* **14**, 1135.

Arpin, M., and Strazielle, C. (1977). *Polymer* **18**, 597.

Asrar, J., Preston, J., Ciferri, A., and Krigbaum, W. R. (1982). *J. Polym. Sci., Polym. Chem. Ed.* **20**, 373.

Bacon, R. (1973). *Chem. Phys. Carbon* **9**, 1.

Capaccio, G., and Ward, I. M. (1974). *Polymer* **15**, 233.

Ciferri, A., and Valenti, B. (1979). *In* "Ultra High Modulus Polymers" (A. Ciferri and I. M. Ward, eds.), p. 203. Appl. Sci., London.

Du Pont de Nemours, E. I. (1974). U.S. Patent 3,817,941.

Flory, P. J. (1956). *Proc. R. Soc. London, Ser. A* **234**, 73.

Flory, P. J. (1978). *Macromolecules* **11**, 1138.

Franks, N., and Varga, j. (1979). U.S. Patent 4,145,532 (to Akzona Co.).

Griffin, B. P., and Cox, M. K. (1980). *Br. Polym. J.* **12**, 147.

Husman, G., Helmimiak, T., Adams, W., Wiff, D., and Benner, C. (1979). *Org. Coat. Plast. Chem.* **40**, 797.

Hwang, W. F., Wiff, D. R., and Helmimiak, T. (1981). *Org. Coat. Plast. Chem.* **44**, 32.

Jackson, W. J., Jr. (1980). *Br. Polym. J.* **12**, 154.

Jackson, W. J., Jr., and Kufhuss, H. F. (1976). *J. Polym. Sci., Polym. Chem. Ed.* **14**, 2043.

Kalb, B., and Pennings, A. J. (1980). *J. Mater. Sci.* **15**, 2584.

Krigbaum, W. R., Salaris, F., Ciferri, A., and Preston, J. (1979). *J. Polym. Sci., Polym. Lett. Ed.* **17**, 601.

Luise, R. R. (1980). U.S. Patent 4,183,895 (to Du Pont Co.).

Marrucci, G., and Sarti, G. C. (1979). *In* "Ultra High Modulus Polymers" (A. Ciferri and I. M. Ward, eds.), p. 137. Appl. Sci., London.

Payet, C. R. (1979). U.S. Patent 4,159,365 (to Du Pont Co.).

Pennings, A. J., and Meihingen, K. E. (1979). *In* "Ultra High Modulus Polymers" (A. Ciferri and I. M. Ward, eds.), p. 117. Appl. Sci., London.

Peterlin, A. (1979). *In* "Ultra High Modulus Polymers" (A. Ciferri and I. M. Ward, eds.), p. 279. Appl. Sci., London.

Preston, J. (1979). *In* "Ultra High Modulus Polymers" (A. Ciferri and I. M. Ward, eds.), p. 155. Appl. Sci., London.

Prevorsek, D. C., Harget, P. J., Sharma, R. K., and Reimschuessel, A. C. (1973). *J. Macromol. Sci., Phys.* **B8**, 127.

Prevorsek, D. C., Kwon, Y. D., and Sharma, R. K. (1977). *J. Mater. Sci.* **12**, 2310.

Reynolds, W. N. (1973). *Chem. Phys. Carbon* **11**, 1.

Schaefgen, J. R. (1978). U.S. Patent 4,118,372 (to Du Pont Co.).

Schaefgen, J. R., Bair, T. I., Ballou, J. W., Kwolek, S. L., Morgan, P. W., Pavar, M., and Zimmerman, J. (1979). *In* "Ultra High Modulus Polymers" (A. Ciferri and I. M. Ward, eds.), p. 203. Appl. Sci., London.

Singer, L. S. (1979). *In* "Ultra High Modulus Polymers" (A. Ciferri and I. M. Ward, eds.), p. 251. Appl. Sci., London.

Smith, P., and Lemstra, P. J. (1979). *Makromol. Chem.* **180**, 2983.

Smith, P., and Lemstra, P. J. (1980a). *Polymer* **21**, 1341.

Smith, P., and Lemstra, P. J. (1980b). *J. Mater. Sci.* **15**, 505.

Smith, P., Lemstra, P. J., Kalb, B., and Pennings, A. J. (1979). *Polym. Bull. (Berlin)* **1**, 733.

Smith, P., Lemstra, P. J., and Booij, H. C. (1981). *J. Polym. Sci., Polym. Phys. Ed.* **19**, 877.

Southern, J. H., and Porter, R. S. (1970). *J. Appl. Polym. Sci.* **14**, 2305.

Takayanagi, M., Ogata, T., Morikawa, M., and Kai, T. (1980). *J. Macromol. Sci., Phys.* **B17**, 591.

Valenti, B., Alfonso, G. C., Ciferri, A., Giordani, P., and Marrucci, G. (1981). *J. Appl. Polym. Sci.* **26**, 3643.

Wilding, M. A., and Ward, I. M. (1978). *Polymer* **19**, 969.

Wissbrun, K. F. (1980). *Br. Polym. J.* **12**, 163.

Wolfe, J. F., and Arnold, F. E. (1981). *Macromolecules* **14**, 909.

Wolfe, J. F., Loo, B. H., and Arnold, F. E. (1981). *Macromolecules* **14**, 915.

Zachariades, A. E., Mead, W. T., and Porter, R. S. (1979). *In* "Ultra High Modulus Polymers" (A. Ciferri and I. M. Ward, eds.), p. 77. Appl. Sci., London.

Index

MATERIALS SCIENCE AND TECHNOLOGY

EDITORS

A. S. NOWICK
Henry Krumb School of Mines
Columbia University
New York, New York

G.G. LIBOWITZ
Solid State Chemistry Department
Materials Research Center
Allied Corporation
Morristown, New Jersey

Manfred von Heimendahl, ELECTRON MICROSCOPY OF MATERIALS: AN INTRO-
DUCTION, 1980

O. Toft Sørensen (editor), NONSTOICHIOMETRIC OXIDES, 1981

M. Stanley Whittingham and Allan J. Jacobson (editors), INTERCALATION
CHEMISTRY, 1982

A. Ciferri, W. R. Krigbaum, and Robert B. Meyer (editors), POLYMER LIQUID
CRYSTALS, 1982